Fruit Flies of Economic Importance XI

Fruit flies (Diptera: Tephritidae) adversely impact the ability of the world's nations to feed a growing world population. Hundreds of scientists work throughout many countries studying economically damaging fruit flies to improve our understanding of these pests and how to manage them. This book contains 19 contributions from many scientists, which is intended to improve the ability of all nations to manage tephritids and to provide good-quality food for their people.

This book starts with a reflection on the career and contribution of Prof. Antigone Zacharopoulou. It continues with three chapters expanding our knowledge and information on the biology, ecology, physiology, and behavior of tephritid flies. The following three chapters explore opportunities for international trade. Aspects of the sterile insect technique (SIT) and natural enemies from different countries are explained in the next five chapters. Finally, the last seven chapters describe a diverse range of action programs and area-wide integrated pest management (IPM).

Key features:

- Provides information on fruit fly diversity, courtship, and components involved in mating success.
- Offers opportunities to improve international trade on fruits susceptible to tephritids.
- Explores methods to optimize the components of SIT and the use of parasitoids.
- Provides new insights into area-wide IPM and action programs.
- Includes contributions from relevant fruit-growing areas of the globe.

PROCEEDINGS OF THE ELEVENTH INTERNATIONAL SYMPOSIUM ON FRUIT FLIES OF ECONOMIC IMPORTANCE (ISFFEI XI)

Fruit Flies of Economic Importance XI

Edited by
Diana Pérez-Staples, Bernard Dominiak,
María Teresa Vera, and Abdeljelil Bakri

CRC Press
Taylor & Francis Group
Boca Raton London New York

CRC Press is an imprint of the
Taylor & Francis Group, an **informa** business

Designed cover image: top left: Sterile Insect Facility photograph provided by Abdeljelil Bakri; top right: Bactrocera dorsalis mating photograph provided by Carlos Cáceres; bottom left: Anastrepha ludens, Mexican fruit fly ovipositing photograph provided by Andrés Díaz Cervantes; bottom right: Queensland fruit fly photograph provided by Polychronis Rempoulakis.

First published 2026
by CRC Press
4 Park Square, Milton Park, Abingdon, Oxon, OX14 4RN

and by CRC Press
2385 NW Executive Center Drive, Suite 320, Boca Raton FL 33431

NSW GOVERNMENT | Department of Primary Industries

Publication of this book has been funded by the project: Biosecurity preparedness for oriental fruit fly (FF18001). This project has been funded by Hort Innovation, using the research and development levies, contributions from the Australian Government and co-investment from NSW Department of Primary Industries, Macquarie University and PIRSA-SARDI. Hort Innovation is the grower-owned, not-for-profit research and development corporation for Australian horticulture.

For Product Safety Concerns and Information please contact our EU representative GPSR@taylorandfrancis.com. Taylor & Francis Verlag GmbH, Kaufingerstraße 24, 80331 München, Germany.

British Library Cataloguing-in-Publication Data
A catalogue record for this book is available from the British Library

ISBN: 978-1-041-08197-5 (hbk)
ISBN: 978-1-041-08219-4 (pbk)
ISBN: 978-1-003-64433-0 (ebk)

DOI: 10.1201/9781003644330

Typeset in Sabon
by codeMantra

Access the Support Material: www.routledge.com/9781041081975

Contents

Section II
Risk assessment, quarantine and post-harvest

Section III
Sterile insect technique

Section IV
Natural enemies and biological control

Section V
Area-wide IPM and action programs

Preface

The 11th International Symposium on Fruit Flies of Economic Importance was held in Sydney, NSW, Australia, from 13 to 18 November 2022. It was co-organized by the New South Wales Department of Primary Industries and the Joint FAO/IAEA Division of Nuclear Techniques in Food and Agriculture. The event took place at Macquarie University's North Ryde Campus and was logistically supported by the Conference Company, a professional event organizer.

For the first time, the symposium was conducted in a hybrid format, offering both in-person and online attendance options. Participants included fruit fly researchers, plant protection officials, fruit industry representatives, students, and exhibitors. A total of 243 participants from 45 countries attended in person, while 109 participants from 33 countries joined virtually. The program featured a plenary talk, 60 oral presentations, and 157 poster presentations.

The symposium covered a wide range of topics, including the following:

1. Biology, Ecology, Physiology, and Behavior
2. Taxonomy and Morphology
3. Genetics and Biotechnology
4. Chemical Ecology and Attractants
5. Risk Assessment, Quarantine, and Post-Harvest Management
6. Sterile Insect Technique (SIT)
7. Natural Enemies and Biological Control
8. Other Control Methods and New Developments
9. Area-Wide Integrated Pest Management (AW-IPM) and Action Programs
10. Social, Economic, and Policy Issues of Action Programs

A significant portion of the symposium focused on the SIT as a key component of AW-IPM strategies against fruit fly pests. Presentations covered aspects such as mass-rearing, field release, the development of genetic sexing strains, quality control, and preventive releases. Several success stories of SIT application were shared, along with oral presentations on emerging and ongoing SIT programs worldwide. These studies highlighted the cost-effectiveness, environmental sustainability, and public acceptance of SIT as a method for integrated fruit fly management. In addition, new developments in SIT were discussed, including next-generation genetic sexing strains, the combined application of SIT and male annihilation techniques for enhanced pest control, and the adaptation of SIT for spotted-wing drosophila (SWD).

Participants also took part in two field trips:

1. A tour of Sydney Markets, an internationally recognized hub for fresh produce and community markets, playing a key role in Australia's horticultural industry and economy.
2. A visit to the Elizabeth Macarthur Agricultural Institute (EMAI), a premier biosecurity facility dedicated to enhancing food and fiber production while protecting the environment.

As a social event, a night boat tour of Sydney Harbour was organized for attendees.

Two sessions were dedicated to honoring the memory of distinguished fruit fly researchers who passed away since the 10th ISFFEI: Roger Vargas (Session 1) and Antigoni Zacharopoulou (Session 3). During the symposium's opening, a historical overview was presented to mark the 40th anniversary of ISFFEI, which was first held in Athens in November 1982. Since then, and over the past four decades, the symposium has been hosted in ten countries across six continents.

Beyond formal sessions, attendees had ample opportunities to share knowledge, establish collaborations, and build friendships during poster sessions, coffee breaks, lunches, the welcome cocktail, the social function, and field trips. Once again, the symposium demonstrated the strong collaboration and camaraderie within the fruit fly research community, which has undoubtedly contributed to the field's many successes.

This book serves as the official proceedings of the 11th ISFFEI, continuing a long-standing tradition as the 11th volume in this series. It includes 19 contributions from 75 authors worldwide. In addition to papers originating from the symposium, this volume also features valuable contributions from participants in two regional meetings held in 2024: TEAM (Mauritius) and TWWH (Jamaica). All manuscripts underwent peer review by members of the Fruit Fly Steering and Local Organizing Committees, as well as experts in the field to whom we are grateful (see Acknowledgments). The editorial work was carried out by Abdeljelil Bakri, Bernard Dominiak, Diana Pérez-Staples, and María Teresa Vera, whose dedication and high-quality contributions are deeply appreciated.

The publication of this volume was financially supported by a Hort Innovation research grant awarded to Macquarie University, NSW Department of Primary Industries, and South Australia's Research Development Institute (Project FF18001: Biosecurity Preparedness for Oriental Fruit Fly *Bactrocera dorsalis*). Thanks to registration fees, donors, and exhibitors, the proceedings book will be made open access and available online through the CRC Press website and Nucleus IAEA website, ensuring broad accessibility to researchers and stakeholders worldwide.

Chronis Rempoulakis
Vienna, Austria
Spring 2025

Acknowledgments

This book *Fruit Flies of Economic Importance XI* is a compilation of the Proceedings of the 11th International Symposium on Fruit Flies of Economic Importance (ISFFEI), held in Sydney, New South Wales, Australia, from 13 to 18 November 2022. We thank Rui Pereira and Chronis Rempoulakis for the invitation to edit the proceedings.

The aim of this book is to highlight the current status of area-wide programs on tephritid flies in different countries and regions as well as other contributions presented during the symposium. This book is organized into the following sections: Biology, Ecology, Physiology, and Behavior; Risk Assessment, Quarantine, and Post-Harvest; Sterile Insect Technique; Natural Enemies and Biological Control; and Area-Wide IPM and Action Programs. We also include a chapter "In Memoriam" of Prof. Antigone Zacharopoulou. It contains contributions from the Tephritid Workers of the Western Hemisphere (TWWH) and the Tephritid Workers of Europe, Africa, and the Middle East (TEAM) regional meetings held in 2024. Since the submission's deadline was in 2024, the authors had the opportunity to update the content of their chapters.

All chapters were peer reviewed by at least two experts in the field. Reviewers were (in alphabetical order):

Abdeljelil Bakri (Morocco), Brian Barnes (South Africa), Bernard Dominiak (Australia), Francisco Devescovi (Argentina), Walther Enkerlin (FAO/IAEA), Paola Fedyszak (Argentina), Yoav Gazit (Israel), Yeudiel Gomez-Simuta (Mexico), Robert Griffin (USA), Nicholas Manoukis (USA), Pablo Montoya (Mexico), Pablo Liedo (Mexico), Diana Pérez-Staples (Mexico), Jesus Reyes Flores (Mexico), Diego Segura (Argentina), Preeaduth Sookar (Mauritius), Gustavo Taret (Argentina), Florencia Vásquez (Argentina), and María Teresa Vera (FAO/IAEA).

Thank you very much to all reviewers!

We are also thankful to Abraham Moreno-Mejía and Dinesh Rao for help with figures, and Helena Ajuria for help during the editorial process and to Anahita Nanda, Leon Bijnsdorp, Randy Brehm, and Kaustav Ghosh of Taylor & Francis.

Finally, we would like to thank our families during this process for their invaluable support, in particular, Dinesh, Maya, and Lila Rao.

Editorial Team
Diana Pérez-Staples
Bernard Dominiak
María Teresa Vera
Abdeljelil Bakri

Editor biographies

Diana Pérez-Staples is a Research Professor at the Universidad Veracruzana in Xalapa, Veracruz, Mexico. Her research focuses on the sexual behavior and physiology of tephritid fruit flies and other insect pests and on improving current control methods, with over 90 published papers in scientific journals. She is a member of the International Fruit Fly Steering Committee.

Bernard Dominiak works for the New South Wales Department of Primary Industries, and he is interested in dispersal, quarantine distances, and improved market access conditions. He has experience in managing trade zones, surveillance, sterile insect rearing, and release, with over 100 published papers in scientific journals. He was part of the organizing committee for the 2022 symposium.

María Teresa Vera is a Research Entomologist at the Insect Pest Control Section, Joint FAO/IAEA Division of Nuclear Techniques in Food and Agriculture, Vienna, Austria. Her research focuses on fruit fly reproductive biology and the assessment of sexual competitiveness for the implementation of sterile insect technique, with over 70 published papers in scientific journals. She is on the Editorial Board of the Revista Agronómica del Noroeste Argentino. She is a member of the International Fruit Fly and the Working Group on Fruit Flies of the Western Hemisphere Steering Committees.

Abdeljelil Bakri is a consultant on the control of tephritid fruit flies to reduce their impact on the production of fresh fruits and vegetables. He has been a professor at Cadi Ayyad University since 1983 and founded the Insect Biological Control Unit, dedicated to the study of insect communication and behavior, as well as the development of chemical attractants and insect control methods. He has worked with the International Atomic Energy Agency in Vienna since 1999 on international projects regarding SIT and developed databases to provide information to IAEA Member States. He has served on various committees and is an active member of the International Fruit Fly Steering Committee. He graduated from several universities, including Southampton and Poitiers.

Contributors

Hans T. Alborn
United States Department of Agriculture
Agricultural Research Service Center for Medical, Agricultural and Veterinary Entomology
Gainesville, Florida, USA

Henrique Martinelli Amaral
Center for Nuclear Energy in Agriculture (CENA)
University of São Paulo (USP)
Piracicaba, Brazil

Yassine Aoutil
Medfly Sterile Insect Unit
Maroc Citrus
Agadir, Morocco

Alexandre Santos Araújo
Department of Entomology and Acarology
Luiz de Queiroz College of Agriculture (ESALQ)
University of São Paulo (USP)
São Paulo, Brazil

Emilio Arévalo Peñaranda
Retired, formerly of Colombian Agricultural Institute (ICA).
Medellín, Colombia

Abdeljelil Bakri
Firdaous
Lot Ibn Khaldoun
Marrakech, Morocco

Rob Baxter
Department of Primary Industries and Regional Development
East Perth, Western Australia, Australia

Siham Boujdig
The National Office of Food Safety (ONSSA)
Rabat, Morocco

Kostas Bourtzis
Insect Pest Control Laboratory
Joint FAO/IAEA Centre of Nuclear Techniques in Food and Agriculture
Seibersdorf, Austria

Iara Sordi Joachim Bravo
Institute of Biology
Federal University of Bahia (UFBA)
Salvador, Brazil

Sonya Broughton
Department of Primary Industries and Regional Development
East Perth, Western Australia, Australia

Robert G. Bruton
United States Department of Agriculture
Agricultural Research Service Center for Medical, Agricultural and Veterinary Entomology
Gainesville, Florida, USA

Carlos Cáceres
Retired, formerly of Joint FAO/IAEA Centre of Nuclear Techniques in Food and Agriculture
IAEA
Vienna, Austria

Rene Cano-Pérez
Servicios Mubarqui S. de R.L. de C.V.
Tamaulipas, Mexico

Jeremías Nitor Carrillo
Servicio Agrícola y Ganadero (SAG)
Ministerio de Agricultura del Gobierno de Chile
Santiago, Chile

Maria de Lourdes Zamboni Costa
Center for Nuclear Energy in Agriculture (CENA)
University of São Paulo (USP)
Piracicaba, Brazil

David A. Cousins
Department of Primary Industries and Regional Development
East Perth, Western Australia, Australia

Jane Larissa de Melo Custódio
Institute of Biology
Federal University of Bahia (UFBA)
Salvador, Brazil

Zhor Dehbi
Regional Directorate of Agriculture Souss Massa
Agadir, Morocco

Hélène Delatte
UMR PVBMT
FOFIFA CENRADERU-DRA
Ambatobe, Madagascar
and
CIRAD
UMR PVBMT
La Reunion, France

Vanessa Simões Dias
Insect Pest Control Laboratory
Joint FAO/IAEA Centre of Nuclear Techniques in Food and Agriculture
IAEA
Vienna, Austria

Bernard Dominiak
NSW Department of Primary Industries
Orange, New South Wales, Australia

Paulina Concha Egaña
Servicio Agrícola y Ganadero (SAG)
Ministerio de Agricultura del Gobierno de Chile
Santiago, Chile

Walther Enkerlin
Insect Pest Control Section
Joint FAO/IAEA Centre of Nuclear Techniques in Food and Agriculture
Department of Nuclear Sciences and Applications
International Atomic Energy Agency
Vienna, Austria

Paola Fedyszak
National Service for Agrifood Health and Quality (SENASA)
National Plant Protection Directorate (DNPV)
Ciudad Autónoma de Buenos Aires, Argentina

Ruth Castro Feria
Servicio Agrícola y Ganadero (SAG)
Ministerio de Agricultura del Gobierno de Chile
Santiago, Chile

Jocelyn Yevenes Flores
Servicio Agrícola y Ganadero (SAG)
Ministerio de Agricultura del Gobierno de Chile
Santiago, Chile

Everett Foreman
United States Department of Agriculture
Agricultural Research Service Center for Medical, Agricultural and Veterinary Entomology
Gainesville, Florida, USA

Esteban Garavelli
National Service for Agrifood Health and Quality (SENASA)
National Plant Protection Directorate (DNPV)
Ciudad Autónoma de Buenos Aires, Argentina

Rubén Luis Leal-de la Garza
Servicios Mubarqui S. de R.L. de C.V.
Tamaulipas, Mexico

Paloma Guazzelli Della Giustina
Center for Nuclear Energy in Agriculture (CENA)
University of São Paulo (USP)
Piracicaba, Brazil

Juliet Goldsmith
CAHFSA
Suriname

Robert L. Griffin
Retired
Garner, North Carolina USA

Daniel A. Hahn
Department of Entomology and Nematology
University of Florida
Gainesville, Florida, USA

Alfred M. Handler
United States Department of Agriculture
Agricultural Research Service Center for Medical
Agricultural and Veterinary Entomology
Gainesville, Florida, USA

Darryl Hardie
Department of Primary Industries and Regional Development
East Perth, Western Australia, Australia

Richard B. Johnston
Department of Primary Industries and Regional Development
East Perth, Western Australia, Australia

Sija Kabota
Department of Crop Science and Horticulture
Sokoine University of Agriculture (SUA)
Morogoro, Tanzania
and
Research, Consultancy and Publication Unit
National Sugar Institute (NSI)
Kidatu-Morogoro, Tanzania

Adalécio Kovaleski
Empresa Brasileira de Pesquisa Agropecuária (EMBRAPA)
Estação Experimental de Fruticultura de Clima Temperado
Vacaria, Brazil

Abdul Kudra
Department of Crop Science and Horticulture
Sokoine University of Agriculture (SUA)
Morogoro, Tanzania

Vincent Lanoiselet
Department of Primary Industries and Regional Development
East Perth, Western Australia, Australia

Amine Leghlid
Medfly Sterile Insect Unit
Agadir, Morocco

Pablo Liedo
Ecología de Artrópodos y Manejo de Plagas
El Colegio de la Frontera Sur
Tapachula, Chiapas, Mexico

Luís Anselmo Lopez
Center for Nuclear Energy in Agriculture (CENA)
University of São Paulo (USP)
Piracicaba, Brazil

Daguang Lu
Insect Pest Control Subprogramme
FAO/IAEA Centre of Nuclear Techniques in Food and Agriculture
IAEA
Vienna, Austria

Yahya Machal
Medfly Sterile Insect Unit
Agadir, Morocco

Ramadhani Majubwa
Department of Crop Science and Horticulture
Sokoine University of Agriculture (SUA)
Morogoro, Tanzania

Richard W. Mankin
United States Department of Agriculture
Agricultural Research Service Center for Medical, Agricultural and Veterinary Entomology
Gainesville, Florida, USA

Aruna Manrakhan
Citrus Research International
Mbombela, South Africa
and
Department of Conservation Ecology and Entomology
Stellenbosch University
Matieland, South Africa

Thiago Mastrangelo
Center for Nuclear Energy in Agriculture (CENA)
University of São Paulo (USP)
Piracicaba, Brazil

Rosalie McCauley
Department of Primary Industries and Regional Development
East Perth, Western Australia, Australia

Marc De Meyer
Invertebrates Section and JEMU
Royal Museum for Central Africa
Tervuren, Belgium

Maulid Mwatawala
Department of Crop Science and Horticulture
Sokoine University of Agriculture (SUA)
Morogoro, Tanzania

Aaishah Ollee
Ministry of Agro Industry, Food Security, Blue Economy and Fisheries.
Réduit, Mauritius

Marcelo Contreras Ortiz
Servicio Agrícola y Ganadero (SAG)
Ministerio de Agricultura del Gobierno de Chile
Santiago, Chile

Nausheen Azhaar Patel
Entomology Division
Agricultural Services
Ministry of Agro Industry & Food Security
Réduit, Mauritius

Rui Pereira
Insect Pest Control Section
Join FAO/IAEA Division of Nuclear Techniques in Food and Agriculture
Vienna, Austria

Diana Pérez-Staples
Instituto de Biotecnología y Ecología Aplicada
Universidad Veracruzana
Xalapa, Mexico

Valeska Contreras Pizarro
Servicio Agrícola y Ganadero (SAG)
Ministerio de Agricultura del Gobierno de Chile
Santiago, Chile

Ana Julia Prestes
Center for Nuclear Energy in Agriculture (CENA)
University of São Paulo (USP)
Piracicaba, Brazil

Touhidur Rahman
Department of Primary Industries and Regional Development
East Perth, Western Australia, Australia

Henriette Rasolofoarivao
FOFIFA CENRADERU-DRA
Antananarivo, Madagascar and Département d'Entomologie, Faculté des Sciences
Université d'Antananarivo, B.P.
Antananarivo, Madagascar

Lala Harivelo Ravaomanarivo
Département d'Entomologie
Faculté des Sciences
Université d'Antananarivo, B.P.
Antananarivo, Madagascar

Ricardo Rodríguez Palomino
Retired, Formerly of Agriculture and Livestock Service (SAG) – Plant Protection Division (PPD)
Santiago, Chile

Alzira Kelly Passos Roriz
Institute of Biology
Federal University of Bahia (UFBA)
Salvador, Brazil

Mayren Sanchez
Ecología de Artrópodos y Manejo de Plagas
El Colegio de la Frontera Sur
Tapachula, Chiapas, Mexico

Alies van Sauers-Muller
Ministry of Agriculture, Animal Husbandry and Fisheries
Carambola Fruit Fly Research
Suriname

Marc F. Schetelig
Department of Insect Biotechnology in Plant Protection
Justus-Liebig-University Gießen
Gießen, Germany

Preeaduth Sookar
Entomology Division
Agricultural Services
Ministry of Agro Industry & Food Security
Réduit, Mauritius

Petronila Tarimo
Department of Crop Science and Horticulture
Sokoine University of Agriculture (SUA)
Morogoro, Tanzania

Nicholas M. Teets
Department of Entomology
University of Kentucky
Lexington, Kentucky, USA

Jorge Toledo
Ecología de Artrópodos y Manejo de Plagas
El Colegio de la Frontera Sur
Tapachula, Chiapas, Mexico

Nayadeth Álvarez Ugalde
Servicio Agrícola y Ganadero (SAG)
Ministerio de Agricultura del Gobierno de Chile
Santiago, Chile

María Florencia Vazquez
National Service for Agrifood Health and Quality (SENASA)
National Plant Protection Directorate (DNPV)
Ciudad Autónoma de Buenos Aires, Argentina

María Teresa Vera
Insect Pest Control Section
Joint FAO/IAEA Centre of Nuclear Techniques in Food and Agriculture
Department of Nuclear Sciences and Applications
International Atomic Energy Agency
Vienna, Austria

Massimiliano Virgilio
Invertebrates Section and JEMU
Royal Museum for Central Africa
Tervuren, Belgium

Jose Luis Zavala-Lopez
Servicios Mubarqui S. de R.L. de C.V.
Tamaulipas, Mexico

Jamaa Zim
Department of Plant Protection
Hassan II Institute of Agronomy and Veterinary Medicine
Agadir, Morocco

Hassan Zouhry
Medfly Sterile Insect Unit
Maroc Citrus
Agadir, Morocco

Chapter 1

In memoriam

Antigone Zacharopoulou

Kostas Bourtzis

Professor Antigone Zacharopoulou was born on 28 October 1945 in Agrinio, a town in Western Greece. She was born to parents who worked hard as farmers to provide for the family. Even though her parents had no educational background, she finished primary and secondary school in her hometown, and she then became the first and only child of the family (four sisters and one brother) who went to university, after having successfully passed competitive entrance exams. As a university student, Antigone studied Science with an emphasis on Biology at the Aristotle University of Thessaloniki, Central Macedonia, between 1965 and 1970. At that time, there was no Biology Department in Greece.

After she graduated, she applied for a teaching assistant position in the first, newly established Biology Department in Greece, in the University of Patras, where she remained for the rest of her career and life, for more than 50 years. From 1970 to 2022, she was elected as Lecturer (1982), Assistant Professor (1987), Associate Professor (1992), Professor (1999) and Emeritus Professor upon her retirement (2011). During her time as a teaching assistant in the Genetics Laboratory of the Department, she naturally got interested in genetics and scientific research. With the encouragement of Prof. Pelecanos, she started studying the evolution and characterization of chromosomal rearrangements in the model organism *Drosophila melanogaster,* and she received her PhD in 1978 (Zacharopoulou 1978, 1987; Zacharopoulou and Pelecanos 1980). She continued using *Drosophila* as a model organism, and in collaboration with Prof. George Yannopoulos, she studied the role of transposable elements in the induction of chromosomal rearrangements in laboratory and natural populations of *D. melanogaster* (Yannopoulos and Zacharopoulou 1980; Yannopoulos et al. 1980, 1982).

During the mid-1980s, she decided to shift from the *Drosophila* research field to a long and successful career working with true fruit flies, and mainly on the tephritid *Ceratitis capitata*, also known as Mediterranean fruit fly (medfly). She pioneered the field of cytogenetics in tephritids by studying the mitotic and polytene chromosomes of many species including C. *capitata* and other *Ceratitis* species (members of the FAR complex), the olive fruit fly *Bactrocera oleae*, the Queensland fruit fly *Bactrocera tryoni*, the oriental fruit fly *Bactrocera dorsalis* and other related *Bactrocera* and *Zeugodacus* species, the Mexican fruit fly *Anastrepha ludens*, the South American fruit fly *Anastrepha fraterculus sp.1* and *Drosophila suzukii* (Zacharopoulou 1987, 1990; Bedo and Zacharopoulou 1988; Zacharopoulou et al. 1991a, 2011a, 2011b; Mavragani-Tsipidou et al. 1992; Semeshin et al. 1995; Zhao et al. 1998; Gariou-Papalexiou et al. 1999, 2001, 2016;

DOI: 10.1201/9781003644330-1

Garcia-Martinez et al. 2009; Drosopoulou et al. 2017, 2019a, 2019b; Gouvi et al. 2022). She also studied the mitotic and polytene chromosomes of several tsetse species of the genus *Glossina*, which are major vectors of African trypanosomes causing human (sleeping sickness) and animal (nagana) trypanosomiasis (Gariou-Papalexiou et al. 2002b, 2007).

Antigone participated in many genomic and genetic studies in several tephritid species in collaboration with many Greek and foreign colleagues. These studies contributed to the detection, isolation and characterization of several markers and genes in the medfly using classical genetic, cytogenetic and molecular approaches, as well as the characterization of the nuclear and mitochondrial genomes of many species including *C. capitata*, *B. dorsalis*, *B. carambolae*, *B. oleae* and *Zeugodacus cucurbitae* (Malacrida et al. 1992; Zacharopoulou et al. 1992; Zwiebel et al. 1995; Papadimitriou et al. 1998; Thimianou et al. 1998; Verras et al. 1999; Zambetaki et al. 1999; Rosetto et al. 2000; Michel et al. 2001; Gariou-Papalexiou et al. 2002a; Stratikopoulos et al. 2002, 2009; Augustinos et al. 2002, 2005, 2008; Gomulski et al. 2004; Drosopoulou et al. 2019a). Her many years of work on the genetic and cytogenetic analysis of *C. capitata* resulted in the development of an integrated genetic and cytogenetic map for this species based on microsatellite, gene and morphological markers, which made a major contribution to the genome assembly of this species (Stratikopoulos et al. 2008; Papanicolaou et al. 2016).

Antigone's outstanding cytogenetic expertise on tephritids was one of the major tools in the integrated multidisciplinary approach employed for the species resolution in the *B. dorsalis* species complex. Through thorough cytogenetic analysis, no chromosomal rearrangements were found among *B. dorsalis*, *B. invadens*, *B. philippinensis* and *B. papayae*. This finding, together with additional experimental data, resulted in the synonymization of the three later "species" with *B. dorsalis*, which had an enormous impact on pest management strategies and global trade (Augustinos et al. 2014, 2015; Schutze et al. 2015). Similarly, cytogenetic analysis of pure lines and of their hybrids contributed to the resolution of two of the morphotypes existing in the *Anastrepha fraterculus* species complex, clearly indicating that the Argentinean (*A. fraterculus sp. 1*) and the Peruvian populations (morphotypes) should be considered distinct species (Cáceres et al. 2009). Also, it is important to note that the extended polytene chromosome analysis of numerous tephritids, accompanied by *in situ* hybridization of genes and other markers, has demonstrated the extended synteny of chromosomes within tephritids and with the proposed Muller elements of *Drosophila*. This pioneering work has been verified by extensive recent data derived from the field of genomics (Gariou-Papalexiou et al. 2002a; Zacharopoulou et al. 2017).

Antigone had an almost four-decade-long collaboration with the Insect Pest Control Subprogramme of the Joint FAO/IAEA Centre of Nuclear Techniques in Food and Agriculture (Figure 1.1).

This collaboration started in the 1980s and included a sabbatical leave at the Insect Pest Control Laboratory. This long-lasting collaboration provided useful tools for genomic and genetic studies, including the construction of mitotic and polytene chromosome maps for several major tephritid pest species. Additionally, this collaboration played a catalytic role in the development and characterization of genetic sexing strains (GSSs), which are used in sterile insect technique applications against them globally. Antigone and her team characterized several versions of the medfly

Figure I.1 Prof. Antigone Zacharopoulou at one of the insect rearing rooms of the Genetics and Molecular Biology Group, Insect Pest Control Laboratory, Seibersdorf, Austria.

VIENNA GSS, including VIENNA 7 and VIENNA 8, and they also developed an inversion (Inv D53), which has successfully suppressed genetic recombination and increased the genetic stability of these strains (Kerremans et al. 1990; Zacharopoulou et al. 1991b; Gariou-Papalexiou et al. 2002a; Augustinos et al. 2017; Zacharopoulou et al. 2017). In addition, Antigone's team and collaborators developed the first and, so far, only balancer chromosome in *C. capitata*, which has been used in several genetic studies including the development of new GSSs in this species (Gubb et al. 1998; Gourzi et al. 2000; Porras et al. 2020). Also, Antigone contributed to the characterization of GSSs developed for other major tephritid pest species such as *A. ludens*, *A. fraterculus*, *B. dorsalis* and *Z. cucurbitae* (Zacharopoulou and Franz 2012; Zepeda-Cisneros et al. 2014; Meza et al. 2020; Ramírez-Santos et al. 2021).

Antigone remained an active researcher even after her retirement. She kept going to the lab, dissecting, looking for nuclei and chromosomes and *in situ* signals under the microscope, taking photos and making polytene chromosome maps! (Figure 1.2)

She published more than 70 peer-reviewed manuscripts. The last one, which focused on the comparative analysis of polytene chromosomes of *Z. curcubitae* and *Z. tau*, was published just one month before she passed away (Gouvi et al. 2022). In recognition of her outstanding, lifelong scientific contributions to the tephritid research field, the Greek community of tephritid workers/researchers, collaborators and ex-students honoured Antigone in a special ceremony in October 2017 and presented her a "plaquette".

Antigone was also a great teacher. Her Genetics 101 course was outstanding and inspired many students, including me, to follow her steps and do our BSc, MSc or PhD thesis under her supervision. Antigone was a charismatic teacher with a unique

Figure 1.2 Prof. Antigone Zacharopoulou sorting photographs of polytene chromosomes of tephritid species at her laboratory in the Department of Biology, University of Patras, Patras, Greece.

personality. You would always see her being surrounded by students not only asking her questions about their experiments but also seeking her advice on both professional and personal matters. She was always available to listen and support.

On a personal note, Antigone was not just my PhD supervisor, she was my mentor! She truly shaped not only my professional career but also my personality as a human being. She loved travelling, music and dancing, but above all she loved being with friends, (ex-)students and collaborators, enjoying nice food and a glass of nice wine or, even better, "tsipouro". She was also an excellent cook! But above all, she was amazingly generous in her friendships and collaborations! Her kindness, warmth and humanity will never be forgotten!

Antigone passed away on Friday, May 27, 2022. She has left a deep empty space among all of us who knew her. We miss her so much...

ACKNOWLEDGEMENTS

I would like to thank Dr. Antonios Augustinos and Dr. Angeliki Gariou-Papalexiou for our discussions about the life and scientific career of our beloved teacher, their comments on the text and providing the photos included in this manuscript.

REFERENCES

Augustinos, A. A., E. Drosopoulou, A. Gariou-Papalexiou, et al. 2014. The *Bactrocera dorsalis* species complex: Comparative cytogenetic analysis in support of sterile insect technique applications. *BMC Genetics* 15(Suppl. 2): S16.

Augustinos, A. A., E. Drosopoulou, A. Gariou-Papalexiou, et al. 2015. Cytogenetic and symbiont analysis of five members of the *B. dorsalis* complex (Diptera, Tephritidae): No evidence of chromosomal or symbiont-based speciation events. *Zookeys* 540:273–298.

Augustinos, A. A., Z. Mamuris, E. Stratikopoulos, et al. 2005. Microsatellite analysis of olive fly populations in the Mediterranean indicates a westward expansion of the species. *Genetica* 125(2–3):231–241.

Augustinos, A. A., E. E. Stratikopoulos, E. Drosopoulou, et al. 2008. Isolation and characterization of microsatellite markers from the olive fly, *Bactrocera oleae*, and their cross-species amplification in the Tephritidae family. *BMC Genomics* 9:618.

Augustinos, A. A., E. Stratikopoulos, A. Zacharopoulou, et al. 2002. Polymorphic microsatellite markers in the olive fly, *Bactrocera oleae*. *Molecular Ecology Notes* 2:278–280.

Augustinos, A. A., A. Targovska, E. Cancio-Martinez, et al. 2017. *Ceratitis capitata* genetic sexing strains: Laboratory evaluation of strains from mass rearing facilities worldwide. *Entomologia Experimentalis et Applicata* 164:305–317.

Bedo, D., and A. Zacharopoulou. 1988. Inter-tissue variability of polytene chromosome banding patterns. *Trends in Genetics* 4:90–91.

Cáceres, C., D. F. Segura, T. Vera, et al. 2009. Incipient speciation revealed by studies on mating compatibility, sex pheromones, hybridization and cytology, between laboratory strains of *Anastrepha fraterculus* from Peru and Argentina. *Biological Journal of the Linnean Society* 97:152–165.

Drosopoulou, E., A. Gariou-Papalexiou, E. Karamoustou, et al. 2019a. The chromosomes of Drosophila suzukii (Diptera: Drosophilidae): Detailed photographic polytene chromosomal maps and in situ hybridization data. *Molecular Genetics and Genomics* 294(6):1535–1546.

Drosopoulou, E., C. Pantelidou, A. Gariou-Papalexiou, et al. 2017. The chromosomes and the mitogenome of Ceratitis fasciventris (Diptera: Tephritidae): Two genetic approaches towards the Ceratitis FAR species complex resolution. *Scientific Reports* 7(1):4877.

Drosopoulou, E., A. Syllas, P. Goutakoli, et al. 2019b. The complete mitochondrial genome of *Bactrocera carambolae* (Diptera: Tephritidae): Genome description and phylogenetic implications. *Insects* 10(12):4.

Garcia-Martinez, V., E. Hernandez-Ortiz, C. S. Zepeta-Cisneros, et al. 2009. Mitotic and polytene chromosome analysis in the Mexican fruit fly, *Anastrepha ludens* (Loew) (Diptera: Tephritidae). *Genome* 52(1):20–30.

Gariou-Papalexiou, A., G. Chrysanthis, A. Mintzas, et al. 1999. Patterns of puffing activity in the salivary gland polytene chromosomes of the medfly *Ceratitis capitata*, during the larval and prepupal development. *Genome* 42:919–929.

Gariou-Papalexiou, A., M. C. Giardini, A. A. Augustinos, et al. 2016. Cytogenetic analysis of the South American fruit fly *Anastrepha fraterculus* (Diptera: Tephritidae) species complex: Construction of detailed photographic polytene chromosome maps of the Argentinian Af. sp.1 member. *PLOS One* 11(6):e0157192.

Gariou-Papalexiou, A., P. Gourzi, A. Delprat, et al. 2002a. Polytene chromosomes as tools in the genetic analysis of the Mediterranean fruit fly *C. capitata*. *Genetica* 116:59–71.

Gariou-Papalexiou, A., A. C. Mintzas, and A. Zacharopoulou. 2001. Ecdysone regulated puffs in the Medfly *Ceratitis capitata*. *Genome* 44:752–763.

Gariou-Papalexiou, A., G. Yannopoulos, A. S. Robinson, et al. 2007. Polytene chromosome maps in four species of tsetse flies Glossina austeni, G. pallidipes, G. morsitans morsitans and G. m. submorsitans (Diptera: Glossinidae): A comparative analysis. *Genetica* 129(3):243–251.

Gariou-Papalexiou, A., G. Yannopoulos, A. Zacharopoulou, et al. 2002b. Photographic polytene chromosome maps for *Glossina morsitans* submorsitans (Diptera: Glossinidae): Cytogenetic analysis of a colony with sex-ratio distortion. *Genome* 45:871–880.

Gomulski, L. M., S. Brogna, A. Babaratsas, et al. 2004. Molecular basis of the size polymorphism of the first intron of the Adh-1 gene of the Mediterranean fruit fly, *Ceratitis capitata*. *Journal of Molecular Evolution* 58(6):732–742.

Gourzi, P., D. Gubb, Y. Livadaras, et al. 2000. The construction of the first balancer chromosome for the Mediterranean fruit fly *Ceratitis capitata*. *Molecular and General Genetics MGG* 264:127–136.

Gouvi, G., A. Gariou-Papalexiou, A. A. Augustinos, et al. 2022. The chromosomes of *Zeugodacus tau* and *Zeugodacus cucurbitae*: A comparative analysis. *Frontiers in Ecology and Evolution* 10:854723.

Gubb, D., A. Zacharopoulou, Y. Livadaras, et al. 1998. Recovery of a marked translocation strain that will facilitate the isolation of balancer chromosomes in the Mediterranean fruit fly *Ceratitis capitata*. *Genome* 41:256–265.

Kerremans, Ph., K. Bourtzis, and A. Zacharopoulou. 1990. Cytogenetic analysis of three genetic sexing strains of *Ceratitis capitata*. *Theoretical and Applied Genetics* 80:177–182.

Malacrida, A., G. Gasperi, A. Zacharopoulou, et al. 1992. Evidence for a genetic duplication involving alcohol dehydrogenase genes in *Ceratitis capitata*. *Biochemical Genetics* 30:35–48.

Mavragani-Tsipidou, P., G. Karamanlidou, A. Zacharopoulou, et al. 1992. Mitotic and polytene chromosome analysis in *Dacus oleae* (Diptera: Tephitidae). *Genome* 35:373–378.

Meza, J. S., K. Bourtzis, A. Zacharopoulou, et al. 2020. Development and characterization of a pupal-colour based genetic sexing strain of *Anastrepha fraterculus* sp. 1 (Diptera: Tephritidae). *BMC Genetics* 21(Suppl 2):134.

Michel, K., A. Stamenova, A. C. Pinkerton, et al. 2001. Hermes-mediated germ line tranformation of the Mediterranean fruit fly *Ceratitis capitata*. *Insect Molecular Biology* 10:155–162.

Papadimitriou, E., D. Kritikou, M. Mavroidis, et al. 1998. The heat shock 70 gene family in the Mediterranean fruit fly Ceratitis capitata. *Insect Molecular Biology* 7:279–290.

Papanicolaou, A., M. F. Schetelig, P. Arensburger, et al. 2016. The Whole Genome Sequence of the Mediterranean fruit fly, *Ceratitis capitata* (Wiedemann), reveals insights into the biology and adaptive evolution of a highly invasive pest species. *Genome Biology* 17(1):192.

Porras, M. F., J. S. Meza, E. G. Rajotte, et al. 2020. Improving the phenotypic properties of the Ceratitis capitata (Diptera: Tephritidae) temperature sensitive lethal genetic strain in support of sterile insect technique applications. *Journal of Economic Entomology* 113:2688–2694.

Ramírez-Santos, E., P. Rendón, G. Gouvi, et al. 2021. A novel genetic sexing strain of *Anastrepha ludens* for cost-effective sterile insect technique applications: Improved genetic stability and rearing efficiency. *Insects* 12:499.

Rosetto, M., T. de Filippis, M. Mandrioli, et al. 2000. Ceratotoxins: Female specific X-linked genes from the medfly, *Ceratitis capitata*. *Genome* 43:707–711.

Schutze, M. K., N. Aketarawong, W. Amornsak, et al. 2015. Synonymization of key pest species within the *Bactrocera dorsalis* species complex (Diptera: Tephritidae): Taxonomic changes based on 20 years of integrative morphological, molecular, cytogenetic, behavioral, and chemoecological data. *Systematic Entomology* 40:456–471.

Semeshin, V. F., D. Kritikou, A. Zacharopoulou, et al. 1995. Electron microscope investigation of polytene chromosomes in the Medfly *Ceratitis capitata*. *Genome* 38:652–660.

Stratikopoulos, E. E., A. A. Augustinos, A. Gariou-Papalexiou, et al. 2002. Identification and partial characterization of a new medfly- specific 44–bp centromeric repeat. *Chromosome Research* 10:287–295.

Stratikopoulos, E. E., A. A. Augustinos, I. D. Pavlopoulos, et al. 2009. Isolation and characterization of microsatellite markers from the Mediterranean fruit fly, *Ceratitis capitata*: Cross-species amplification in other Tephritidae species reveals a varying degree of transferability. *Molecular Genetics and Genomics* 282(3):283–306.

Stratikopoulos, E. E., A. A. Augustinos, Y. G. Petalas, et al. 2008. An integrated genetic and cytogenetic map for the Mediterranean fruit fly, *Ceratitis capitata*, based on microsatellite and morphological markers. *Genetica* 133(2):147–157.

Thimianou, S., M. Mavroidis, G. Kokolakis, et al. 1998. Cloning and characterization of a cDNA encoding a male specific serum protein of the Mediterranean fruit fly *Ceratitis capitata* with sequence similarity to odorant-binding proteins. *Insect Molecular Biology* 7:345–353.

Verras, M., M. Mavroidis, G. Kokolakis, et al. 1999. Cloning and characterization of CcEcR. An edysone receptor homolog from the Mediterranean fruit fly *Ceratitis capitata. European Journal of Biochemistry* 265:798–808.

Yannopoulos, G., M. Pelecanos, and A. Zacharopoulou. 1980. Combined action of diethyl-sulphate (DES) and of the male recombination factor 31,1 MRF in *Drosophila melanogaster. Genetica* 12:41–48.

Yannopoulos, G., N. Stamatis, and A. Zacharopoulou. 1982. Genetic studies of a cytoplasmic suppressor which inactivates the male recombination from 31.1 MRF in *Drosophila melanogaster. Heredity* 49:235–241.

Yannopoulos, G. and A. Zacharopoulou. 1980. Studies on the chromosomal rearrangements induced by the male recombination factor 31.1 MRF in *Drosophila melanogaster. Mutation Research* 73:81–92.

Zacharopoulou, A. 1978. Study of the polymorphism of chromosomal inversions in a Greek population of *Drosophila melanogaster*. PhD thesis, University of Patras, Greece (in Greek).

Zacharopoulou, A. 1987. Cytogenetic analysis of mitonic and salivary gland chromosomes in the Medfly *Ceratitis capitata. Genome* 29:67–71.

Zacharopoulou, A. 1990. Polytene chromosome maps in *C. capitata. Genome* 33:184–197.

Zacharopoulou, A., A. A. Augustinos, E. Drosopoulou, et al. 2017. More than thirty years of cytogenetic studies of Tephritidae in support of sterile insect technique and global trade. *Entomologia Experimentalis et Applicata* 164:204–225.

Zacharopoulou, A., A. A. Augustinos, W. A. A. Sayed, et al. 2011a. Mitotic and polytene chromosomes analysis of the oriental fruit fly, *Bactrocera dorsalis* (Hendel) (Diptera: Tephritidae). *Genetica* 139(1):79–90.

Zacharopoulou, A., K. Bourtzis, and Ph. Kerremans. 1991a. A comparison of polytene chromosomes in salivary glands and orbital bristle trichogen cells in *Ceratitis capitata. Genome* 34:215–219.

Zacharopoulou, A., and G. Franz. 2012. Genetic and cytogenetic characterization of genetic sexing strains of *Bactrocera dorsalis* and *Bactrocera cucurbitae* (Diptera: Tephritidae). *Journal of Economic Entomology* 106 (2):995–1003.

Zacharopoulou, A., M. Frisardi, C. Savakis, et al. 1992. The genome of the Mediterranean fruit fly Ceratitis capitata: Localization of molecular markets by in situ hybridization to salivary-gland polytene chromosomes. *Chromosoma* 101:448–455.

Zacharopoulou, A., and M. Pelecanos. 1980. Seasonal and year-to year inversion polymorphism in a Southern Greek Drosophila melanogaster wild population. *Genetica* 54:105–111.

Zacharopoulou, A., M. E. Riva, A. Malacrida, et al. 1991b. Cytogenetic characterization of a genetic sexing strain of *Ceratitis capitata. Genome* 34:606–611.

Zacharopoulou, A., W. A. A. Sayed, A. A. Augustinos, et al. 2011b. Analysis of mitotic and polytene chromosomes and photographic polytene chromosome maps in *Bactrocera cucurbitae* (Diptera: Tephritidae). *Annals of the Entomological Society of America* 104(2):306–318.

Zambetaki, A., A. Zacharopoulou, Z. Scouras, et al. 1999. The genome of the olive fruit fly *Bactrocera oleae*: Localization of molecular markers by in situ to the salivary gland polytene chromosomes. *Genome* 42:744–751.

Zepeda-Cisneros, C. S., J. S. M. Hernández, V. García-Martínez, et al. 2014. Development, genetic and cytogenetic analyses of genetic sexing strains of the Mexican fruit fly, *Anastrepha ludens* Loew (Diptera: Tephritidae). *BMC Genetics* 15: S1.

Zhao, J. T., M. Frommer, J. A. Sved, et al. 1998. Mitotic and polytene chromosome analysis in the Queensland fruit fly, *Bactrocera tryoni* (Diptera: Tephritidae). *Genome* 41:510–526.

Zwiebel, J., G. Saccone, A. Zacharopoulou, et al. 1995. The white gene of *Ceratitis capitata*: A phenotype marker for germline transformation. *Science* 270:2005–2008.

Section I

Biology, ecology, physiology and behavior

Chapter 2

Diversity and abundance of fruit flies (Diptera: Tephritidae) infesting cucurbit crops in the Morogoro region of Tanzania

Petronila Tarimo, Sija Kabota, Maulid Mwatawala, Ramadhani Majubwa, Abdul Kudra, Massimiliano Virgilio, and Marc De Meyer

2.1 INTRODUCTION

The "true" fruit flies are members of the Tephritidae family (Diptera) and include agricultural pests, that cause economic losses to farmers worldwide. There are approximately 5,000 known species of fruit flies worldwide, of which 200 species are the most destructive to fruit and vegetable crops (Papadopoulos et al. 2024). The intensity of fruit fly damage varies considerably among fruit fly species and across different regions of the world (Mwatawala et al. 2009; Papadopoulos et al. 2024). For instance, in Sub-Saharan Africa (SSA), fruit flies are responsible for up to 53% of fruit and vegetable losses (Ekesi et al. 2016; Gnanvossou et al. 2017; Zida et al. 2020). SSA is home to over 1,000 fruit fly species, of which more than 400 can develop in wild and cultivated fruits (Zida et al. 2020). While most of these species are indigenous to SSA, a few have exotic origins (White and Elson-Harris 1992; Papadopoulos et al. 2024).

Fruit fly infestations remain a significant challenge for farmers in SSA, as they cause extensive damage to horticultural crops (Vayssieres et al. 2008a; Nankinga et al. 2014; Ekesi et al. 2016). Several biodiversity monitoring studies have been conducted, which have played a crucial role in developing effective management strategies (Mwatawala et al. 2006; Ekesi et al. 2016; Gnanvossou et al. 2017; Kambura et al. 2018; Mokam et al. 2018). Previous research on tephritid pests in SSA largely focused on pests of fleshy fruits (see Mwatawala et al. 2006; Vayssieres et al. 2008a; Mkiga and Mwatawala 2015; Tanga and Rwomushana 2016; Gnanvossou et al. 2017) and described their species richness in different agroecological zones, using attractants such as para-pheromones, food baits, or collected samples from reared fruits (Geurts et al. 2012; Mokam et al. 2014; Kambura et al. 2018; Hanna et al. 2020). Only a few studies have focused on the diversity of fruit flies infesting cucurbit crops in SSA (Kambura et al. 2018). Consequently, information on these pests remains scarce (Tanga and Rwomushana 2016), limiting the development of comprehensive management plans to control them in the region.

Besides other factors, altitudinal differences between sites play a major role in influencing the diversity of fruit flies (Mwatawala et al. 2006; Geurts et al. 2012;

DOI: 10.1201/9781003644330-3

Salazar-Mendoza et al. 2021), as they determine the types of host plants that fruit flies can access and the necessary microclimatic conditions for their survival and breeding (Kounatidis et al. 2008; Vayssières et al. 2009; Hassani et al. 2022). Seasonal variation in climatic conditions is another factor that plays a significant role in shaping the diversity of fruit flies (Zida et al. 2020), as it directly impacts host availability, growth and development, and the distribution patterns of these insects (Vayssières et al. 2009; Hassani et al. 2016; Amin et al. 2019). Often, climatic factors such as temperature, rainfall and humidity dictate the phenology of host plants (Laskar and Chatterjee 2010; Ganie et al. 2013), creating temporal windows of resource abundance or scarcity that affect fruit fly populations and diversity (Zida et al. 2020). Despite the well-established presence of cucurbit-infesting fruit flies at all altitudes in the Morogoro region (Mwatawala et al. 2006), their spatial and temporal dynamics across mountainous and plateau agroecological zones remain largely unknown. Understanding these dynamics is crucial for developing effective and sustainable control strategies (Mwatawala et al. 2006). While fruit flies infesting fleshy fruits have been extensively studied in the region, there is a significant knowledge gap concerning the diversity patterns and dynamics of fruit flies that infest cucurbits, which needs to be addressed.

Proper assessment of diversity patterns of fruit flies infesting cucurbit crops requires sampling completeness and a combination of procedures for estimating species richness (Colwell et al. 2012; Mokam et al. 2014; Callaghan et al. 2019). According to Chao et al. (2009) and Chao and Shen (2004), sampling completeness can be determined using species accumulation curves (SACs) and nonparametric estimators of species richness. SACs allow the quantification of within-inventory efficacy, sampling completeness and estimation of the minimum sampling effort required (asymptotic species richness) (Chao et al. 2000, 2009; Mokam et al. 2014). In this study, we describe the diversity patterns of fruit flies infesting cucurbit crops in the Morogoro region of Tanzania, using diversity indices and nonparametric estimates of species richness in order to develop *ad hoc* and bespok management programmes for fruit flies infesting cucurbits in this region.

2.2 MATERIALS AND METHODS

2.2.1 Study site

Studies on the diversity and abundance of pestiferous fruit flies attacking cucurbit crops were conducted over two seasons in each year: May to June season (Season I) and October to November season (Season II) in 2020 and 2021 in the Morogoro Region, Eastern-Central Tanzania. These months were selected to coincide with the two cucurbit seasons per year in the region. The Morogoro region is in the transition zone between the bimodal and unimodal rainfall belts at S5°58′–S10°0′ and E35°25′–E38°30′ (United Republic of Tanzania (URT) 2002; Mwatawala et al. 2006). Study sites (plateau and mountainous) were selected based on their differences in climatic conditions (Table 2.1).

Five experimental plots located at least 1 km apart were identified in each agroecological zone (Table 2.2). Three cucurbitaceous crops were established per plot, viz, cucumber (*Cucumis sativus* L.) variety "Ashley," watermelon (*Citrullus lanatus* (Thunb.) Matsum. and Nakai) variety "Sugar baby," and squash (*Cucurbita moschata* D.)

Table 2.1 Agroecological zones of Morogoro region

Agroecological zone	*Characteristics*
Mountainous zone	Altitude: >600 m asl; average rainfall: 800 mm–2,500 mm p.a.; annual average temperature of 29°C.
Plateau zone	Altitude: 300–600 m asl; average rainfall: 700 mm–1,200 mm p.a.; annual average temperature of ~24°C

Source: Mwatawala et al. (2006) and United Republic of Tanzania (URT) (2002).

Table 2.2 Locations of the experimental plots established in two agroecological zones of Morogoro

Agroecological zones	*Location*	*Coordinates*	*Altitude*
Plateau	SUA Horticulture Unit (HT)	S06°50′41.4″ E37°39′43.3″	524 m
	SUA Crop Museum (CM)	S06°51′00.53″ E37°39′17.90″	528 m
	SUGECO (SG)	S06°50′22″ E37°38′42.2″	511 m
	SUA Mazimbu (MZ)	S06°47′26.208″ E37°38′7.926″	486 m
	SUA Mafiga (MF)	S06°50′22.764″ E37°37′53.46″	503 m
Mountainous	Morning Site (MS)	S06°53′17.9″ E37°40′14.93″	1274 m
	Mkumbulu (MK)	S06°52′24.2″ E37°40′21.5″	1105 m
	Ruvuma (RV)	S06°52′34.6″ E37°40′3.7″	995 m
	Kifuru (KF)	S06°53′32.1″ E37°40′9.5″	1418 m
	Mgola (MG)	S06°51′41.4″ E37°40′4.3″	1084 m

SUGECO stands for Sokoine University Graduate Entrepreneurs Cooperative.

variety "Waltham." Cucumber, watermelon and squash were each planted on an area of 1 120 m^2 (70 m×16 m) at a spacing of 50 cm×60 cm, 1 m×1.5 m and 1 m×1.5 m, respectively.

2.2.2 Sample collection

Sampling of fruit flies was carried out using traps baited with lures and by collecting and incubating infested fruits. Trapping was conducted using clusters of three Tephri-Traps (Sorygar, Spain), each baited with commercially available Cuelure (Scentry Biologicals, Billings, Montana, USA), Zingerone (ZN) (Sigma-Aldrich, St. Louis, Missouri, USA) and BioLure (BL) (Chempac (Pty) Ltd, Suterra LLC, USA), for eight weeks in each season. Traps were set following the guidelines recommended by the International Atomic Energy Agency (IAEA and FAO 2011). The traps were set in the ten established cucurbit plots. To minimize lure interference, within each plot traps were placed at a distance of at least 20 m (IAEA and FAO 2011). The lures were selected to attract taxa from the genera *Dacus* and *Zeugodacus*. Cuelure and Zingerone are male-specific each with a specific range of species that can be attracted (Manrakhan et al. 2017), while BioLure is a general attractant for both sexes and predominantly catches female fruit flies.

A set of three traps, each baited with one of the attractants, was placed in each experimental plot. Each trap was baited with one of the attractants placed on a tree branch or held on a wooden pole. A strip of DDVP (Vaportape II; Hercon Environmental, Emigsville, PA, USA) containing the insecticide dichlorvos was placed at the bottom of each trap to kill and retain trapped insects. Traps were inspected once a week and were randomly replaced within the plot after each inspection to minimise location bias. Attractants and insecticide strips were replaced every four weeks. Collected specimens were preserved in 70% ethanol. Trap catches were recorded as the number of fruit flies per trap per week.

In addition, ten fruits per crop species were sampled weekly for eight consecutive weeks throughout the two cucurbit seasons in each year from 2020 to 2021. A sample set of ten fruits per crop species was collected by randomly harvesting a mixture of immature and semimatured fruits still attached to the plants. Then, each sample was weighed and placed in a container to rear the infesting fruit flies following the method described by Copeland et al. (2002). The containers were fitted with mesh lids prevent emerging flies from escaping. They were kept at room temperature maintained between 23°C and 25°C. Sterilized sand was placed at the bottom of the container as pupation media.

Trap flies were recorded, and fruits were collected from May to June and October to November in 2020 and 2021. Collected fruits were incubated at the Sokoine University of Agriculture (SUA) rearing facility in Morogoro, utilizing (24.8×14.6×11 cm) plastic rearing containers. These containers were inspected daily for adult fruit flies until no more fruit flies emerged. At the end of the process, the samples were sieved, and if pupae were found, they were returned to the incubation containers until all pupae had transformed into adult flies (Mkiga and Mwatawala 2015; Mwatawala et al. 2010). Unmerged pupae were not recorded because it was impossible to confirm their species identity. The emerged fruit flies were preserved in absolute ethanol for subsequent morphological identification.

Fruit fly identification was conducted at the SUA entomology laboratory with the aid of a binocular stereo microscope using keys and characters described by White and Elson-Harris (1992), Billah et al. (2006) and Virgilio et al. (2014). Specimens were sent to the Royal Museum for Central Africa (RMCA) for confirmation of identification by Dr Marc De Meyer. Representative images of all collected species are publicly available at the website of the RMCA: https://virtualcol.africamuseum.be/ (Royal Museum for Central Africa (RMCA) 2020).

2.2.3 Data analysis

Sampling effort in each agroecological zone was evaluated using three non-parametric abundance-based estimators (ABEs) of species richness (Abundance-based Coverage Estimator (ACE), Chao 1 and Jackknife 1) following Chao et al. (2000), Chao and Shen (2004), and Colwell et al. (2012), and rarefaction (interpolation) and prediction (extrapolation) curves as described by Dove and Cribb (2006), Colwell et al. (2012) and Chao et al. (2014). Moreover, species diversity between the two agroecological zones was compared through both α and β diversity indices. Only catches from traps baited with Biolure and adults emerged from fruits were used, as this is a specific lure more suitable for comparisons of the relative abundance of species (Pogue et al. 2024).

The α-diversity between the two agroecological zones was assessed using the following indices (*Simpson 1949; Margalef 1958; Shannon and Weaver 1963; Pielou 1969*); -

$$\text{Shannon}\left(H' = \sum_{i=1}^{S} \frac{ni}{N} \ln \frac{ni}{N}\right) \tag{2.1}$$

$$\text{Inverse Simpson}\left(1/D = 1\bigg/\left(\sum_{i=1}^{S}\frac{ni}{N}\left(\frac{ni-1}{N-1}\right)\right)\right) \tag{2.2}$$

$$\text{Species richness index of Margalef}\left(\text{DMg} = \left(\frac{S-1}{\ln\ln N}\right)\right) \tag{2.3}$$

$$\text{Evenness index of Pielou}\left(J = \frac{H}{\ln\ln S}\right) \tag{2.4}$$

The computations of diversity indices were done using the Vegan package in R (Oksanen et al. 2018).

The normality of α-diversity data was preliminarily checked via the Shapiro–Wilk test. A two-way analysis of variance (ANOVA) was performed to determine the effects of season and agroecological zone on the α-diversity of dominant species. The homogeneity of variance was preliminarily checked using Cochran's C tests, and *post hoc* comparisons were performed using the Student–Newman–Keuls (SNK) method as implemented in the GAD package in R (Underwood et al. 2000; Sandrini-Neto and Camargo 2012). β-diversity was determined using the Sorensen and Jaccard indices (Magurran 1988). The seasonal abundances of dominant species were also investigated using the cumulative trapping data from Zingerone, Cuelure and Biolure baits.

2.3 RESULTS

2.3.1 Species abundances

The sampling effort allowed the trapping of 99.9% and 98.8% of all fruit fly species infesting cucurbit crops in the mountainous and plateau zones (Table 2.3). The asymptote of species richness was reached in all agroecological zones at 4,375 and 1,800 individuals for the mountainous and plateau zones, respectively (Figure 2.1).

Table 2.3 Estimators of species richness of fruit flies infesting cucurbit crops in two agroecological zones of Morogoro, Tanzania

Richness estimators	*Mountainous zone*	*Plateau zone*
Sobs	11.0	8.0
ACE	11.2	8.3
Chao1	11.0	8.0
Jack 1	11.0	8.0
Mean of the three ABE	11.0	8.1
Sampling efforts (%)	99.9	98.8

KEY: Sobs: number of species observed generated per Mao Tau.
ABE: Abundance-based estimators.
ACE: Abundance-based coverage estimator.

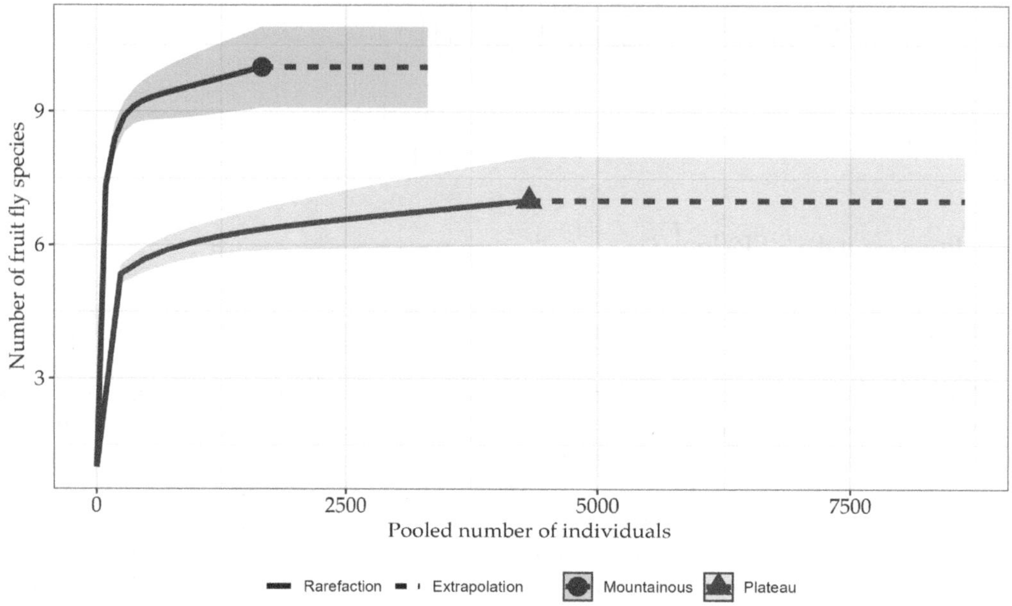

Figure 2.1 Rarefaction and extrapolation curves based on the number of fruit fly individuals collected in two of the agroecological zones of Morogoro, from May 2020 to November 2021.

The accumulation curves further indicated that the fruit fly species trapped in each of the agroecological zones represent more than 98% of the estimated fruit fly species in the sampling area.

In total, 60,906 fruit fly specimens from Biolure baits and fruits were collected from the ten cucurbit plots across the two agroecological zones (see Supplementary material 2.1). Specimens belonged to the three genera *Dacus, Zeugodacus* and *Bactrocera*. Of the total specimens collected, 61.73% were from the mountainous zone, and the remaining 38.27% were from the plateau zone. Species from the genus *Zeugodacus* were the most abundant and accounted for 71.08% of the total of specimens, while species from the *Dacus* and *Bactrocera* genera constituted the remaining 28.92% (see Supplementary material 2.1; Figure 2.2).

Furthermore, a total of 97,008 specimens were collected from all attractants (Biolure, Cuelure and Zingerone) plus reared fruits. Of these, 58,407 (60.21%) were from the mountainous zone and 38,601 (39.79%) were from the plateau zone. Among the fruit fly species caught, *Z. cucurbitae* (Coquillet) was the most abundant species, followed by *D. vertebratus* Bezzi and *D. bivittatus* (Bigot) (see Supplementary material 2.2). 24,807 (25.6% of the total catch) specimens were from lure-baited traps, and 74,201 (74.4%) were from collected fruits. Of the trapped specimens, 20,128 (81.14%) were from Cuelure, 3,166 (12.76%) were from Zingerone and 1,513 (6.1%) were from Biolure. In Zingerone, *Z. cucurbitae* was the most abundant, followed by *D. humeralis* (Bezzi) and *D. vertebratus* (Figure 2.3a). In Cuelure, *Zeugodacus cucurbitae* was the

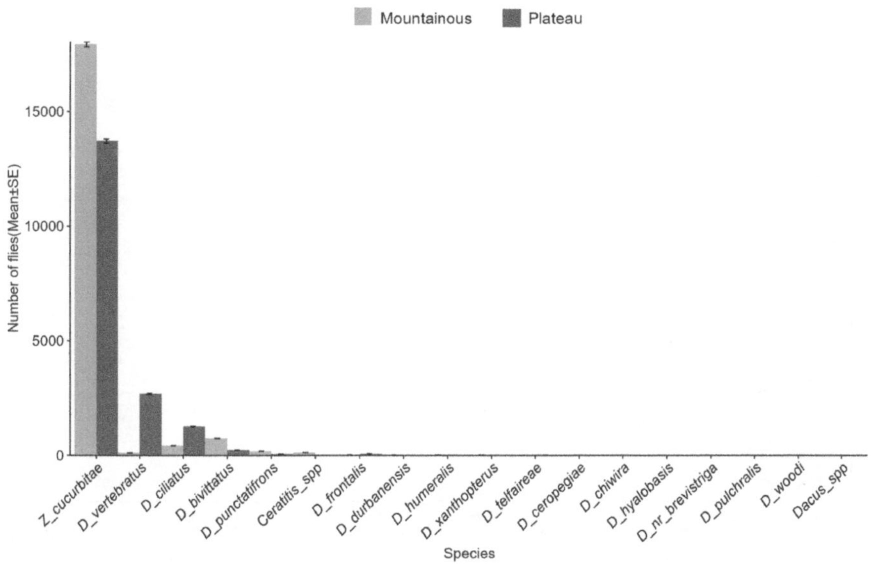

Figure 2.2 Abundance of fruit fly species infesting cucurbit crops in the mountainous zone of Morogoro, Tanzania, from May 2020 to November 2021.

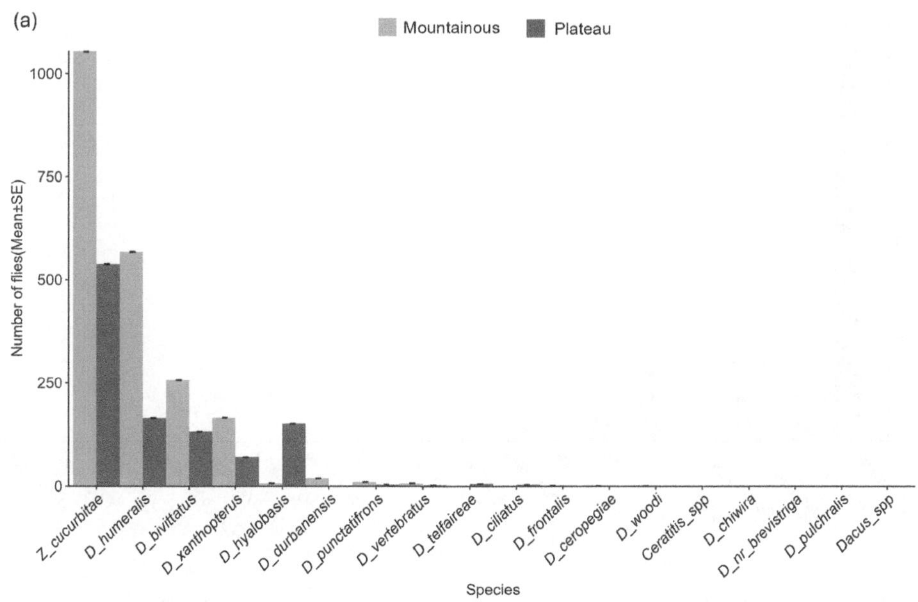

Figure 2.3 (a) Abundance of cucurbit infester species trapped by Zingerone in the two agroecological zones of Morogoro, Tanzania, from May 2020 to November 2021. (b) Abundance of cucurbit infester species trapped by Cuelure in the two agroecological zones of Morogoro, Tanzania, from May 2020 to November 2021. (c) Abundance of cucurbit infester species trapped by Biolure in the two agroecological zones of Morogoro, Tanzania, from May 2020 to November 2021.

(Continued)

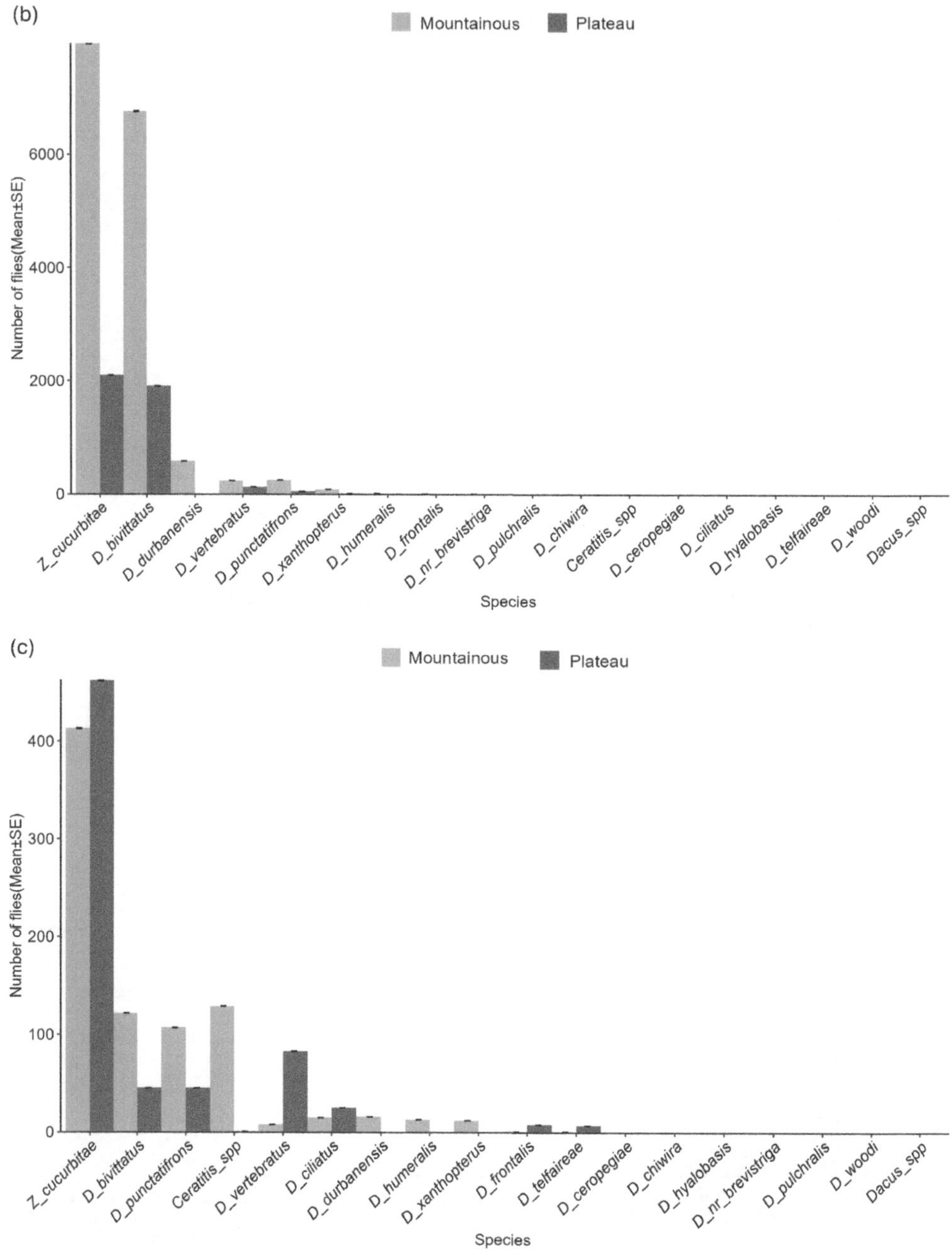

Figure 2.3 (Continued) (a) Abundance of cucurbit infester species trapped by Zingerone in the two agroecological zones of Morogoro, Tanzania, from May 2020 to November 2021. (b) Abundance of cucurbit infester species trapped by Cuelure in the two agroecological zones of Morogoro, Tanzania, from May 2020 to November 2021. (c) Abundance of cucurbit infester species trapped by Biolure in the two agroecological zones of Morogoro, Tanzania, from May 2020 to November 2021.

most abundant, followed by *D. vertebratus and D. durbanensis* Munro (Figure 2.3b), and in Biolure, *Z. cucurbitae* was followed by *D. vertebratus* and *D. ciliatus* Loew (Figure 2.3c).

Among the fruit flies reared out from collected fruits, 25,530 (35.36%) were recovered from squash, 17,512 (24.25%) from watermelon and 29,162 (40.39%) from cucumber (see Supplementary material 2.2).

2.3.2 Species diversity

The results demonstrated that fruit fly species richness, evenness and diversity, as measured by Shannon's and Inverse Simpson's indices, were higher in the plateau zone than in the mountainous zone (Table 2.4). These differences were statistically significant across agroecological zones and seasons (Table 2.5). Notably, the Shannon and Inverse Simpson diversities were significantly higher in the plateau zone during the June–August season (Season I) than in the mountainous zone (Figures 2.4 and 2.5). Additionally, species evenness was significantly higher in the plateau zone compared

Table 2.4 Mean alpha diversity index values

Agroecological zone	*Richness*	*Evenness*	*Shannon*	*Inverse Simpson*
Mountainous	2.756±0.112	0.314±0.022	0.334±0.023	1.208±0.039
Plateau	2.925±0.122	0.419±0.023	0.498±0.028	1.418±0.054

Table 2.5 Analysis of variance (ANOVA) of the effects of agroecological zone and season on species richness, evenness and diversity of fruit flies

		Statistics	
Variables	*Source of variation*	*F value*	*P-value*
Richness	Agroecological zone	0.083	0.774 ns
	Season	16.955	**<0.0001***
	Agroecological zone×season	1.499	0.2217 ns
Evenness	Agroecological zone	11.074	**0.0009***
	Season	0.001	0.9794 ns
	Agroecological zone×season	3.137	0.0775 ns
Shannon	Agroecological zone	10.549	**0.0013***
	Season	3.887	**0.0495***
	Agroecological zone×season	4.339	**0.0380***
Inverse Simpson	Agroecological zone	10.152	**0.0015***
	Season	5.665	0.0179*
	Agroecological zone×season	7.219	0.0076*

Note: values with * are significant at $P<0.05$; **ns**=not significant.

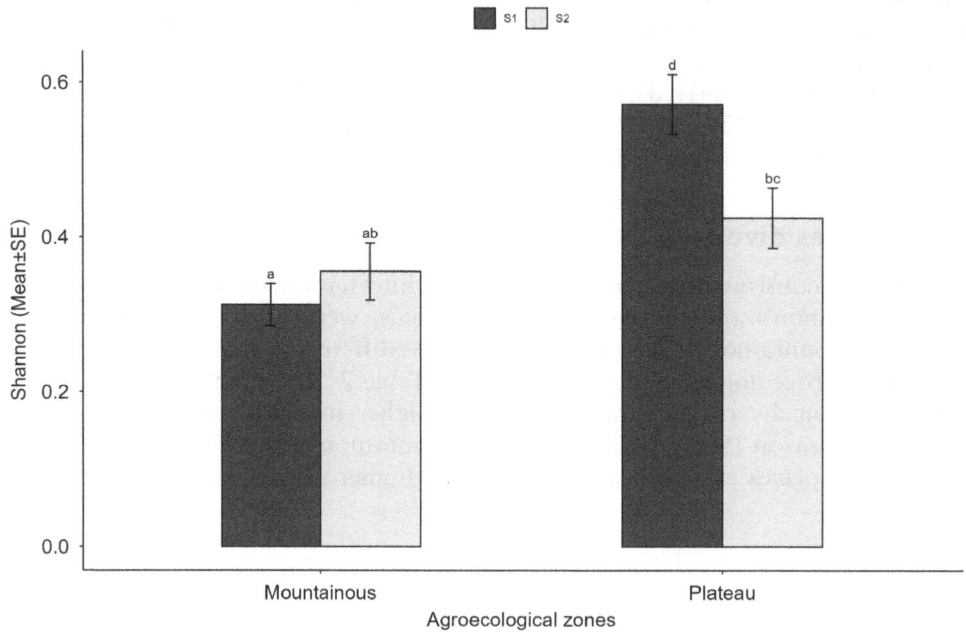

Figure 2.4 Shannon diversity of cucurbit-infesting species across the agroecological zones of Morogoro, Tanzania, from May 2020 to November 2021. Significance letters and standard errors are indicated (SI = season I, S2 = season II).

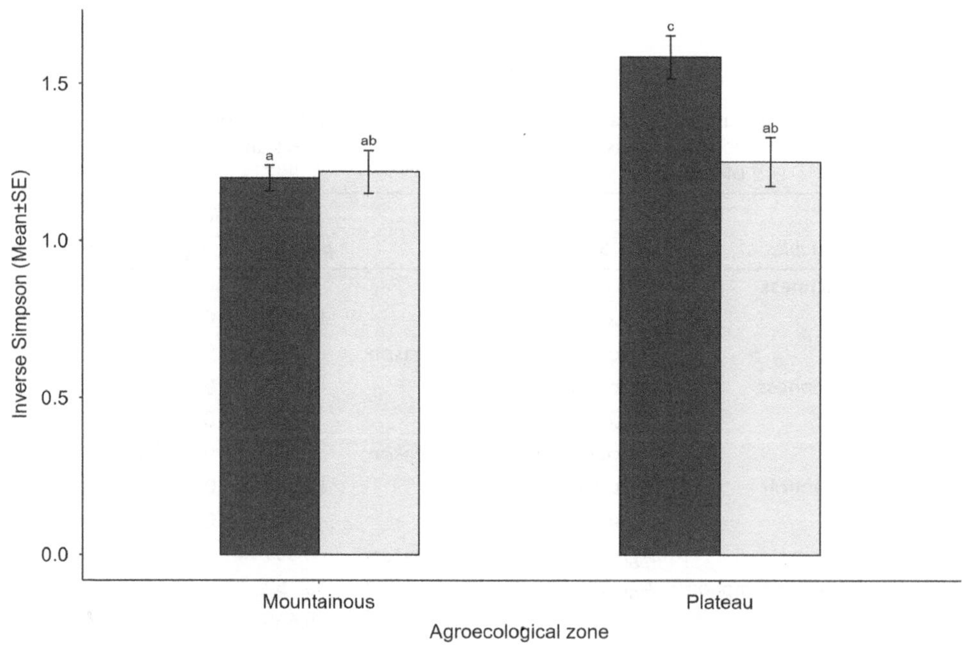

Figure 2.5 Inverse Simpson diversity of fruit fly species infesting cucurbit crops across the agroecological zones of Morogoro, Tanzania, from May 2020 to November 2021. Significance letters and standard errors are indicated.

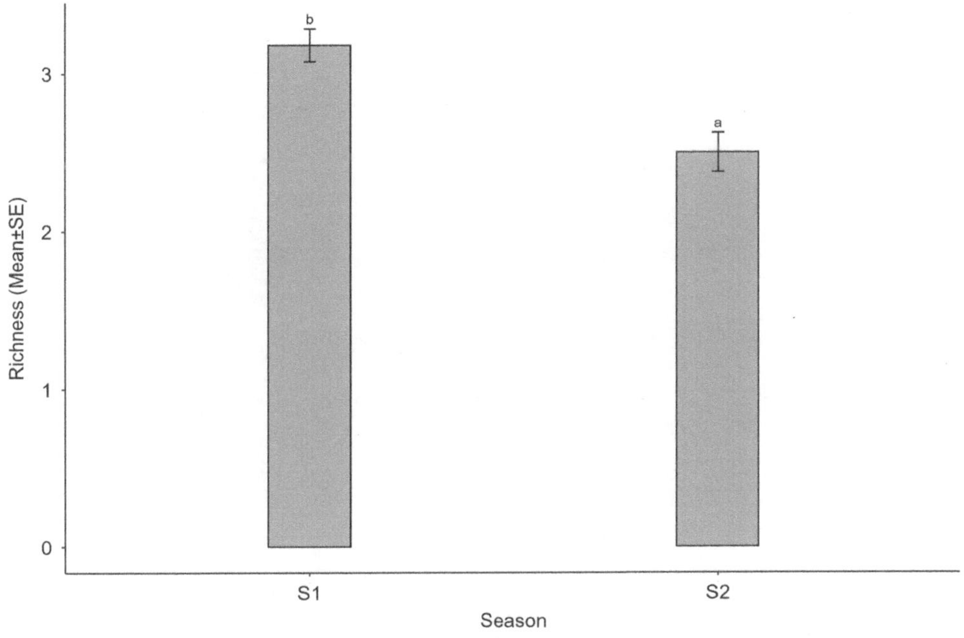

Figure 2.6 Species richness of fruit flies infesting cucurbit crops across the agroecological zones of Morogoro, Tanzania, from May 2020 to November 2021. Significance letters and standard errors are indicated (S1 = season I, S2 = season II).

to the mountainous zone (Figure 2.6). Furthermore, we observed significantly higher species richness of fruit flies during the June–August season (Season I) compared to the September–November season (Season II) (Figure 2.7).

Both Sorensen's and Jaccard's indices had high values (Table 2.6), indicating a high similarity in terms of species composition between the plateau and the mountainous zones (Table 2.6).

2.3.3 Spatiotemporal patterns of the dominant species

Zeugodacus cucurbitae, D. bivittatus and *D. punctatifrons* were the three most abundant species out of the ten species trapped and are major prominent pests of cucurbit fruits known in the Morogoro region (Mwatawala et al. 2006, 2009, 2010; Geurts et al. 2012). Results from Figure 2.8a–c show that *Z. cucurbitae* dominated the Morogoro region with one to two distinct population peaks throughout the two cucurbit cropping seasons for 2020. Moreover, a similar trend was observed in *Z. cucurbitae* in 2021 (see Figure 2.8a). The other two species followed similar trends. *Z. cucurbitae* was recorded in higher numbers in traps baited with Biolure and Zingerone in October and November (Figures 2.8a and 2.8b). For traps baited with Cuelure, a higher number of *Z. cucurbitae* and *D. bivittatus* per week were recorded at the end of May and early November (Figure 2.8c). The other two species had similar fluctuation trends but with relatively low numbers (Figure 2.8a–c).

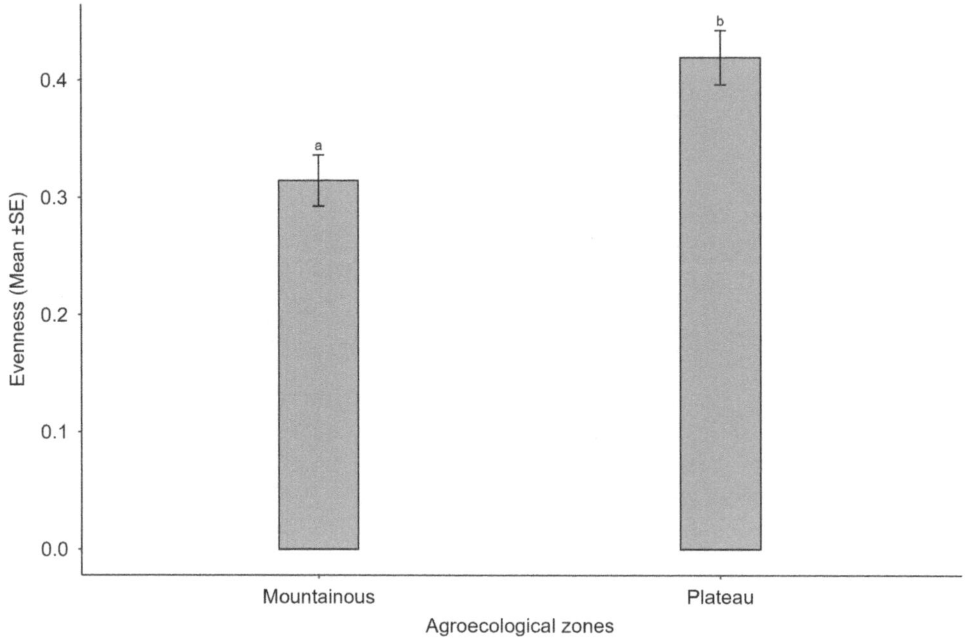

Figure 2.7 Species evenness of fruit flies infesting cucurbit crops across the agroecological zones of Morogoro, Tanzania, from May 2020 to November 2021. Significance letters and standard errors are indicated.

Table 2.6 Beta diversity indices comparing mountainous and plateau zones

Index	*Value*
Sorensen (S)	0.764
Jaccard (J)	0.613

The ANOVA on the effect of agroecological zone and cropping season is presented in Table 2.7. The results showed that the abundance of a species varied significantly between seasons. Further examination of the results showed that the abundance of fruit flies was significantly higher in Season I (May–June) than in Season II (October–November) (Figure 2.9) (*post hoc* test=Student–Newman–Keuls). The effect of the interaction between season and agroecological zone was not significant.

2.4 DISCUSSION

Optimal sampling effort is highly important when generating inventory lists of pest species of importance in agricultural management (Mokam et al. 2014; Lengyel et al. 2018). The results of the three non-parametric estimators of species richness revealed that the sampling effort of this study allowed for the recovery of more than 98% of

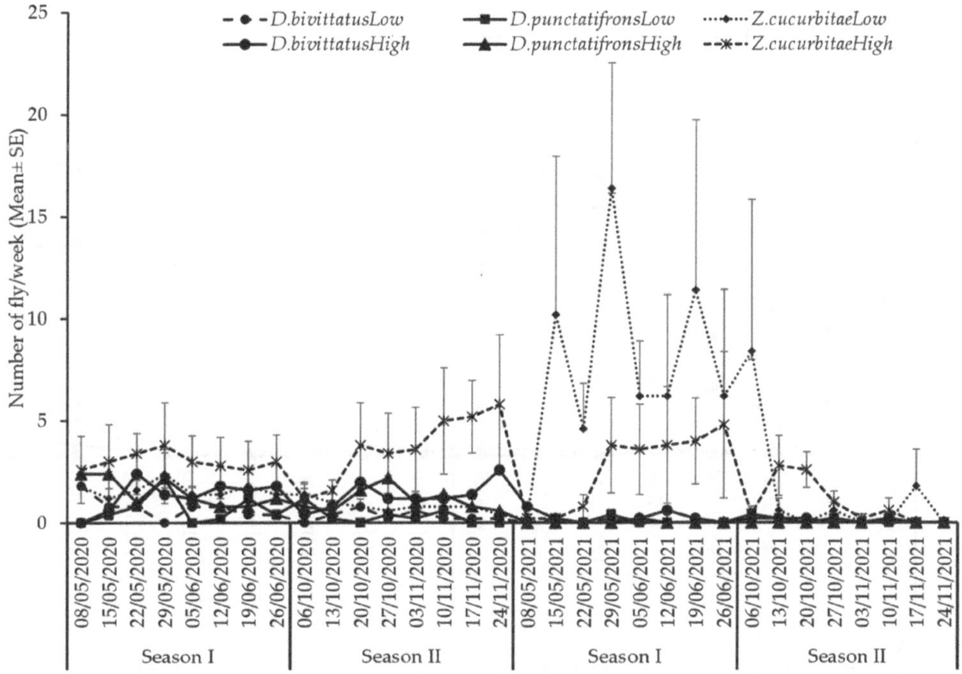

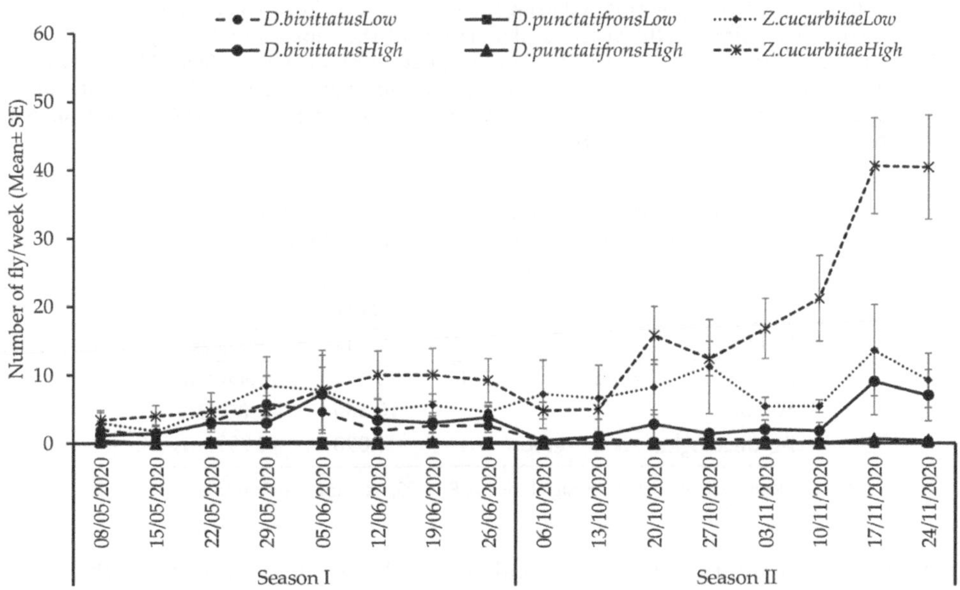

Figure 2.8 (a) Seasonal fluctuation of the most abundant fly species infesting cucurbit crops in the Morogoro region trapped using Biolure from May 2020 to November 2021. Standard errors are indicated. (b) Seasonal fluctuation of the most abundant fly species infesting cucurbit crops in the Morogoro region trapped using Zingerone from May to November 2020. Standard errors are indicated. (c) Seasonal fluctuation of the most abundant fly species infesting cucurbit crops in the Morogoro region trapped using Cuelure from May to November 2020. Standard errors are indicated.

(Continued)

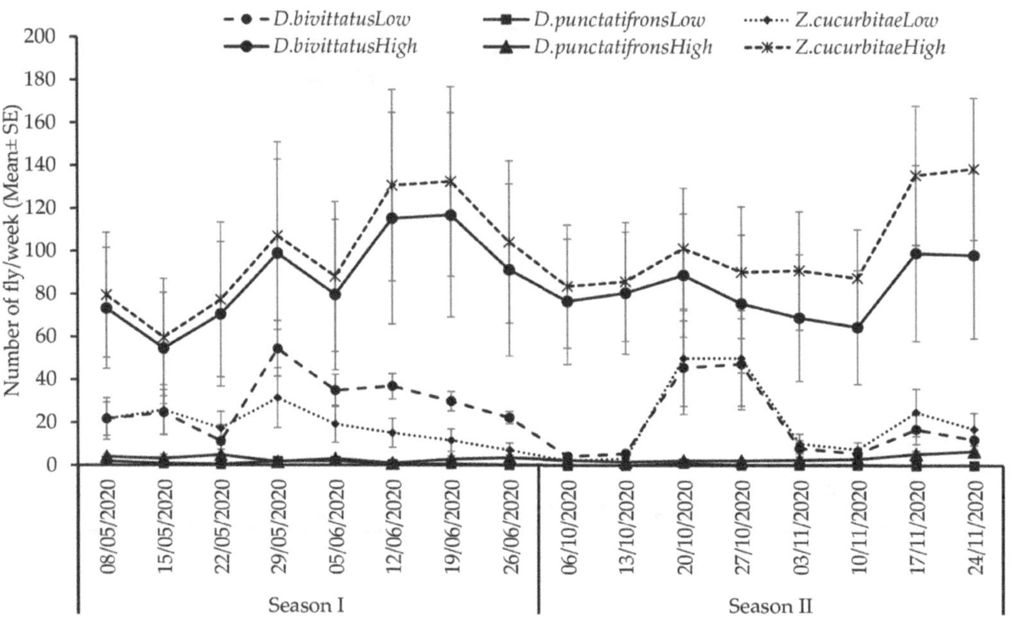

Figure 2.8 (*Continued*) (a) Seasonal fluctuation of the most abundant fly species infesting cucurbit crops in the Morogoro region trapped using Biolure from May 2020 to November 2021. Standard errors are indicated. (b) Seasonal fluctuation of the most abundant fly species infesting cucurbit crops in the Morogoro region trapped using Zingerone from May to November 2020. Standard errors are indicated. (c) Seasonal fluctuation of the most abundant fly species infesting cucurbit crops in the Morogoro region trapped using Cuelure from May to November 2020. Standard errors are indicated.

Table 2.7 Analysis of variance (ANOVA) of the effects of agroecological zone and season on the weekly abundance of fruit flies

	Statistics		
Factors	*df*	*F value*	*P-values*
Agroecological zone	1	0.476	0.4906 ns
Season	1	10.585	**0.0013***
Agroecological zone×season	1	1.090	0.2973 ns

Note: values with * are significant at $P<0.05$; **ns**=not significant.

fruit fly species in the two agroecological zones. Based on the results of the SACs, it was determined that a minimum of 4,375 individuals in the mountainous zone and 1,800 individuals in the plateau zone were necessary to recover over 98% of the fruit fly species from a total of 720 samples per zone. According to De Meyer et al. (2010), the minimum sample size accepted for optimal sampling and efficient comparisons of ecological communities is 20 samples. Therefore, we consider that the sampling effort implemented in this study was sufficient to obtain a satisfactory representation of the fruit fly communities of the Morogoro area.

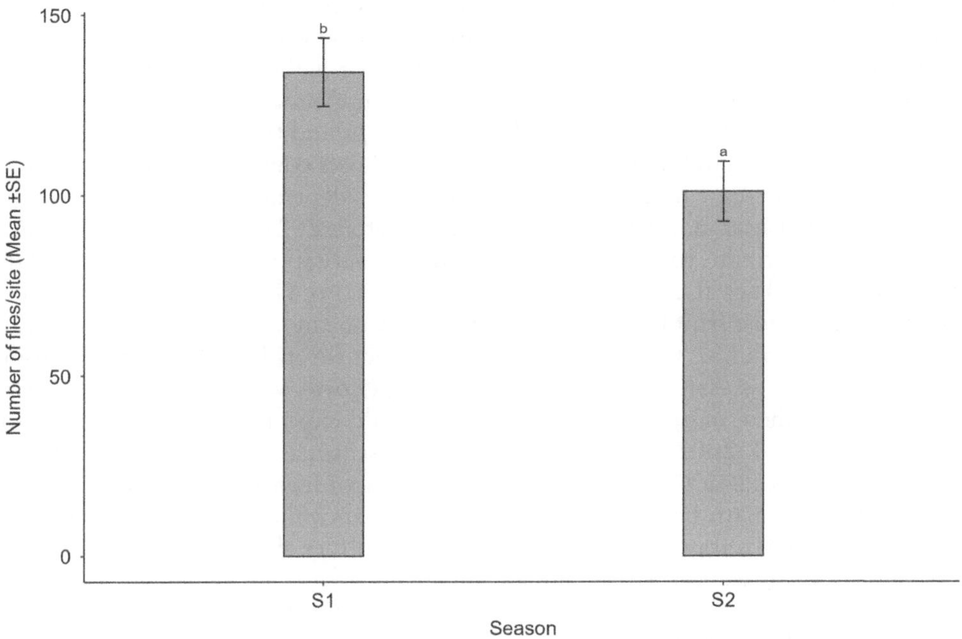

Figure 2.9 Abundance of fruit flies between the cucurbit cropping seasons. Significance letters and standard errors are indicated.

The results revealed significant differences in fruit fly species richness, evenness and diversity between the two agroecological zones of the Morogoro area as well as between the two cropping seasons. The plateau zone was more diverse and species-rich (17 species) compared to the mountainous zone (12 species). Compared to the plateau, the mountainous zone harboured various known cucurbit infesters (*Z. cucurbitae*, *D. bivittatus*, *D. punctatifrons*, *D. ciliatus*, *D. telfaireae* and D. frontalis) and other non-cucurbit infesters (*D. durbanensis*, *D. humeralis*, *D. xanthopterus*, *D. hyalobasis*, *D. woodi*, *D. ceropegiae*, *D. chiwira*, *D. pulchralis*, *D. nr. brevistriga* and *Ceratitis* spp.).

The observed higher species richness and diversity of fruit flies in the plateau zone of the Morogoro region can be explained by a confluence of ecological and environmental factors. Climatically, the plateau zone typically experiences a warmer climate, which is conducive to fruit fly development and survival, thereby supporting a higher diversity of species (Mwatawala et al. 2009; Nyasasi and Kisetu 2014; Bwire et al. 2022; Rweyemamu et al. 2024). Additionally, the wide diversity of cultivated fruits in the plateau zone offers ample resources for various fruit fly species, with different species exhibiting specific host preferences for oviposition and larval development (Ekesi et al. 2006; Mwatawala et al. 2009; Bwire et al. 2022). The agricultural practices adopted by most small-scale farmers (Bwire et al. 2022; Bakengesa et al. 2023) are largely dominated by low use of pesticides, which further likely contributed to the higher species richness and diversity observed in the plateau zone than in the mountainous zone. Thus, the microhabitats of the plateau zone provide a mosaic of ecological niches that can support many fruit fly species (Bwire et al. 2022; Bakengesa et al. 2023).

Such a microhabitat allows for the coexistence of diverse fruit fly species, contributing to the observed differences in species richness and diversity between the plateau and the mountainous zones. Conversely, the mountainous zone had a cooler climate, fewer types of cultivated fruits, less favourable agroecological practices for fruit fly development and fewer microhabitats for fruit fly development and thriving.

Also, seasonal shifts to higher availability of fruit hosts could result in higher fruit fly abundance or an increase in the number of rare species depending on the season (Raghu et al. 2000; Valadão et al. 2010; Salazar-Mendoza et al. 2021). In this case, host availability might contribute to the observed seasonal fluctuations in fruit fly population density (Mwatawala et al. 2009; Geurts et al. 2012, 2014). Given the fact that the complete life cycle of most fruit flies takes approximately one month when the temperatures are between 20°C and 32°C (see biology details in fact sheets for all major Afrotropical fruit fly pests at https://fruitflies.africamuseum.be/) (Royal Museum for Central Africa (RMCA) 2022), then, in optimal conditions, we could expect up to 13 generations per year (Copeland et al. 2002). In the tropics, fruit fly seasonality is often linked to other hosts that may serve as a bridge allowing populations of fruit flies to persist at times of the year when the main hosts are not available (Mwatawala et al. 2009; Geurts et al. 2014; Mkiga and Mwatawala 2015), resulting in fruit flies shifting hosts depending on the season (Salazar-Mendoza et al. 2021).

The frequent encounter of *Z. cucurbitae, D. punctatifrons* and *D. bivittatus* as dominant agricultural pests across the ecological zones has also been reported in different parts of Africa (Vayssieres et al. 2008b; Geurts et al. 2012; Ekesi et al. 2016; Gnanvossou et al. 2017; Kambura et al. 2018; Amevoin et al. 2021).

The populations of cucurbit-infesting flies in the Morogoro region was significantly influenced by the seasons. Generally, fruit flies were more abundant in the May–June season than in the October–November season. This difference in abundance can be attributed to several interrelated factors, primarily the availability of fruits and climatic conditions during these periods. The May–June season coincides with the end of the main rainy season in Tanzania, which typically spans from March to May (Mwatawala et al. 2010; Rweyemamu et al. 2024), which results resulted in a surge in fruit production, providing an abundant food source for fruit flies (Mwatawala et al. 2009). Many fruiting plants, including mangoes, papayas and citrus fruits, reach their peak ripeness during these months. The increase in available food resources supports a larger fruit fly population, as the flies have more opportunities to feed, reproduce and lay eggs (Mwatawala et al. 2006). In contrast, the October–November period follows the short rainy season, which is less intense and produces fewer fruits, thereby limiting the food supply for the flies (Mwatawala et al. 2009; Bwire et al. 2022). Moreover, the climatic conditions during the May–June season are generally more favourable for the survival and reproduction of fruit flies. During this time, the temperature and humidity levels create an optimal environment for fruit flies to thrive. Studies indicate that higher humidity, in particular, is crucial for the development of fruit fly eggs and larvae (Duyck et al. 2006; Vayssières et al. 2008a, 2009; Amin et al. 2019; Zida et al. 2020). In contrast, the October–November season may experience more variable weather patterns, including lower humidity and cooler temperatures, which can negatively impact fruit fly populations. These less favourable conditions can reduce the reproductive rates and survival of the flies, leading to a lower abundance. This suggests that multiple interacting factors, including climatic conditions and plant host availability, modulate the population abundances and dynamics of fruit flies (Mwatawala et al. 2006, 2009; Sarwar et al. 2013; Mkiga and Mwatawala 2015).

2.5 CONCLUSION

This study characterizes the fruit fly communities in the Morogoro region of Tanzania, both in the plateau zone and in the mountainous zone. It highlights the importance of considering the agroecological zones, hosts availability and fruit fly seasonality when designing sustainable control strategies against fruit flies. Programmes for managing cucurbit pests in the Morogoro area should focus on *Z. cucurbitae, D. punctatifrons* and *D. bivittatus,* as these are the major fruit fly species infesting cucurbit crops in this region, regardless of the agroecological zones. Furthermore, both short and long fruiting seasons of the host plants need to be considered, since the findings of this study revealed a higher abundance of fruit flies at the end of the long rainy (May–June) and short dry (October–November) seasons.

ACKNOWLEDGEMENTS

Thanks are due to Cesilia Kijalo, Evance Matowo, Frida Mangu, Christian Barnabas and Tabitha Fussi (Sokoine University of Agriculture) for their assistance in the field and Laboratory work. Also, we extend our thanks to Wouter Hendrycks and Lore Esselens for assisting with the analysis of data. This study was financed by Belgian Development Cooperation through the AGROVEG project within the framework agreement with the Royal Museum for Central Africa.

REFERENCES

Amevoin, K., L. K. Agboyi, M. Gomina, et al. 2021. Fruit fly surveillance in Togo (West Africa): state of diversity and prevalence of species. *International Journal of Tropical Insect Science* 41 (4): 3105–3119. https://doi.org/10.1007/s42690-021-00504-9

Amin, M. R., N. P. Nancy, M. R. U. Miah, et al. 2019. Fluctuations in fruit fly abundance and infestation in sweet gourd fields in relation to varied meteorological factors. *Entomological Research* 49 (5): 223–228. https://doi.org/10.1111/1748-5967.12351

Bakengesa, J., A. B. Kudra, M. De Meyer, et al. 2023. Agroecological farming practices against fruit flies in Central Eastern Tanzania. *Fruits* 78: 1–12. https://doi.org/10.17660/th2023/010

Billah, M., M. Mansell, M. De Meyer, et al. 2006. Fruit fly taxonomy and identification. In *A Field Guide to the Management of Economically Important Tephritid Fruit Flies in Africa*, ed. S. Ekesi and M. K. Billah, H1–H19. Nairobi, Kenya: ICIPE Science Press.

Bwire, P. M., J. Backengesa, S. Kabota, et al. 2022. Comparison of agroecological and conventional on management of fruit flies (Diptera: Tephritidae) infesting cucurbit crops in Morogoro, Tanzania. *Poster Presented at the International Symposium on Fruit Flies of Economic Importance (ISFFEI 2022)*, Sydney, Australia. https://agritrop.cirad.fr/602996 (Accessed July 7 2024).

Callaghan, C. T., A. G. B. Poore, R. E. Major, et al. 2019. Optimizing future biodiversity sampling by citizen scientists. *Proceedings of the Royal Society B: Biological Sciences* 286 (1912): 20191487. https://doi.org/10.1098/rspb.2019.1487

Chao, A., R. K. Colwell, C. W. Lin, et al. 2009. Sufficient sampling for asymptotic minimum species richness estimators. *Ecology* 90 (4): 1125–1133.

Chao, A., N. J. Gotelli, T. C. Hsieh, et al. 2014. Rarefaction and extrapolation with Hill numbers: a framework for sampling and estimation in species diversity studies. *Ecological Monographs* 84 (1): 45–67. https://doi.org/10.1098/rspb.2019.1487

Chao, A., W. H. Hwang, Y. C. Chen, et al. 2000. Estimating the number of shared species in two communities. *Statistica Sinica* 10:227–246.

Chao, A., and T. J. Shen. 2004. Nonparametric prediction in species sampling. *Journal of Agricultural, Biological, and Environmental Statistics* 9 (3): 253–269.

Colwell, R. K., A. Chao, N. J. Gotelli, et al. 2012. Models and estimators linking individual-based and sample-based rarefaction, extrapolation and comparison of assemblages. *Journal of Plant Ecology* 5 (1): 3–21.

Copeland, R. S., R. A. Wharton, Q. Luke, et al. 2002. Indigenous hosts of *Ceratitis capitata* (Diptera: Tephritidae) in Kenya. *Annals of the Entomological Society of America* 95 (6): 672–694.https://doi.org/10.1603/0013-8746(2002)095[0672:IHOCCD]2.0.CO;2

De Meyer, M., M. P. Robertson, M. W. Mansell, et al. 2010. Ecological niche and potential geographic distribution of the invasive fruit fly *Bactrocera invadens* (Diptera, Tephritidae). *Bulletin of Entomological Research* 100 (1): 35–48. https://doi.org/10.1017/S0007485309006713

Dove, A. D., and T. H. Cribb. 2006. Species accumulation curves and their applications in parasite ecology. *Trends in Parasitology* 22 (12): 568–574.

Drew, R. A. I., and G. H. S. Hooper. 1983. Population studies of fruit flies (Diptera: Tephritidae) in south-east Queensland. *Oecologia* 56 (2–3): 153–159.

Duyck, P. F., P. David, and S. Quilici. 2006. Climatic niche partitioning following successive invasions by fruit flies in La Réunion. *Journal of Animal Ecology* 75 (2): 518–526. https://doi.org/10.1111/j.1365-2656.2006.01072.x

Ekesi, S., M. De Meyer, S. A. Mohamed, et al. 2016. Taxonomy, ecology, and management of native and exotic fruit fly species in Africa. *Annual Review of Entomology* 61 (1): 219–238. https://doi.org/10.1146/annurev-ento-010715-023603

Ekesi, S., P. W. Nderitu, and I. Rwomushana. 2006. Field infestation, life history and demographic parameters of the fruit fly *Bactrocera invadens* (Diptera: Tephritidae) in Africa. *Bulletin of Entomological Research* 96 (4): 379–386. https://doi.org/10.1079/BER2006442

Ganie, S. A., Z. H. Khan, R. A. Ahangar, et al. 2013. Population dynamics, distribution, and species diversity of fruit flies on cucurbits in Kashmir Valley, India. *Journal of Insect Science* 13 (1): 1–7. https://doi.org/10.1673/031.013.6501

Geurts, K., M. Mwatawala, and M. De Meyer. 2012. Indigenous and invasive fruit fly diversity along an altitudinal transect in Eastern Central Tanzania. *Journal of Insect Science* 12 (1):1–18. https://doi.org/10.1673/031.012.1201

Geurts, K., M. W. Mwatawala, and M. De Meyer. 2014. Dominance of an invasive fruit fly species, *Bactrocera invadens*, along an altitudinal transect in Morogoro, Eastern Central Tanzania. *Bulletin of Entomological Research* 104 (3): 288–294. https://doi.org/10.1017/S0007485313000722

Gnanvossou, D., R. Hanna, G. Goergen, et al. 2017. Diversity and seasonal abundance of tephritid fruit flies in three agro-ecosystems in Benin, West Africa. *Journal of Applied Entomology* 141 (10): 798–809. https://doi.org/10.1111/jen.12429

Hanna, R., D. Gnanvossou, G. Goergen, et al. 2020. Efficiency of food-based attractants for monitoring tephritid fruit flies diversity and abundance in mango systems across three West African agro-ecological zones. *Journal of Economic Entomology* 113 (2): 860–871. https://doi.org/10.1093/jee/toz338

Hassani, I. M., H. Delatte, L. H. R. Ravaomanarivo, et al. 2022. Niche partitioning via host plants and altitude among fruit flies following the invasion of *Bactrocera dorsalis*. *Agricultural and Forest Entomology* 24 (4): 575–585. https://doi.org/10.1111/afe.12522

Hassani, I. M., L. H. Raveloson-Ravaomanarivo, H. Delatte, et al. 2016. Invasion by *Bactrocera dorsalis* and niche partitioning among tephritid species in Comoros. *Bulletin of Entomological Research* 106 (6): 749–758. https://doi.org/10.1017/S0007485316000456

International Atomic Energy Agency (IAEA) and FAO. 2011. FAO/IAEA guidelines for implementing systems approaches for P.|INIS [online]. https://inis.iaea.org/search/search.aspx?orig_q=RN:43033077

Kambura, C., C. M. Tanga, D. Kilalo, et al. 2018. Composition, host range and host suitability of vegetable-infesting tephritids on cucurbits cultivated in Kenya. *African Entomology* 26 (2): 379–397.

Kounatidis, I., N. T. Papadopoulos, P. Mavragani-Tsipidou, et al. 2008. Effect of elevation on spatio-temporal patterns of olive fly (*Bactrocera oleae*) populations in northern Greece. *Journal of Applied Entomology* 132 (9–10): 722–733. https://doi.org/10.1111/j.1439-0418.2008.01349.x

Laskar, N., and H. Chatterjee. 2010. The effect of meteorological factors on the population dynamics of melon fly, *Bactrocera cucurbitae* (Coq.) (Diptera: Tephritidae) in the foot hills of Himalaya. *Journal of Applied Sciences and Environmental Management* 14(3):53–58. https://doi.org/10.4314/jasem.v14i3.61466

Lengyel, S., B. Kosztyi, D. S. Schmeller, et al. 2018. Evaluating and benchmarking biodiversity monitoring: Metadata-based indicators for sampling design, sampling effort and data analysis. *Ecological Indicators* 85: 624–633. https://doi.org/10.1016/j.ecolind.2017.11.012

Magurran, A. E. 1988. *Ecological Diversity and Its Measurement*. Princeton: Princeton University Press.

Manrakhan, A., J. H. Daneel, R. Beck, et al. 2017. Efficacy of trapping systems for monitoring of Afrotropical fruit flies. *Journal of Applied Entomology* 141 (10): 825–840. https://doi.org/10.1111/jen.12373

Margalef, R. 1958. *Temporal Succession and Spatial Heterogeneity in Natural Phytoplankton*. Berkeley: University of California Press. https://doi.org/10.1525/9780520350281-024

Mkiga, A. M., and M. W. Mwatawala. 2015. Developmental biology of *Zeugodacus cucurbitae* (Diptera: Tephritidae) in three cucurbitaceous hosts at different temperature regimes. *Journal of Insect Science* 15 (1): 1–5. https://doi.org/10.1093/jisesa/iev141

Mokam, D. G., C. Djiéto-Lordon, and C. F. Bilong Bilong. 2014. Patterns of species richness and diversity of insects associated with cucurbit fruits in the southern part of Cameroon. *Journal of Insect Science* 14 (1):1–9. https://doi.org/10.1093/jisesa/ieu110

Mokam, D. G., C. Djiéto-Lordon, C. F. Bilong Bilong, et al. 2018. Host susceptibility and pest status of fruit flies (Diptera: Tephritidae) attacking cucurbits in two agroecological zones of Cameroon, Central Africa. *African Entomology* 26 (2): 317–332.

Mwatawala, M. W., M. De Meyer, R. H. Makundi, et al. 2006. Biodiversity of fruit flies (Diptera, Tephritidae) in orchards in different agro-ecological zones of the Morogoro region, Tanzania. *Fruits* 61 (5): 321–332. https://doi.org/10.1051/fruits:2006031

Mwatawala, M. W., M. De Meyer, R. H. Makundi, et al. 2009. Host range and distribution of fruit-infesting pestiferous fruit flies (Diptera, Tephritidae) in selected areas of Central Tanzania. *Bulletin of Entomological Research* 99 (6): 629–641. https://doi.org/10.1017/S0007485309006695

Mwatawala, M., A. P. Maerere, R. Makundi, et al. 2010. Incidence and host range of the melon fruit fly *Bactrocera cucurbitae* (Coquillett) (Diptera: Tephritidae) in Central Tanzania. *International Journal of Pest Management* 56 (3): 265–273. https://doi.org/10.1080/09670871003596792

Nankinga, C. M., B. E. Isabirye, H. Muyinza, et al. 2014. Fruit fly infestation in mango: A threat to the Horticultural sector in Uganda. *Uganda Journal of Agricultural Sciences* 15 (1): 1–14.

Nyasasi, B. T., and E. Kisetu. 2014. Determination of land productivity under maize-cowpea intercropping system in agro-ecological zone of Mount Uluguru in Morogoro, Tanzania. *Global Journal of Agricultural Sciences* 2 (2): 147–157.

Oksanen, J., F. G. Blanchet, R. Kindt, et al. 2018. Community ecology package. *R Package Version* 2: 5–2.

Papadopoulos, N. T., M. De Meyer, J. S. Terblanche, et al. 2024. Fruit flies: Challenges and opportunities to stem the tide of global invasions. *Annual Review of Entomology* 69 (1): 022723–103200. https://doi.org/10.1146/annurev-ento-022723-103200

Pielou, E. C. 1969. *An Introduction to Mathematical Ecology.* New York: Wiley-Inter-science.

Pogue, T., K. Malod, and C. W. Weldon. 2024. Effects of physiological status and environmental factors on the lure responses of three pest fruit fly species (Diptera: Tephritidae). *Journal of Chemical Ecology* 50 (11): 679–700.

Raghu, S., A. R. Clarke, R. A. I. Drew, et al. 2000. Impact of habitat modification on the distribution and abundance of fruit flies (Diptera: Tephritidae) in Southeast Queensland. *Population Ecology* 42 (2): 153–160.

Royal Museum for Central Africa (RMCA). 2020. Virtual collections: Search objects: collection_id:43 [online]. https://virtualcol.africamuseum.be/providence/pawtucket/index.php/Search/objects/key/72d1c128168c46edff116c4d2956baf8/facet/Family_facet/id/40260/view/images

Royal Museum for Central Africa (RMCA). 2022. Fact sheets | Fruit flies [online]. https://fruitflies.africamuseum.be/outputs/factsheets

Rweyemamu, E. W., M. W. Mwatawala, G. M. Tryphone, et al. 2024. Impact of cucurbit crop management techniques on the foraging behavior of honeybees and hoverflies in Morogoro, Tanzania. *BMC Ecology and Evolution* 24 (1): 1–13.

Salazar-Mendoza, P., I. Peralta-Aragón, L. Romero-Rivas, et al. 2021. The abundance and diversity of fruit flies and their parasitoids change with elevation in guava orchards in a tropical Andean forest of Peru, independent of seasonality. *PLOS One* 16 (4): 1–24. https://doi.org/10.1371/journal.pone.0250731

Sandrini-Neto, L., and M. G. Camargo. 2012. GAD: Analysis of variance from general principles. R package version 1.1.1. https://cran.rproject.org/web/packages/GAD/index.html

Sarwar, M., M. Hamed, B. Rasool, et al. 2013. Host preference and performance of fruit flies *Bactrocera zonata* (Saunders) and *Bactrocera cucurbitae* (Coquillett) (Diptera: Tephritidae) for various fruits and vegetables. *International Journal of Scientific Research in Environmental Sciences* 1 (8): 188–194. https://doi.org/10.12983/ijsres-2013-p188-194

Simpson, E. H. 1949. Measurement of diversity. *Nature* 163: 688. https://doi.org/10.1038/163688a0

Shannon, C. E., and Weaver, W. 1963. *The Mathematical Theory of Communication.* Champaign, IL: University of Illinois Press.

Tanga, C. M., and I. Rwomushana. 2016. Fruit fly species composition, distribution and host plants with emphasis on vegetable-infesting species. In *Fruit Fly Research and Development in Africa - Towards a Sustainable Management Strategy to Improve Horticulture*, ed. S. Ekesi, S. A. Mohamed, and M. De Meyer, 107–126. Cham: Springer International Publishing. https://doi.org/10.1007/978-3-319-43226-7_6

Underwood, A. J., M. G. Chapman, and S. D. Connell. 2000. Observations in ecology: you can't make progress on processes without understanding the patterns. *Journal of Experimental Marine Biology and Ecology* 250 (1–2): 97–115. https://doi.org/10.1016/S0022-0981(00)00181-7

United Republic of Tanzania (URT). 2002. *Morogoro Regional Socio-Economic Profile.* Morogoro: National Bureau of Statistics, Morogoro Regional Commissioner's office, 229p.

Valadão, H., J. D. V. Hay, and R. Tidon. 2010. Temporal dynamics and resource availability for drosophilid fruit flies (Insecta, Diptera) in a gallery forest in the Brazilian Savanna. *International Journal of Ecology* 2010 (e152437): 1–7. https://doi.org/10.1155/2010/152437

Vayssières, J. F., Y. Carel, M. Coubès, et al. 2008a. Development of immature stages and comparative demography of two cucurbit-attacking fruit flies in Reunion Island: *Bactrocera cucurbitae* and *Dacus ciliatus* (Diptera Tephritidae). *Environmental Entomology* 37 (2): 307–314. https://doi.org/10.1093/ee/37.2.307

Vayssières, J. F., S. Korie, and D. Ayegnon. 2009. Correlation of fruit fly (Diptera Tephritidae) infestation of major mango cultivars in Borgou (Benin) with abiotic and biotic factors and assessment of damage. *Crop Protection* 28 (6): 477–488. https://doi.org/10.1016/j.cropro.2009.01.010

Vayssieres, J. F., S. Korie, O. Coulibaly, et al. 2008b. The mango tree in central and northern Benin: cultivar inventory, yield assessment, infested stages and loss due to fruit flies (Diptera Tephritidae). *Fruits* 63 (6): 335–348. https://doi.org/10.1051/fruits:2008035

Virgilio, M., I. White, and M. De Meyer. 2014. A set of multi-entry identification keys to African frugivorous flies (Diptera, Tephritidae). *ZooKeys* 428: 97–108. https://doi.org/10.3897/zookeys.428.7366

White, I. M., and M. M. Elson-Harris. 1992. *Fruit Flies of Economic Significance: Their Identification and Bionomics*. Wallingford, Oxon: CAB International.

Zida, I., S. Nacro, R. Dabiré, et al. 2020. Seasonal abundance and diversity of fruit flies (Diptera: Tephritidae) in three types of plant formations in Western Burkina Faso, West Africa. *Annals of the Entomological Society of America* 113 (5): 343–354. https://doi.org/10.1093/aesa/saaa004

Chapter 3

Assessment of courtship behavior and mating success in *Ceratitis capitata* (Diptera: Tephritidae)

A comparison between laboratory-reared and wild flies

Jane Larissa de Melo Custódio, Alexandre Santos Araújo, Alzira Kelly Passos Roriz, Vanessa Simões Dias, and Iara Sordi Joachim Bravo

3.1 INTRODUCTION

Domestication is an evolutionary process that arises from a mutualistic ecological interaction, in which one species actively controls the fitness components (survival and reproduction) of another species (Purugganan 2022). In human-associated domestication, populations are reared in human-controlled environments, and their genetic, physiological, and behavioral traits are modified across generations from their wild ancestors, allowing them to evolve in new directions (Hoffmann and Ross 2018; Lecocq 2018).

Several insects undergo domestication through artificial rearing in laboratories or insect production facilities to serve as laboratory organisms or for pest control strategies. Over succeeding generations, artificial rearing allows the selection of individuals that reproduce early (Meats et al. 2004; Diamantidis et al. 2011) and develop fast (Economopoulos 1992; Gaire et al. 2022). Additionally, behavioral modifications have been reported in laboratory-reared insects, including shortened courtship (Briceño and Eberhard 1998), faster copulations (Cayol et al. 1999), impaired flight ability (Nakamori and Simizu 1983), and reduced predator avoidance skills (Rao et al. 2014; Dor and Liedo 2019). These characteristics enable a rapid turnover of generations and ease of handling by humans, favoring the maintenance of colonies in artificial environments (Cayol 1999).

Although some modifications favor increased fitness under artificial conditions (e.g., rapid development and early reproduction) (Hoffmann and Ross 2018), they can be deleterious when laboratory populations are compared to their wild counterparts (Rössler 1975; Economopoulos 1992; Liimatainen et al. 1997). Artificial environments lack the variation and complexity found in natural ecosystems, and therefore, selective pressures related to vulnerability and variation are relaxed (Cayol 1999; Hoffmann and Ross 2018). This can compromise the ability of laboratory populations to survive, compete, escape predators, and reproduce effectively in the wild (Hoffmann and Ross 2018). These negative consequences of domestication could affect the efficiency of insect pest control techniques that rely on the interaction between laboratory-reared and wild individuals, such as the sterile insect technique (SIT) (Hendrichs et al. 2002).

DOI: 10.1201/9781003644330-4

In the SIT strategy, mass-reared males of a target insect species are sterilized by ionizing radiation and released in the field to compete with their wild counterparts for sexual mates (Hendrichs et al. 2002; Dunn and Follett 2017; Pereira et al. 2023). Sterile males must be able to survive, compete, court, and mate in nature with wild females (Hendrichs and Robinson 2009). Therefore, SIT success relies significantly on understanding the sexual behavior of the target species. However, achieving the ideal balance between the logistical efficiency of artificially rearing populations and preserving their behavioral traits at suitable levels that allow good performance in the field is challenging. Studies investigating the domestication effects on insect sexual behavior are important to address these issues.

Due to its elaborate courtship and importance for SIT application, the Mediterranean fruit fly *Ceratitis capitata* (Wiedemann) (Diptera: Tephritidae) is a suitable study species to investigate the effects of laboratory domestication on sexual behavior. In *C. capitata*, the primary mating system for acquiring sexual partners involves forming leks (Eberhard 2000; Benelli et al. 2014; Dias et al. 2023). The leks are small aggregations of males in arenas, usually gathered in the undersurface of leaves of host trees (Prokopy and Hendrichs 1979; Shelly 2018). The males initiate an elaborate behavior involving chemical, acoustic, and visual signals to attract females to the exhibition site (Prokopy and Hendrichs 1979; Sugayama and Malavasi 2000; Dias et al. 2023). In the leks, males defend small territories and compete for mating with visiting females (Eberhard 2000; Shelly 2018), performing a series of aggressive agonistic behaviors while defending these territories (Briceño et al. 1999; Benelli 2015). Male courtship begins after a female arrives on the leks (Eberhard 2000).

The primary sequence of the courtship performed by the males of *C. capitata* was first detailed by Féron (1962) and revisited in later studies (Briceño et al. 1996; Liimatainen et al. 1997; Briceño and Eberhard 2000, 2002). The courtship comprises the primary behavioral steps: emission of sex pheromone (calling behavior), emission of an acoustic signal produced by the continuous vibration of the wings (wing vibration), emission of an intermittent buzzing produced by rhythmic movement back and forth of the wings (wing buzzing), and execution of rapid head movements (head rock) that typically occur simultaneously with the wing movements. During the execution of the head rock, the male often stimulates the female mechanically, establishing contact by touching the female's aristae and antennae (Briceño and Eberhard 2002). After performing these steps, the male jumps onto the female to attempt copulation (Briceño et al. 1996; Eberhard 2000).

Many studies have shown that the courtship behavior and mating success of *C. capitata* can be affected by laboratory rearing (Economopoulos 1992; Briceño et al. 1996; Liimatainen et al. 1997; Briceño and Eberhard 1998; Calcagno et al. 2002; Lux et al. 2002; Pereira et al. 2007). One of these effects is the reduction of the courtship duration in crowded conditions, which is probably advantageous since it decreases the probability of the courtship being interrupted by other flies (Briceño and Eberhard 1998). Additionally, transitions between courtship behavioral units were either absent or less elaborate in mass-reared males (Liimatainen et al. 1997). As a result, wild females that often select a mate based on a long and detailed courtship tend to reject the courtship of laboratory-reared males (Cayol 1999), which decreases their mating success in comparison to wild males (Lance et al. 2000; Pereira et al. 2007; De Aquino and Joachim-Bravo 2014).

Although the current knowledge about the sexual behavior of *C. capitata* is extensive, some parameters have yet to be assessed. For instance, only a few studies have compared the temporal pattern of calling behavior between laboratory-reared and wild males (Papadopoulos et al. 1998; Diamantidis et al. 2008). Furthermore, there is a lack of quantification of behaviors related to courtship other than those that comprise the primary sequence described for *C. capitata*.

In this study, we investigated the effects of laboratory rearing on the sexual behavior of *C. capitata*. We hypothesize that domestication through laboratory rearing negatively affects the courtship behavior and mating success of *C. capitata* males. To test this hypothesis, comparisons were made between laboratory-reared and wild males, evaluating the following aspects of sexual behavior: temporal pattern of calling behavior, description and quantification of courtship behavioral units, and mating success.

3.2 MATERIALS AND METHODS

3.2.1 Biological material and rearing protocol

Experiments were carried out using sexually mature *C. capitata* adult flies from laboratory-reared and wild populations maintained in the Laboratório de Ecologia Comportamental de Insetos at the Universidade Federal da Bahia (Salvador, Bahia, Brazil). The laboratory-reared population was collected on infested almonds (*Terminalia catappa* L.) in Ribeirão Preto, São Paulo, Brazil (21°10′22.9″S 47°48′48.6″W) and reared for 35 years under controlled conditions. In the larval stage, the laboratory-reared flies were fed an artificial diet of Brewer's yeast, soybean bran, and oat bran, adapted from Carvalho et al. (1998). The wild population was obtained from infested almonds (*T. catappa*) collected in Salvador, Bahia, Brazil (12°55′36.6″S 38°19′15.8″W; 12°56′13.4″S 38°19′53.5″W). The almonds were taken to the laboratory and placed in plastic trays containing vermiculite (Brasil Minérios ®, Goiânia, Goiás, Brazil). The fruits were monitored every day for the presence of pupae. Subsequently, the pupae were sieved and transferred into plastic containers (11×11×3.5 cm) that were held inside plastic cages (20×24×33 cm) and maintained in the dark until the adults emerged. Immature stages from the wild population were reared on guava fruits (*Psidium guajava* L.) infested by wild females. Only wild flies from the F0 and F1 generations were used to perform the experiments. Adult flies from both populations were kept in plastic cages (20×24×33 cm) under controlled environmental conditions (25±1°C temperature, 65±10% relative humidity, 12L:12D photoperiod). Flies received water and a diet based on hydrolyzed protein (Bionis Biorigin ®, Lençóis Paulista, São Paulo, Brazil) and sugar (1:3) *ad libitum* (Silva Neto et al. 2012).

3.2.2 Temporal pattern of calling behavior

To evaluate the temporal pattern of calling behavior, ten males of the same population, individually marked on the mesonotum with different colors of nontoxic paint, were released in an acrylic box (20×15×12 cm). A glass box (9×7×9 cm) containing ten wild females was placed inside each acrylic box to visually stimulate the males (Roriz et al. 2019). The number of males performing the calling behavior was counted every

30 minutes from 7:30 to 18:30. The calling behavior was identified by visual observation of a bright, translucent droplet of pheromone on the distal portion of the anal epithelium with or without the distension of the lateral abdominal glands (Levinson et al. 1987). Ten replicates were performed for each population using flies aged 10–12 days.

3.2.3 Courtship recording

To record the courtship, two males of the same population and one wild female aged 10–12 days were released into a glass box (9×7×9 cm) containing a leaf of *Eugenia uniflora* L. to provide an exhibition site for sexual interactions. Recording began with the release of the female and lasted up to 30 minutes (the recording was interrupted if there was confirmation of mating within this interval). Despite releasing two males per box, only the successful male was analyzed. A digital video camera (Geovision-GV-BX220D-3) was positioned in front of the glass box to record the courtship. The software Eagle Vision Pro version 4.1 and GeoVision 800 were used to capture the images and save the videos in AVI format, respectively. Twenty videos that resulted in a successful copulation were obtained for each population. Ten videos were randomly chosen to construct an ethogram, and the other ten recordings were used to analyze the data.

3.2.4 Video analysis

To construct the ethogram, we relied on previous studies on the sexual behavior of *C. capitata* (Levinson et al. 1987; Briceño et al. 1996, 1999; Briceño and Eberhard 2000, 2002) and the fruit fly *Anastrepha fraterculus* (Wiedemann) (Roriz et al. 2018). Once the ethogram was constructed, the videos selected for the data analysis were analyzed. The behavioral units exhibited by the males were identified and sequenced, second by second, following the ethogram descriptions (see Table 3.2). The following parameters were analyzed: number of behavioral units (total number of behavioral units exhibited in a courtship event), courtship duration (total duration, in seconds, of the courtship), behavioral sequence (probabilistic occurrence of behavioral routines), and relative frequency of behavioral units (number of occurrences of the behavioral unit divided by the total duration of the courtship) (Martin and Bateson 2007).

The behavioral sequence was analyzed using the EthoSeq software (Japyassu et al. 2006), built to extract probabilistic behavioral sequences (tree-generated sequences, or TGSs) from observational data. EthoSeq uses graph theory to produce hierarchical representations in directed trees (DiTree), where each behavioral action appears only once. The DiTree maximizes the sum of the probabilities of transitions between all the behavioral units so that the behavioral routines contained in the tree are the most likely to occur within the dataset (Japyassu et al. 2006). The DiTrees, whose roots indicated mating success (i.e., ended in the *Mating* behavioral unit), were selected to assess the behavioral sequences that most contributed to mating occurrence.

3.2.5 Mating success in laboratory-reared and wild flies

To investigate male mating success, mate choice experiments were performed in laboratory cages (89×69×69 cm). A small potted plant (*E. uniflora*) approximately 60 cm tall and with a 40 cm radius was placed in the center of each cage. At least 15 minutes before the test, 20 males (ten of each population), marked on the mesonotum with non-toxic paint, were released inside the cage. Fifteen minutes later, ten wild females were released inside the cage. Observations of the occurrence of mating took place from 7:00 to 14:00. Mating pairs were removed from the cage using a small glass tube. The population of the male, the time until mating (latency), and mating duration were registered. Ten experimental cages were set up using flies from two cohorts aged 7–11 days.

3.2.6 Statistical analysis

The temporal pattern of calling behavior was analyzed using a generalized linear model (GLM) with a Poisson distribution, considering the number of males calling as the response variable. Population, time, and the interaction between population and time were included as explanatory variables. To account for temporal dependence between observations, time was treated as a factor reflecting the temporal structure, with the correlation between five consecutive data collection times incorporated, after which temporal independence was assumed for the time factor. The number of behavioral units, courtship duration, and the relative frequencies of behavioral units were compared between laboratory and wild flies through GLMs, assuming a Gaussian or Poisson distribution. The models' significances were assessed using F-tests or Chi-squared tests. Generalized linear mixed models (GLMMs) were used to analyze the influence of the population (explanatory variable) on the response variables (proportion of matings, mating latency, and duration), assuming either a Binomial or Gaussian distribution. The cohort was treated as a random effect in the models. The Akaike Information Criterion was used to identify the model that best fitted the data. The models' significance was assessed using the Type-3 Wald Chi-square test. All statistical analyses were performed in R Studio (R Core Team 2024) using the following packages: *ggplot2* (Wickham et al. 2024), *emmeans* (Lenth et al. 2024), *multcomp* (Hothorn et al. 2023), *lme4* (Bates et al. 2024), and *car* (Fox et al. 2023).

3.3 RESULTS

3.3.1 Temporal pattern of calling behavior

The populations did not differ statistically in the temporal pattern of calling behavior, and the Time×Population interaction was not significant (Table 3.1). However, the number of males calling varied significantly with time (Table 3.1). Males from both

Table 3.1 Generalized linear model for the temporal pattern of calling behavior of *C. capitata* males

	df	χ^2	*P*	*Significance*
Population	1	1178.72	0.3648	ns
Time	1	395.13	<0.001	***
Population×time	1	391.59	0.0597	ns

Significance codes: ***$p<0.001$; ns=not significant.

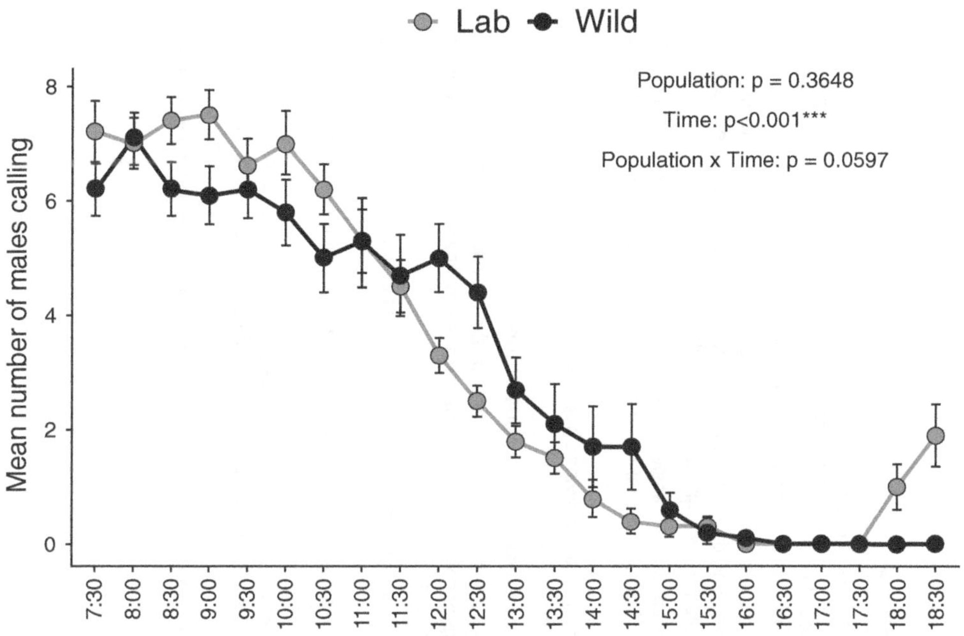

Figure 3.1 Temporal pattern of male calling behavior in *Ceratitis capitata* wild and laboratory-reared (Lab) populations. Points and error bars represent mean values and ±SE, respectively ($p<0.05$, generalized linear model—GLM).

populations began pheromone emission early in the morning and continued for at least 8 hours (Figure 3.1). The highest activity occurred at 8:00 for wild males and 9:00 for laboratory-reared males. The laboratory-reared population maintained a relatively high mean number of males calling until 10:00 when activity decreased exponentially. Conversely, the mean number of wild males calling was roughly constant, with some fluctuations until 12:30, when activity began to decline. In both populations, the males called until around 16:00, but only laboratory-reared males resumed calling behavior at 18:00.

3.3.2 Courtship behavior analysis

3.3.2.1 Ethogram

Sixteen behavioral units were identified for the courtship of *C. capitata*, which were divided into eight behavioral categories: pheromone emission, wing movements or positions, head movements, body movements or positions, cleaning, agonistic behaviors, male–female interactions, and copulation (Table 3.2). The behavioral units *Oscillation*, *Stationary*, *Grooming*, and *Falling*, previously described for *A. fraterculus* (Roriz et al. 2018; Cardoso 2021), were observed in *C. capitata*. The description of the *Oscillation* behavior was modified, as the movement exhibited by *C. capitata* was less elaborate than that of *A. fraterculus*. We described the behavioral unit *Reorientation* during our video analysis. Laboratory-reared and wild males exhibited all behavioral units described here and performed the behaviors similarly.

Table 3.2 Descriptions of the behavioral units performed by *C. capitata* males during courtship

Behavioral categories	*Behavioral units*	*Descriptions of the behavioral units*
Pheromone emission	**CALL**: Calling	Emission of a bright, translucent droplet of pheromone on the distal portion of the anal epithelium with or without distension of the lateral abdominal glands (Levinson et al. 1987).
Wing movements or positions	**WVb:** Wing vibration	The wings, partially extended laterally, move rhythmically in a dorsoventral direction, producing a continuous vibration. During the movement, the male's everted anal epithelium is directed ventrally, with the pheromone droplet still visible (Briceño and Eberhard 2000).
	WBz: Wing buzzing	The wings, extended laterally at approximately 60° angle in relation to the thorax, move rhythmically forward and then return to their initial position, resulting in a 'back-and-forth' movement (Briceño and Eberhard 2000).
Head movements	**HRk:** Head rock	The male rapidly shakes its head, often in rotation swings, but lateral deflections or dorsoventral movements may also occur (Briceño et al. 1996).
Body movements or positions	**FL:** Flying	The fly flies, beating its wings, and moves toward a landing point (Roriz et al. 2018).
	OC: Oscillation	The fly moves its legs laterally, walking, creating unilateral arcs with the movement of the body. With each oscillation, the fly performs wing opening and alignment movements at around 90° in relation to the thorax (modified from Roriz et al. 2018).
	MO: Mobile	The fly walks, moving its legs synchronously, usually maintaining its wings at a 45° angle (Roriz et al. 2018).
	RE: Reorientation	The male turns its body to reorient itself toward the female after detecting some movement made by the female—usually the landing after a flight event.
	ST: Stationary	The male remains stationary with the wings relaxed, near the body, or held almost transversely, with the proximal portion of the wings positioned downwards and the distal portion upwards (Roriz et al. 2018).
	FG: Falling	The male falls with its legs facing upwards after an unsuccessful mating attempt or failed flight landing (Cardoso 2021).

(*Continued*)

Table 3.2 (Continued) Descriptions of the behavioral units performed by *C. capitata* males during courtship

Behavioral categories	*Behavioral units*	*Descriptions of the behavioral units*
Cleaning	**GR:** Grooming	While standing still, the fly performs body cleaning movements, rubbing its legs against the wings, head, thorax, abdomen, proboscis, and antennae (Roriz et al. 2018).
Agonistic behavior	**FI:** Fight	Aggressive interactions between males or between males and females, in which the attacking fly pushes its opponent or hits its opponent with the head, or the attacking fly flies over its opponent, hitting it sharply on the back (modified from Briceño et al. 1999).
Male–female interactions	**AL:** Alignment	The male and female position themselves opposite each other, at short distances, looking directly at each other (Briceño and Eberhard 2002).
	CO: Contact	The male's aristae repeatedly strike the female's aristae, and/or the male takes small steps forward and touches the female's antennae with its dimorphic pair of bristles (modified from Briceño and Eberhard 2002).
	AT: Attempt	The male jumps and lands onto the female, with the anterior end of its body facing toward the posterior end of the female's body. Then, the male turns about 180° around the female's body to attempt copulation (Briceño et al. 1996).
Copulation	**MT:** Mating	The male holds the female with its three pairs of legs and introduces its copulatory organ into the female, beginning the copulation.

3.3.2.2 Number of behavioral units and courtship duration

The populations differed statistically in the number of behavioral units exhibited during courtship (*df*=1, *F*=7.3702, *p*=0.0147). The number of behavioral units was significantly lower in the laboratory-reared population (Figure 3.2a). However, courtship duration did not differ statistically between the populations (*df*=1, χ^2=62743, *p*=0.4604, Figure 3.2b).

3.3.2.3 Behavioral sequence

The TGSs showed similarities between the two populations regarding the behavioral units that most contributed to mating success. The populations differed more in terms of the probability of occurrence of the routines (Figure 3.3). The behavioral unit *Mating*

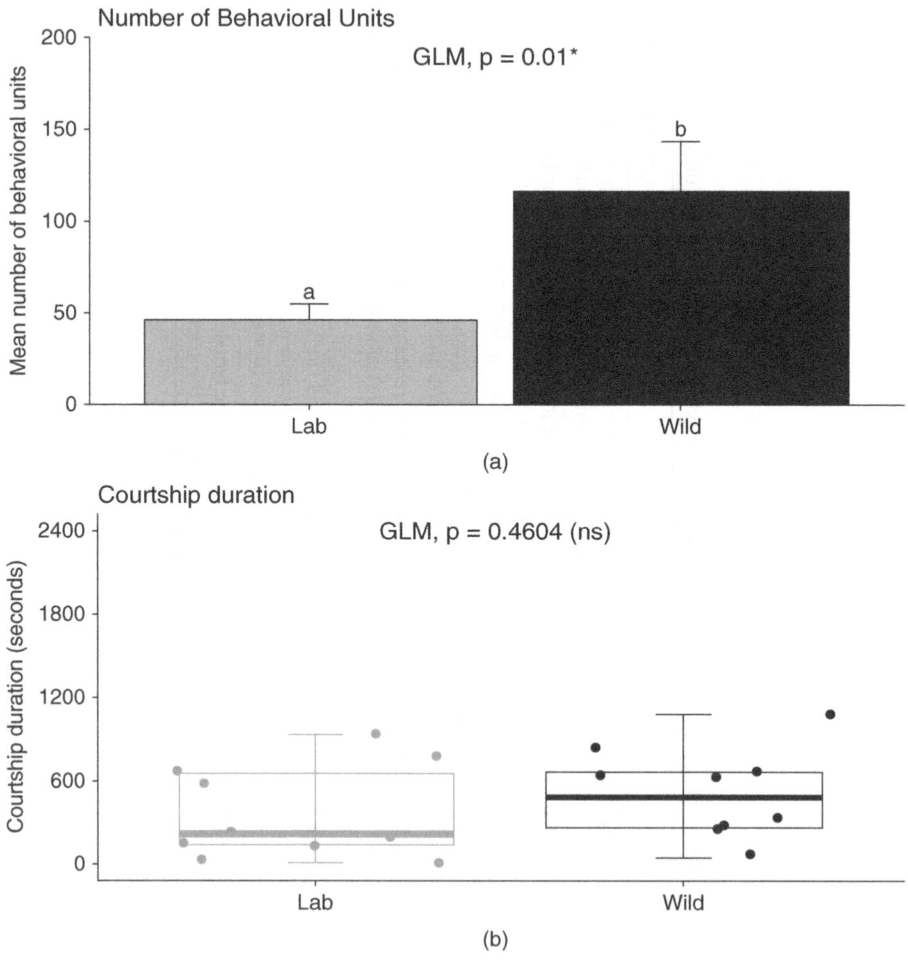

Figure 3.2 Mean number of behavioral units performed by the males during courtship (a) and courtship duration in seconds (b) for the laboratory-reared (Lab) and wild populations of *C. capitata*. Graph bars represent mean values with standard errors (±SE). Bars followed by different letters indicate statistical differences between the populations (*p*<0.05, GLM).

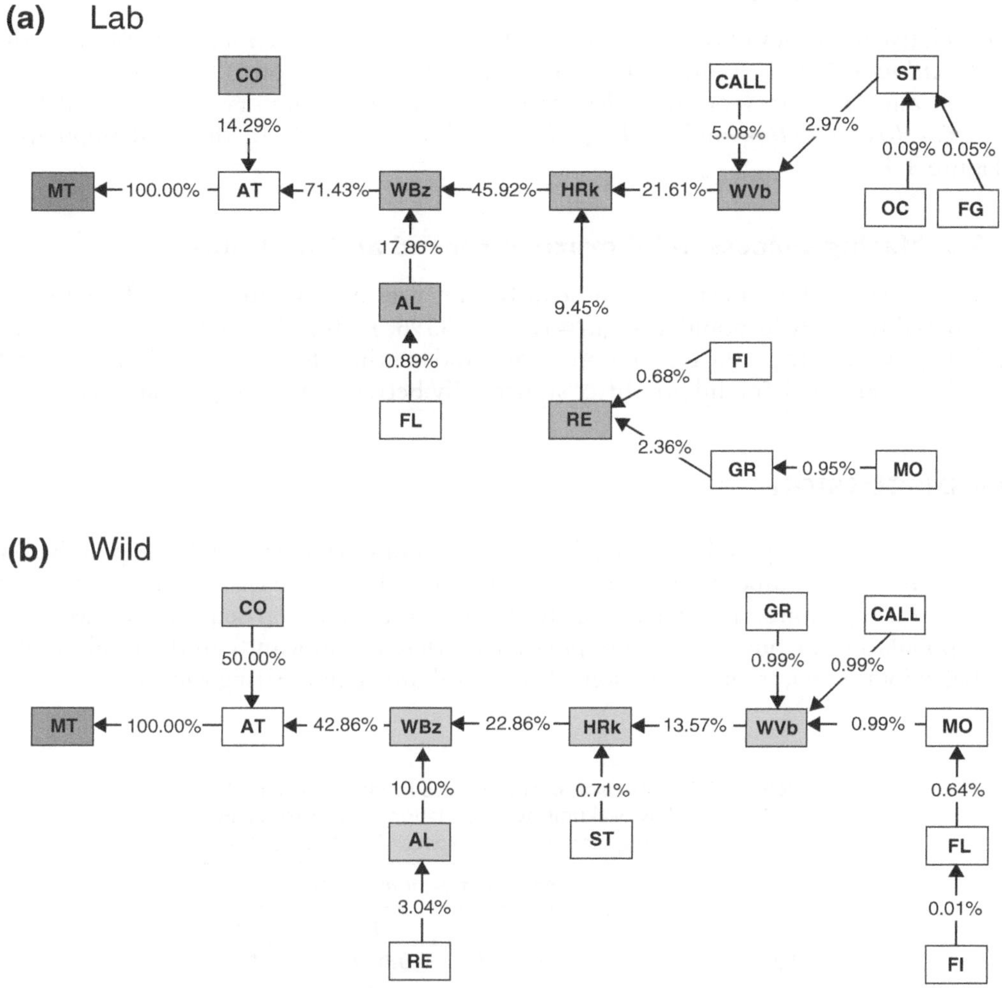

Figure 3.3 Tree-generated sequences (TGSs) that resulted in successful copulation (ended in MT) for laboratory-reared (Lab) (a) and wild (b) populations of *C. capitata*. The arrows indicate the behavioral routines, and the percentage values represent the probability of each path. Example: in the laboratory population, the probability of the path FL/AL/WBz/AT/MT is 0.89%, while that of AL/WBz/AT/MT is 17.86%. The highlighted behavioral units and percentages represent behavioral routines that most contributed to mating. The abbreviations of the behavioral units are described in Table 3.2.

(MT) was always preceded by *Attempt* (AT) (AT>MT: 100.00%). In both populations, only *Contact* (CO) and *Wing buzzing* (WBz) preceded AT. The sequence *Wing vibration* (WVb)>*Head rock* (HRk)>*Wing buzzing* (WBz)>*Attempt* (AT)>*Mating* (MT) was confirmed in both populations and presented a higher probability of occurrence in the laboratory-reared population (21.61%) than in the wild population (13.57%). *Reorientation* (RE) preceded *Head rock* (HRk) in the laboratory-reared population, whereas in the wild population, *Reorientation* (RE) was more likely to precede *Alignment* (AL).

3.3.2.4 *Relative frequency of behavioral units*

The relative frequency of the behavioral units *Mobile*, *Flying*, *Stationary*, *Reorientation*, and *Calling* differed significantly between the two populations (Table 3.3). *Mobile*, *Flying*, and *Stationary* occurred less frequently in the laboratory-reared population, whereas *Reorientation* and *Calling* occurred less frequently in the wild population (Figure 3.4).

3.3.3 Mating success in laboratory-reared and wild flies

The proportion of matings was significantly higher in the laboratory-reared population compared to the wild population (df=1, χ^2=5.8396, p=0.0156, Figure 3.5a). Latency (df=1, χ^2=0.2541, p=0.6142, Figure 3.5b) and mating duration (df=1, χ^2=0.0688, p=0.7931, Figure 3.5c) did not differ statistically between the two populations.

3.4 DISCUSSION

In this study, we compared the sexual behavior of laboratory-reared and wild populations of *C. capitata*. Our findings indicate similarities in calling patterns, courtship duration, behavioral repertoire, and behavioral routines of the courtship sequence between the two populations. Conversely, the populations differed significantly in the total number of behavioral units, frequency of some behavioral units, and mating success.

Table 3.3 Comparison of the relative frequencies of each behavioral unit between laboratory-reared and wild populations of *C. capitata* (p<0.05, GLMs)

	Generalized linear model (GLM)			
Behavioral unit	*df*	χ^2	p	*Significance*
MO	1	0.0477	**0.0028**	**
FL	1	0.0487	**0.0103**	*
WVb	1	0.0100	0.1154	ns
ST	1	0.0037	**0.0023**	**
GR	1	0.0023	0.7705	ns
RE	1	0.0035	**0.0351**	*
HRk	1	0.0043	0.2276	ns
WBz	1	0.0046	0.5189	ns
AL	1	0.0034	0.4897	ns
CALL	1	0.0011	**0.0248**	*
AT	1	0.0011	0.6237	ns
CO	1	0.0004	0.1605	ns
FI	1	0.0001	0.8853	ns
OC	1	< 0.001	0.8383	ns
FG	1	0.0120	0.9751	ns

Significance codes: bold values are significant at *p<0.05 and **p<0.01; ns=not significant

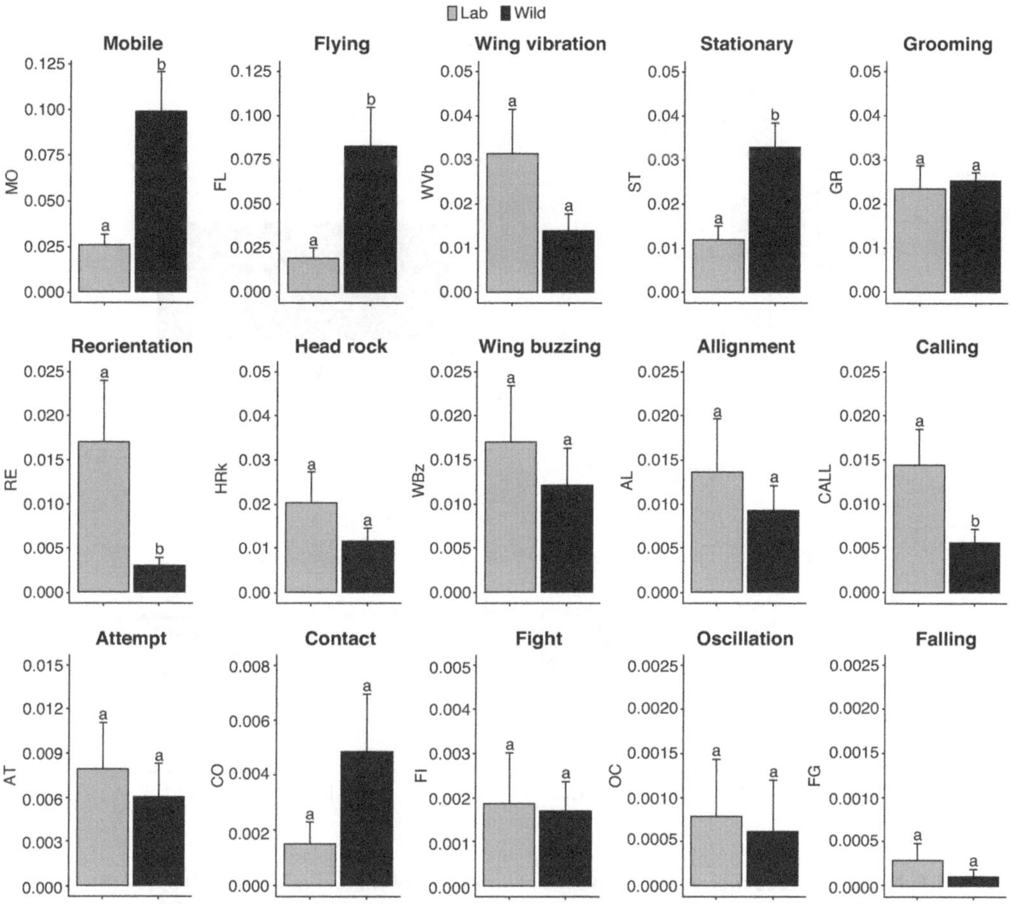

Figure 3.4 Relative frequencies of behavioral units for laboratory-reared (Lab) and wild populations, calculated as the number of occurrences divided by the total duration of the courtship. Graph bars represent mean values with standard errors (±SE). Bars followed by different letters indicate statistical differences between the populations ($p<0.05$, GLMs).

Calling behavior has been identified as a crucial step in a male's effort to ensure mating success (Whittier et al. 1994; Silva Neto et al. 2009). Our findings showed that the temporal pattern of calling behavior of the laboratory-reared males did not differ significantly from that of the wild males, indicating the quality of our laboratory population regarding this parameter. Moreover, we found a significant variation in the number of calling males from the two populations at different hours of the day. For both populations, we observed an unimodal pattern of calling behavior, with only one calling peak occurring during the day. This pattern was observed in wild *C. capitata* males evaluated in the laboratory under a similar photoperiod regime as ours (12L:12D) (Landolt et al. 1992). However, wild flies tested in the laboratory and maintained in a different photoperiod (14L:10D) showed an additional peak of calling in the afternoon (Papadopoulos et al. 1998; Diamantidis et al. 2008). Also, these patterns were found in

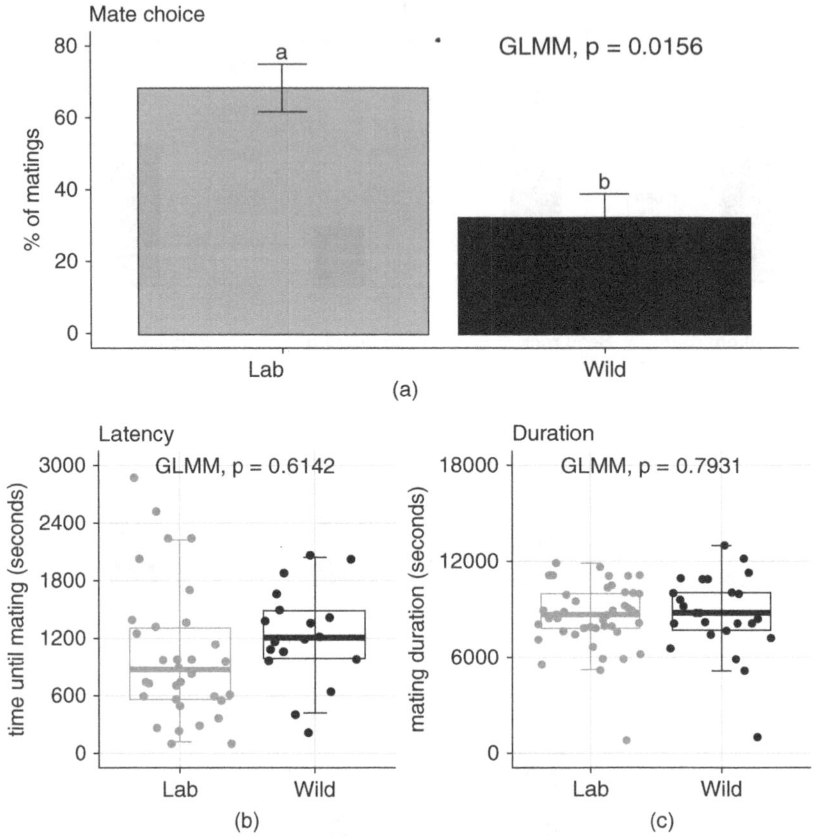

Figure 3.5 Mating success measured by the percentage (%) of mating (a), mating latency (b), and mating duration (c) for the laboratory-reared (Lab) and wild populations of *C. capitata*. Graph bars represent mean values with standard errors (±SE). Bars followed by different letters indicate statistical differences between the populations ($p<0.05$, GLMMs).

wild populations of *C. capitata* evaluated in field cages (Prokopy and Hendrichs 1979; Hendrichs and Hendrichs 1990).

Prokopy and Hendrichs (1979) observed only one peak of calling behavior in flies from Guatemala around 12:30 when the temperature was suitable enough to allow high activity levels. In a field experiment conducted in Egypt, Hendrichs and Hendrichs (1990) observed two peaks of calling behavior, one in the morning before the hottest time of the day and another small peak in the afternoon. Therefore, abiotic environmental conditions, such as temperature and photoperiod, seem to play a role in regulating the daily rhythm of sexual activity in *C. capitata*. Here, wild males presented a high incidence of calling behavior from early morning until 12:30, a range of time in which the temperature in Salvador, Bahia, Brazil (12°55′36.6″S 38°19′15.8″W; 12°56′13.4″S 38°19′53.5″W) is commonly warm enough for this pattern to occur. Alternatively, the absence of fluctuations in the temperature during the test could explain the concentration of elevated levels of calling behavior between the early hours of the day and early afternoon, as well as the lack of a second peak of calling. The males of the laboratory-reared population resumed calling in the late afternoon (18:00 and 18:30), possibly in response to domestication effects in an environment where the shift between light and dark is abrupt.

The primary courtship behaviors previously described for *C. capitata* (i.e., *Wing vibration*, *Wing buzzing*, and *Head rock*) (Briceño et al. 1996; Briceño and Eberhard 2000, 2002) occurred in both populations. The probability of the routine *Wing vibration* (WVb)>*Head rock* (HRk)>*Wing buzzing* (WBz) contributing to female acceptance was higher in the laboratory-reared population. This result reinforces the quality of our laboratory males in attracting the wild females and being successfully chosen through this courtship sequence. *Wing vibration* and *Wing buzzing* are critical courtship stages, producing acoustic signals to attract the female (Briceño and Eberhard 2000; Dias et al. 2023). As previously observed in *C. capitata* (Briceño et al. 1996), the males analyzed in this work continue to release pheromones during *Wing vibration* by everting the rectal epithelium ventrally. Furthermore, previous studies have reported that the pheromone droplet is either fully or partially reabsorbed during *Wing buzzing* and *Head rock* (Briceño et al. 1996; Lux et al. 2002). In this work, the pheromone droplet remained visible during these behaviors. However, due to the visual limitations of some footage, we cannot confirm whether this was the case in all occurrences of *Wing buzzing* and *Head rock*.

Head rock occurred in two stages in our study, as previously observed in *C. capitata* (Briceño and Eberhard 2002; Lux et al. 2002). In the first stage, *Head rock* begins during the final moment of *Wing vibration* and immediately before *Wing buzzing*. In the second stage, the male performs *Head rock* when *Wing buzzing* starts. The contact between male and female aristae occurs during the second stage of *Head rock* (Briceño and Eberhard 2002). *Ceratitis capitata* males have sexually dimorphic orbital bristles on the anterior portion of the head that were possibly selected to promote tactile stimulation and induce mating acceptance in the female (Miranda 2000; Briceño and Eberhard 2002). Here, the behavioral routine *Contact* (CO)>*Attempt* (AT) contributed substantially to female acceptance in both populations, with a higher probability in the wild population.

Laboratory-reared and wild males exhibited similar courtship durations. Compared with mass rearing in insect production facilities or large-scale laboratory rearing, our artificial environment provides more relaxed rearing conditions, such as low fly density and sufficient feeding surfaces. These conditions may have favored the preservation of courtship duration in laboratory-reared males since faster courtships are often associated with crowded environments where the risk of interruption by other flies is high (Liimatainen et al. 1997; Briceño and Eberhard 1998; Calcagno et al. 1999). Similarly, Gomes et al. (2024) found that rearing *Anastrepha obliqua* (Macquart) under relaxed conditions provided a suitable environment for preserving some aspects of courtship behavior, including courtship duration. Nevertheless, we also found that laboratory-reared males performed significantly fewer behavioral units during courtship. This indicates that these males performed fewer behavioral transitions, resulting in a less frenetic courtship compared to wild males.

Calling and *Reorientation* behaviors occurred more frequently in the laboratory-reared population. The behavioral unit described here as *Reorientation* was briefly addressed in previous studies on *C. capitata* (Arita and Kaneshiro 1983; Briceño et al.

1996). Also, a behavior of orientation toward the female was described more recently in the fruit fly *Anastrepha curvicauda* (Gerstaecker) (Salmerón-Muñiz et al. 2021). Here, we consider this behavior to be the male's precision in reorientating itself toward the female, usually after detecting that the female has repositioned or moved from one spot to another. The increased frequency of *Calling* and *Reorientation* in the laboratory-reared males could reflect an adaptation to the artificial environment, particularly to the recording cages, as these behaviors facilitate sexual interactions in confined spaces.

The behaviors *Mobile*, *Flying,* and *Stationary* occurred less frequently in the laboratory-reared population. These behaviors are not directly aimed at attracting females and are related to the general levels of activity. The lower frequency of *Mobile* and *Flying* could indicate that our rearing environment exerts a selective force that favors less-active flies in terms of mobility. The selection of less-active flies could reduce the occurrence of physical damage through collisions in the rearing cages (FAO/IAEA/USDA 2019). In addition, food and water sources are always nearby, and the flies do not have to travel long distances in search of these resources (Cayol 1999). Although the frequency of *Stationary* behavior was also lower, this result does not reflect higher overall activity levels. Instead, it suggests that the rearing environment possibly selected an increase in behaviors more related to female attraction and sexual interactions (e.g., *Calling* and *Reorientation* discussed above), which are typically low-mobility behaviors.

The proportion of matings is a measure that indicates the suitability of the flies and the environment for mating, providing an assessment of mating propensity (FAO/IAEA/USDA 2019). In the mate choice tests, the laboratory-reared males showed a higher proportion of matings compared to their wild counterparts. On the one hand, this result could indicate that not all wild males were fully sexually mature, as the wild flies used for the mate choice tests were at an age on the borderline of sexual maturity. On the other hand, the greater mating success of the laboratory-reared males suggests that the environment was more suitable for their sexual performance. Although wild and laboratory flies were kept under similar rearing conditions before the experiments and the mate choice tests were conducted under specific settings (i.e., larger cages than standard rearing cages, low fly density, and the presence of a small potted plant), the controlled and limited environment seems to have affected the performance of wild males. Indeed, it has been previously shown that wild flies exhibited lower sexual performance when tested under laboratory conditions than in field cages (Meza-Hernández and Díaz-Fleischer 2006). To determine if the laboratory-reared males were favored under specific rearing conditions or if they inherently performed better, we recommend conducting complementary studies in an environment that includes natural stimuli.

Mating latency and mating duration were similar between wild and laboratory-reared flies. Latency is a valuable proxy of readiness to mate (FAO/IAEA/USDA 2019). The similarity of this parameter between the two populations is desirable to allow the match between the courtship display of laboratory-reared males and the time at which wild females are receptive to mating and visiting the leks in nature. Regarding mating duration, fast copulations may be associated with the female's possibility of remating with fertile males (FAO/IAEA/USDA 2019). However, previous studies on *C. capitata* did not find a correlation between mating duration and the amount of sperm transferred to the female (Taylor et al. 2000; Costa et al. 2012). Furthermore, in *C. capitata*, it is still unclear whether it is the male or the female that exerts more control over mating duration, and, therefore, it remains difficult to define the contributions of each sex in controlling this parameter (Eberhard 2000).

Although laboratory domestication has quantitatively affected some parameters of sexual behavior in *C. capitata*, the relaxed conditions of the rearing environment, as compared to mass rearing conditions, seem to have favored the preservation of the temporal pattern of calling behavior and courtship behavioral repertoire, as well as the ability of laboratory-reared males to perform these behaviors. Within the context of the SIT, further investigations are needed to determine whether mass-reared flies exhibit similar behavioral patterns regarding the parameters evaluated here. Such studies will contribute to a comprehensive understanding of the relationship between the intensity of rearing efforts and the behavioral quality expected in the insects. Furthermore, given the economic importance of *C. capitata* as a major agricultural pest, our findings provide new insights into the effects of laboratory domestication on the sexual behavior of this species.

ACKNOWLEDGEMENTS

We thank the Centro Tecnológico Agropecuário do Estado da Bahia (CETAB) and the Universidade Federal da Bahia (UFBA) for providing support in maintaining the laboratory-reared and wild flies, respectively. We thank UFBA and the Programa de Pós-graduação em Ecologia (PPG ECOTAV) for their academic and scientific support. We thank the Coordenação de Aperfeiçoamento de Pessoal de Nível Superior (CAPES) for providing the first author with a scholarship.

REFERENCES

Arita, L. H., and K. Y. Kaneshiro. 1983. Pseudomale courtship behavior of the female Mediterranean fruit fly, *Ceratitis capitata* (Wiedemann). *Proceedings of the Hawaiian Entomological Society* 24:205–210.

Bates, D., M. Maechler, B. Bolker, et al. 2024. Lme4: Linear mixed-effects models using "Eigen" and S4. https://cran.r-project.org/web/packages/lme4/index.html.

Benelli, G. 2015. Aggression in Tephritidae flies: where, when, why? Future directions for research in integrated pest management. *Insects* 6 (1):38–53.

Benelli, G., K. M. Daane, A. Canale, et al. 2014. Sexual communication and related behaviours in Tephritidae: current knowledge and potential applications for integrated pest management. *Journal of Pest Science* 87 (3):385–405.

Briceño, R. D., and W. G. Eberhard. 1998. Medfly courtship duration: a sexually selected reaction norm changed by crowding. *Ethology Ecology & Evolution* 10 (4):369–382.

Briceño, R. D., and W. G. Eberhard. 2000. Male wing positions during courtship by Mediterranean fruit flies (*Ceratitis capitata*) (Diptera: Tephritidae). *Journal of the Kansas Entomological Society* 73 (3):143–147.

Briceño, R. D., and W. G. Eberhard. 2002. Courtship in the medfly, *Ceratitis capitata*, includes tactile stimulation with the male's aristae. *Entomologia Experimentalis et Applicata* 102 (3):221–228.

Briceño, R. D., D. Ramos, and W. G. Eberhard. 1996. Courtship behavior of male *Ceratitis capitata* (Diptera: Tephritidae) in captivity. *The Florida Entomologist* 79 (2):130.

Briceño, R. D., D. Ramos, and W. G. Eberhard. 1999. Aggressive behavior in medflies (*Ceratitis capitata*) and its modification by mass rearing (Diptera: Tephritidae). *Journal of the Kansas Entomological Society* 72 (1):17–27.

Calcagno, G. E., F. Manso, and J. C. Vilardi. 2002. Comparison of mating performance of medfly (Diptera: Tephritidae) genetic sexing and wild type strains: field cage and video recording experiments. *Florida Entomologist* 85 (1):41–50.

Calcagno, G. E., M. T. Vera, F. Manso, et al. 1999. Courtship behavior of wild and mass-reared Mediterranean Fruit Fly (Diptera: Tephritidae) males from Argentina. *Journal of Economic Entomology* 92 (2):373–379.

Cardoso, A. A. S. 2021. *Interações sociais podem alterar o comportamento sexual de Anastrepha fraterculus (Diptera: Tephritidae)?* Repositório Institucional da UFBA.

Carvalho, R. S., A. S. Nascimento, and W. J. R. Matrangolo. 1998. Metodologia de criação do parasitóide exótico: *Diachasmimorpha longicaudata* (Hymenoptera: braconidae), visando estudos em laboratório e em campo. *Embrapa Mandioca e Fruticultura. Circular Técnica* 30.

Cayol, J. P. 1999. Changes in sexual behavior and life history traits of Tephritid species caused by mass-rearing processes. In *Fruit Flies (Tephritidae): Phylogeny and Evolution of Behavior*, ed. M. Aluja, and A. L. Norrbom, 843–860. Boca Raton: CRC Press.

Cayol, J. P., J. Vilardi, E. Rial, et al. 1999. New indices and method to measure the sexual compatibility and mating performance of *Ceratitis capitata* (Diptera: Tephritidae) laboratory-reared strains under field cage conditions. *Journal of Economic Entomology* 92 (1):140–145.

Costa, A. M., C. S. Anjos-Duarte, A. K. P. Roriz, et al. 2012. Male diet and age influence to inhibit female remating in *Ceratitis capitata* (Diptera: Tephritidae). *Journal of Applied Entomology* 136 (6):456–463.

De Aquino, J. C., and I. S. Joachim-Bravo. 2014. Relevance of male size to female mate choice in *Ceratitis capitata* (Diptera: Tephritidae): investigations with wild and laboratory-reared flies. *Journal of Insect Behavior* 27 (2):162–176.

Diamantidis, A. D., J. R. Carey, C. T. Nakas, et al. 2011. Ancestral populations perform better in a novel environment: domestication of Mediterranean fruit fly populations from five global regions: adaptation of different *C. capitata* populations to captivity. *Biological Journal of the Linnean Society* 102 (2):334–345.

Diamantidis, A. D., N. T. Papadopoulos, and J. R. Carey. 2008. Medfly populations differ in diel and age patterns of sexual signalling. *Entomologia Experimentalis et Applicata* 128(3):389–397.

Dias, V. S., I. S. Joachim-Bravo, and B. A. J. Paranhos. 2023. Comportamento das moscas-das-frutas. In *Moscas-das-Frutas no Brasil: conhecimento básico e aplicado*, ed. R. A. Zucchi, A. Malavasi, R. Adaime, and D. E. Nava, 227–259. Piracicaba: Fealq.

Dor, A., and P. Liedo. 2019. Survival ability of Mexican Fruit Fly males from different strains in presence of the predatory Orb-Weaving Spider *Argiope argentata* (Araneae: Araneidae). *Bulletin of Entomological Research* 109 (3):279–286. https://doi.org/10.1017/S0007485318000275.

Dunn, D. W., and P. A. Follett. 2017. The sterile insect technique (SIT) – an introduction. *Entomologia Experimentalis et Applicata* 164 (3):151–154.

Eberhard, W. G. 2000. Sexual behavior and sexual selection in the Mediterranean fruit fly, *Ceratitis capitata* (Dacinae: Ceratitidini). In Fruit Flies (Tephritidae): Phylogeny and Evolution of Behavior, ed. M. Aluja, and A. L. Norrbom, 459–490. Boca Raton: CRC Press.

Economopoulos, A. P. 1992. Adaptation of the Mediterranean fruit fly (Diptera: Tephritidae) to artificial rearing. *Journal of Economic Entomology* 85(3):753–758.

FAO/IAEA/USDA. 2019. *Product Quality Control for Sterile Mass-Reared and Released Tephritid Fruit Flies, Version 7.0*. Vienna, Austria: International Atomic Energy Agency.

Féron, M. 1962. *L'Instinct de reproduction chez la mouche mediterraneenne des fruits Ceratitis capitata Wied (Dipt. Trypetidae) comportement sexuel*. Revue de Pathologie Végétale et d'Entomologie Agricole de France.

Fox, J., S. Weisberg, B. Price, et al. 2023. Car: companion to applied regression. https://cran.r-project.org/web/packages/car/index.html.

Gaire, S. K., S. Pokhrel, M. J. H. Biswas, et al. 2022. Effects of domestication on quality control parameters of the Queensland Fruit Fly *Bactrocera tryoni* (Diptera: Tephritidae). *Journal of Applied Entomology* 146(6):687–699. https://doi.org/10.1111/jen.12998.

Gomes, I. V., A. K. P. Roriz, A. S. Araújo, et al. 2024. Sexual behaviour of *Anastrepha obliqua* (Macquart) (Diptera: Tephritidae): do laboratory domestication conditions influence male courtship behaviour? *Physiological Entomology* 49 (2):99–109.

Hendrichs, J., and M. A. Hendrichs. 1990. Mediterranean fruit fly (Diptera: Tephritidae) in nature: location and diel pattern of feeding and other activities on fruiting and nonfruiting hosts and nonhosts. *Annals of the Entomological Society of America* 83 (3):632–641.

Hendrichs, J., and A. Robinson. 2009. Chapter 243 - Sterile insect technique. In *Encyclopedia of Insects (Second Edition)*, ed. V. H. Resh, and R. T. Cardé, 953–957. San Diego: Academic Press.

Hendrichs, J., A. S. Robinson, J. P. Cayol, et al. 2002. Medfly areawide sterile insect technique programmes for prevention, suppression or eradication: the importance of mating behavior studies. *Florida Entomologist* 85 (1):1–13.

Hoffmann, A. A., and P. A. Ross. 2018. Rates and patterns of laboratory adaptation in (mostly) insects. *Journal of Economic Entomology* 111(2):501–509.

Hothorn, T., F. Bretz, P. Westfall, et al. 2023. Multcomp: simultaneous inference in general parametric models. https://cran.r-project.org/web/packages/multcomp/index.html.

Japyassu, H. F., C. C. Alberts, P. Izar, et al. 2006. EthoSeq: a tool for phylogenetic analysis and data mining in behavioral sequences. *Behavior Research Methods* 38:549–556.

Lance, D. R., D. O. McInnis, P. Rendon, et al. 2000. Courtship among sterile and wild *Ceratitis capitata* (Diptera: Tephritidae) in field cages in Hawaii and Guatemala. *Annals of the Entomological Society of America* 93 (5):1179–1185.

Landolt, P. J., R. R. Heath, and D. L. Chambers. 1992. Oriented flight responses of female Mediterranean fruit flies to calling males, odor of calling males, and a synthetic pheromone blend. *Entomologia Experimentalis et Applicata* 65 (3):259–266.

Lecocq, T. 2018. Insects: the disregarded domestication histories. In *Animal Domestication*, ed. F. Teletchea, 35–68. London: IntechOpen.

Lenth, R. V., B. Bolker, P. Buerkner, et al. 2024. Emmeans: estimated marginal means, aka least-squares means. https://cran.r-project.org/web/packages/emmeans/index.html.

Levinson, H. Z., A. R. Levinson, and K. Schäfer. 1987. Pheromone biology of the Mediterranean fruit fly (*Ceratitis capitata* Wied.) with emphasis on the functional anatomy of the pheromone glands and antennae as well as mating behaviour. *Journal of Applied Entomology* 104 (1–5):448–461.

Liimatainen, J., A. Hoikkala, and T. Shelly. 1997. Courtship behavior in *Ceratitis capitata* (Diptera: Tephritidae): comparison of wild and mass-reared males. *Annals of the Entomological Society of America* 90 (6):836–843.

Lux, S. A., F. N. Munyiri, J. C. Vilardi, et al. 2002. Consistency in courtship pattern among populations of medfly (Diptera: Tephritidae): comparisons among wild strains and strains mass reared for SIT operations. *Florida Entomologist* 85 (1):113–125.

Martin, P., and P. Bateson. 2007. *Measuring Behaviour: An Introductory Guide*. Cambridge: Cambridge University Press.

Meats, A., H. M. Holmes, and G. L. Kelly. 2004. Laboratory adaptation of *Bactrocera tryoni* (Diptera: Tephritidae) decreases mating age and increases protein consumption and number of eggs produced per milligram of protein. *Bulletin of Entomological Research* 94 (6):517–524.

Meza-Hernández, J. S., and F. Díaz-Fleischer. 2006. Comparison of sexual compatibility between laboratory and wild Mexican fruit flies under laboratory and field conditions. *Journal of Economic Entomology* 99 (6):1979–1086.

Miranda, X. 2000. Sexual dimorphism in the aristae of *Ceratitis capitata* (Diptera, Tephritidae) and its possible importance in courtship. *Journal of the New York Entomological Society* 108 (3/4):339–348.

Nakamori, H., and K. Simizu. 1983. Comparison of flight ability between wild and mass-reared Melon Fly, *Dacus Cucurditae* Coquillett (Diptera: Tephritidae), using a flight mill. *Applied Entomology and Zoology* 18 (3):371–381. https://doi.org/10.1303/aez.18.371.

Papadopoulos, N. T., B. I. Katsoyannos, N. A. Kouloussis, et al. 1998. Effect of adult age, food, and time of day on sexual calling incidence of wild and mass-reared *Ceratitis capitata* males. *Entomologia Experimentalis et Applicata* 89 (2):175–182.

Pereira, R., J. Hendrichs, and A. Malavasi. 2023. Técnica do inseto estéril. In *Moscas-das-Frutas no Brasil: conhecimento básico e aplicado*, ed. R. A. Zucchi, A. Malavasi, R. Adaime, and D. E. Nava, 419–427. Piracicaba: Fealq.

Pereira, R., N. Silva, C. Quintal, et al. 2007. Sexual performance of mass reared and wild Mediterranean fruit flies (Diptera: Tephritidae) from various origins of the Madeira Islands. *Florida Entomologist* 90 (1):10–14.

Prokopy, R. J., and J. Hendrichs. 1979. Mating behavior of *Ceratitis capitata* on a field-caged host tree. *Annals of the Entomological Society of America* 72 (5):642–648.

Purugganan, M. D. 2022. What is domestication? *Trends in Ecology & Evolution* 37 (8):663–671.

R Core Team. 2024. R: *A Language and Environment for Statistical Computing*. Vienna, Austria: R Foundation for Statistical Computing. https://www.R-project.org/.

Rao, D., S. Aguilar-Argüello, P. Montoya, et al. 2014. The effect of irradiation and mass rearing on the anti-predator behaviour of the Mexican Fruit Fly, *Anastrepha ludens* (Diptera: Tephritidae). *Bulletin of Entomological Research* 104 (2):176–181. https://doi.org/10.1017/S0007485313000643.

Roriz, A. K. P., H. F. Japyassú, C. Cáceres, et al. 2019. Pheromone emission patterns and courtship sequences across distinct populations within *Anastrepha fraterculus* (Diptera-Tephritidae) cryptic species complex. *Bulletin of Entomological Research* 109 (3):408–417.

Roriz, A. K. P., H. F. Japyassu, and I. S. Joachim-Bravo. 2018. Courtship in two morphotypes of the *Anastrepha fraterculus* (Diptera: Tephritidae) cryptic species complex and their implications for understanding mate recognition. *Journal of Insect Behavior* 31 (5):535–551.

Rössler, Y. 1975. Reproductive differences between laboratory-reared and field-collected populations of the Mediterranean Fruitfly, *Ceratitis capitata*. *Annals of the Entomological Society of America* 68 (6):987–991.

Salmerón-Muñiz, N. N., R. Arzuffi, N. Robledo-Quintos, et al. 2021. The influence of male dominance in female *Anastrepha curvicauda* mate selection. *Scientific Reports* 11 (1):6311.

Shelly, T. E. 2018. Sexual selection on leks: a fruit fly primer. *Journal of Insect Science* 18 (3):9.

Silva Neto, A. M., V. S. Dias, and I. S. Joachim-Bravo. 2009. Escolha de parceiro para acasalamento em *Ceratitis capitata* (Wiedemann) (Diptera: Tephritidae): influência do envelhecimento dos machos no sucesso de cópula. *Neotropical Entomology* 38 (5):571–577.

Silva Neto, A. M., T. R. O. Santos, V. S. Dias, et al. 2012. Mass-rearing of Mediterranean fruit fly using low-cost yeast products produced in Brazil. *Scientia Agricola* 69:364–369.

Sugayama, R. L., and A. Malavasi. 2000. Ecologia comportamental. In *Moscas-das-frutas de importância econômica no Brasil: conhecimento básico e aplicado*, ed. A. Malavasi, and R. A. Zucchi, 103–108. Ribeirão Preto: Holos.

Taylor, P. W., R. Kaspi, and B. Yuval. 2000. Copula duration and sperm storage in Mediterranean fruit flies from a wild population. *Physiological Entomology* 25 (1):94–99.

Whittier, T. S., F. Y. Nam, T. E. Shelly, et al. 1994. Male courtship success and female discrimination in the Mediterranean fruit fly (Diptera: Tephritidae). *Journal of Insect Behavior* 7 (2):159–170.

Wickham, H., W. Chang, L. Henry, et al. 2024. ggplot2: *Create Elegant Data Visualizations Using the Grammar of Graphics*. https://ggplot2.tidyverse.org/

Chapter 4

Nice songs, good smells

Overexpression of antioxidants in transgenic male fruit flies blunts radiation-induced oxidative stress and improves sexual signaling

Vanessa Simões Dias, Robert G. Bruton, Everett Foreman, Richard W. Mankin, Hans T. Alborn, Nicholas M. Teets, Marc F. Schetelig, Alfred M. Handler, and Daniel A. Hahn

4.1 INTRODUCTION

Mate choice can drive the evolution of elaborate traits that reliably signal an individual's condition (Andersson 1994; Andersson and Simmons 2006; Kotiaho and Puurtinen 2007; Jones and Ratterman 2009; Kuijper et al. 2012). Some sexual signals are costly, constituting a potentially reliable source to convey honest information for mate choice (Kotiaho 2001). This cost ensures that only high-quality individuals can afford to display strong signals, maintaining their honesty – a concept central to the Zahavi–Grafen handicap principle. Hence, the cost of sexual signals may function as a handicap (strategic cost) that can be used as an honest indicator of the signaler's quality (Zahavi 1975; 1977; Grafen 1990).

However, the handicap hypothesis has been criticized for lacking strong empirical support (Lachmann et al. 2001; Grose 2011; Hill 2011; Számadó 2011; Higham and Hebets 2013; Számadó and Penn 2015). Alternative models suggest that honesty in sexual signals is maintained instead by the risk of cheating being too high or by evolutionary constraints (Lachmann et al. 2001; Számadó 2011; Holman 2012). Low-quality individuals cannot effectively mimic high-quality signals due to physiological limitations (Higham and Hebets 2013), but the exact mechanisms preventing dishonesty remain unclear (Galvan et al. 2015). Despite ongoing debate, the handicap hypothesis has been expanded to include production and maintenance costs as key factors in signal honesty (Weaver et al. 2017).

A less controversial alternative to the handicap hypothesis is the **index hypothesis** (Maynard Smith and Harper 1995, 2003). Index signals are constrained by physiological or developmental mechanisms, making them inherently honest indicators of condition (Ruxton and Schaefer 2011; Laidre and Johnstone 2013). Examples of index signals include courtship displays in *Drosophila obscura* (Maynard Smith 1956), the calling song of the male black-horned tree cricket *Oecanthus nigricornis* (Brown et al. 1996), and pheromone production in the cockroach *Nauphoeta cinerea* (South et al. 2011). Unlike handicap signals, index signals do not incur unnecessary costs but require sufficient physiological capacity to be effectively produced (Maynard Smith and Harper 1995, 2003). This means that individuals with intrinsic limitations cannot fake them (Biernaskie et al. 2014; Higham 2014; Wyatt 2014).

DOI: 10.1201/9781003644330-5

A key mechanistic explanation for honest condition-dependent signals is their sensitivity to **oxidative stress** (von Schantz et al. 1999). Sexual traits may reflect an individual's ability to manage oxidative damage, which arises from an imbalance between pro-oxidants (e.g., reactive oxygen species, ROS) and antioxidant defenses (Monaghan et al. 2009; Brieger et al. 2012). Females may prefer males with strong antioxidant defenses, as indicated by their high-quality sexual signals (Velando et al. 2008; Metcalfe and Alonso-Alvarez 2010; Garratt and Brooks 2012). **Mitochondrial function**, which regulates cellular energy production and ROS generation, plays a central role in this process (Hill 2014, 2015). Disruptions in mitochondrial efficiency can impair sexual signaling, making mitochondria reliable indicators of individual condition (Koch et al. 2017; Weaver et al. 2017; Koch and Hill 2018).

4.1.1 *Anastrepha suspensa* as an important pest and suitable model system

The Caribbean fruit fly *Anastrepha suspensa* is a polyphagous species with 89 suitable host plants recorded under natural field conditions (Hanlin et al. 2020). Widespread in native habitats of the Caribbean region, *A. suspensa* is invasive to the United States, where it is restricted to a naturalized range of Southern and Central Florida (Norrbom 2022). Despite the low potential for the establishment of *A. suspensa* in temperate regions, a major invasive risk exists for novel tropical and subtropical areas (Santana et al. 2023; EPPO 2025). Besides its economic importance, *A. suspensa* serves as a suitable model for examining how mitochondrial function influences sexual signaling and mate choice. Like several other important tephritid fruit fly pests, males of *A. suspensa* establish leks where they compete for the attention of a limited number of females through complex courtship displays that include calling songs and pheromone production (Burk 1981; Sivinski and Webb 1985; Sivinski 1989; Webb et al. 1975, 1976, 1984). Nutritional studies reveal that only well-fed males can afford to produce these costly signals (Yuval et al. 1998; Kaspi et al. 2000). Furthermore, irradiation studies indicate that oxidative stress hampers both acoustic and chemical signaling, thereby reducing male mating success (Sharp and Webb 1977; Webb et al. 1987; Walder and Calkins 1993).

4.1.2 Relevance of the study on sexual signals for pest management

Besides providing results that can contribute substantially to our understanding of the physiological basis of fruit fly female mate choice and the role of antioxidants in sexual selection, this study offers practical benefits for pest management, including the sterile insect technique (SIT). The SIT is an environmentally friendly method for reducing a pest population through mating between sterile males and wild females, thus minimizing the use of pesticides (Knipling 1955; Vreysen et al. 2006). Insect sterilization using ionizing radiation is an essential operational component for SIT application (FAO/IAEA 2017). Unfortunately, irradiation in air reduces the sexual competitiveness and quality (e.g., adult emergence and flight ability) of fruit flies (reviews in Calkins and Parker 2005; Lance and McInnis 2005). One of the causes of detrimental effects in irradiated insects involves increasing oxidative stress in living tissues via free radical production during and/or after irradiation (Calkins and Parker 2005; López-Martínez and Hahn 2012). This study examined whether enhancing antioxidant capacity can reduce oxidative stress and improve sexual signaling in irradiated *A. suspensa* males. Specifically, we explored whether overexpressing mitochondrial superoxide dismutase (SOD2) – a key antioxidant enzyme – decreases oxidative damage and enhances calling song and pheromone production. SOD2 is essential for neutralizing mitochondrial ROS, potentially boosting cellular respiration efficiency under stress (Murphy 2009; Candas and Li 2014; Zhou et al. 2017). Based on the mitochondrial function hypothesis, we anticipate that irradiated males overexpressing SOD2 will exhibit reduced oxidative stress and enhanced sexual signals compared to wild-type (WT) irradiated males. Furthermore, testing the extent to which overexpression of SOD2 can improve sexual performance in irradiated males may provide new perspectives that can be applied to improve the efficacy or economy of operational SIT programs.

4.2 MATERIALS AND METHODS

4.2.1 Insect strains and rearing protocol

We used two lines of *A. suspensa* in our experiments, a WT strain and a transgenic mitochondrial SOD (SOD2 5.2) strain created in the WT genetic background. The transgenic SOD2 5.2 line expressed a Y-linked male-only insertion resulting in an additional genomic copy of the *A. suspensa* SOD2 coding sequence (Teets et al. 2019). SOD activity in the SOD2 5.2 line was approximately 50% higher than WT (Teets et al. 2019). Both lines were kept at 26±1°C, 60% relative humidity, and 14L:10D photoperiod following a rearing protocol that consisted of routinely collecting and seeding eggs onto an artificial diet for larval development, followed by pupation, adult emergence, and maintenance. The artificial larval diet used to maintain both lines was composed of water, torula yeast (75.3 g/L), wheat germ (75.3 g/L), dextrose (75.3 g/L), cholesterol (0.6 g/L), sodium benzoate (1.3 g/L), methyl 4-hydroxybenzoate (1.3 g/L), butyl 4-hydroxybenzoate (1.0 g/L), and hydrochloric acid (5.3 mL/L). Six days after egg seeding, Petri dishes containing diet and larvae were placed into containers with vermiculite for pupation. Nearly one week after pupation, pupae were sifted and kept in cylindrical plastic cages made for from food storage containers (5 cm diameter×10 cm height) until adult emergence. Adults were fed an artificial adult diet consisting of a three-part extra fine-granulated cane sugar (Publix®) and a one-part yeast hydrolysate enzymatic (MP Biomedicals®). Before the experiments and within 48 hours after adult emergence, a period in which females remained unmated, flies were separated by sex and placed in standard cages made of food storage canisters (15 cm length×13 cm width×26 cm height) with a density of up to 100 flies/cage with unlimited access to water and adult diet.

4.2.2 Irradiation treatment

Two days before adult emergence, WT and SOD2 5.2 late-stage pupae (pharate adults) were irradiated in normal air and room temperature at 70 Gy, the dose recommended for *A. suspensa* SIT programs (FAO/IAEA/USDA 2014). Briefly, pupae were placed in Petri dishes (6 cm diameter×1.5 cm height) that fit into a cardboard cylinder (7 cm diameter × 16 cm height). This cylinder was placed into a Cesium-137 Gammacell 1000 Irradiator, and pupae were irradiated at a dose rate of ~8 Gy/min until exposure to 70 Gy was achieved (8 minutes 45 seconds). Dose accuracy was routinely verified using an HD-V2 Gafchromic film (uncertainty 5.6% at 95% confidence level) and alanine pellet dosimeters (3.8% at 95% confidence level) (FWT®). The HD-V2 film was read at ~26°C through an optical density meter (DoseReader 4, RadGen®) between 24 and 36 hours after radiation exposure. Alanine pellets were read at the National Center for Electron Beam Research facility on the Texas A&M University campus. Untreated (0 Gy) and irradiated treated (70 Gy) males from both WT and SOD2 5.2 lines were evaluated across all experiments totaling four groups: WT-0 Gy, WT-70 Gy, SOD2 5.2-0 Gy, and SOD2 5.2-70 Gy.

4.2.3 Acoustic communication

4.2.3.1 Laboratory bioassays

Sexual acoustic signals of nonirradiated and irradiated males were compared to test the extent to which SOD2 overexpression could improve the courtship songs of irradiated males. Before the recording started, one focal and three muted sexually mature (11 ± 2 days old) males from the same origin and treatment were released into a screen-mesh cage (7 cm diameter × 15 cm height). The focal male was nonmuted and, therefore, the only one being recorded. The screen-mesh cage used in the experiments was made of 1.6 mm copper silver-coated wire (two circular bases and six pillars), tulle mesh fabric (lateral face), galvanized wire mesh (roof and floor), and a polypropylene translucent podium (3 cm length × 1.5 cm width) glued internally in the center-top of the cage. The podium was fixed onto the cage roof to increase adhesion stability to the substrate while wing-fanning males were calling. Males were muted by chilling them on ice for 3 minutes (Figure 4.1a) and cutting their wings (Figure 4.1b) with iris microdissection scissors. Wing removal procedures were done one day before the experiments

Figure 4.1 Muted *Anastrepha suspensa* males used in the acoustic communication bioassays. (a) Winged males were first chilled on ice for 3 minutes. (b) The wings of chilled and motionless males were cut with iris microdissection scissors.

to allow males to recover and be able to sexually interact with their counterparts by emitting chemical signals (pheromones) but not acoustic signals (calls). We recognize, however, that wing removal may have limited the ability of muted males to disperse sexual pheromones. Preliminary experiments showed that the presence of muted males increased the calling effort of the focal males. Five minutes after the males were released into the screen-mesh cage, one untreated WT female was exposed to the four males, and the recording started. Acoustic signals produced by focal males were recorded either for up to 30 minutes or until the end of the copulatory songs produced right after the male jumped over the female during mating. All recordings were performed in the afternoon, from 14:00 to 19:00 hours, the peak time of *A. suspensa* mating (Burk 1983; Landolt and Sivinski 1992). At least 20 replicate recordings, each from a different focal male, were performed for each treatment.

4.2.3.2 Acoustic signal recordings

Male songs produced by wing fanning were recorded in a sound-insulated room (Webb et al. 1976; Mankin et al. 1996) to reduce background noise and improve signal detectability and quality. Briefly, a microphone (model 4145, Brüel and Kjær, Naerum, Denmark) was placed 3–5 mm above the screen-mesh cage containing the insects and was used as the signal transducer. Acoustic signals detected by the microphone were conditioned with a bandpass filter (model 3100, Krohn-Hite, Avon, MA) between 0.1 and 10 kHz. The filtered signals were preamplified (model 2639, Brüel and Kjær [B&K], Naerum, Denmark), then amplified (model 4370, B&K) 20–40 dB, and stored on a digital recorder (Model HDP2, Tascam, TEAC, Montebello, CA) at a 44.1 kHz sampling rate for further analyses. Simultaneously with the sound recordings, an infrared-sensitive digital camera (model 951130, CCS-Sony-HR N266, Sony, Tokyo, Japan) was used to video record each male evaluated in this study. The videos were later used to confirm whether the sounds were related to sexual signals or other general activities, such as flight within the cage.

4.2.3.3 Acoustic signal processing

Acoustic signals were processed following the classification proposed by Sivinski et al. (1984). Briefly, sexual signals were categorized as calling song, consisting of repeated trains (bursts) of wing-beat sound pulses produced during pheromone release, and copulatory song (initially named precopulatory song), a continuous train of wing-beat sound pulses produced by mounted males right before and during the beginning of copulation. In this study, the calling song was defined as a repeated sound containing at least three pulse train bursts separated from each other by no more than 1 second. Only the first bout of calling and copulatory songs was analyzed from each recording. The calling and copulatory songs were analyzed using Raven Lite 2.0 software (Cornell Lab of Ornithology, Ithaca, NY).

Structural, temporal, and spectral parameters of calling and copulatory songs were evaluated and compared among untreated and treated males. Structural parameters were defined as the general arrangement of the calling song, including the number of pulse train bursts in a song and the length of each pulse train, by which it is classified as a short (<3 seconds) or long (≥3 seconds) burst (Figure 4.2a). Because the copulatory

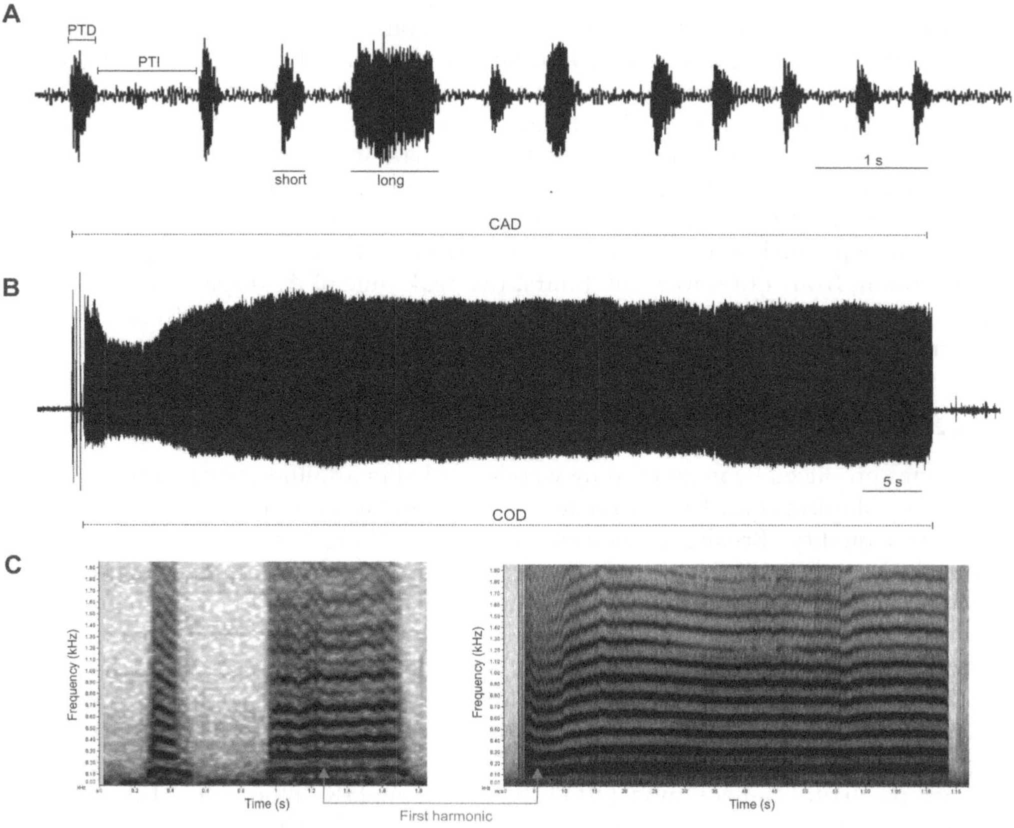

Figure 4.2 Example of common calling and copulatory songs of *Anastrepha suspensa* males. (a) Oscillogram shows the structure of a calling song with its short- and long-pulse trains. The temporal parameters of the calling song consisted of pulse train duration (PTD), pulse train interval (PTI), and calling duration (CAD). (b) Oscillogram of a copulatory song showing its temporal measure, the copulatory duration (COD). (c) Spectrograms display the harmonics of a calling song with two pulse trains (left) and a copulatory song (right). The first harmonic was used to measure the fundamental frequency. The relative power level was measured at the start, midpoint, and end of each pulse train and copulatory song. The first harmonic is also the wing-beat frequency.

song is continuous, no structural parameter was associated with copulatory songs. Temporal parameters consisted of pulse train duration, pulse train interval (PTI), calling duration (Figure 4.2a), and copulatory duration (Figure 4.2b). The fundamental frequency (FF) and relative power level (power) spectral parameters were measured from spectrograms (2048-focus) generated by Raven Lite 2.0 software (Figure 4.2c). The FF is considered the wing-beat frequency (Webb et al. 1976), and the power measured in this study was used to determine the signal intensity directly associated with the FF measured in each pulse train. Both measures of FF and power were obtained by vertically zooming into the spectrograms (11×) and positioning the mouse pointer at the middle of the first harmonic (Figure 4.2c, first wave showed in the spectrogram) in a time axis of 0.5-second major tick interval to obtain and record the values specific to the selected point in each time. Specifically, FF and relative spectrum levels were measured at the start, midpoint, and end of each pulse train and copulatory song. The use of multiple measures at different time points of the same acoustic signal accounts for the intrinsic dynamic properties of the calling and copulatory songs. This multiple time point assessment approach is particularly important because it is known that the FF of the *A. suspensa* trains tends to decrease over time within the same song (Webb et al. 1987; Mankin et al. 2000).

4.2.4 Sexual communication with pheromones

4.2.4.1 Pheromone collection

The sexual pheromones of *A. suspensa* males were collected and analyzed to determine whether SOD2 overexpression can affect pheromone production in irradiated transgenic insects compared to irradiated WT males. Pheromone was collected in the first hours of the sexual activity period, from 14:30 to 17:00 hours (Burk 1983; Heath et al. 1993). Briefly, ten sexually mature virgin males (11±2 days old) from the same origin and treatment were acclimated in a glass volatile collection chamber (10 cm diameter×30 cm height) for 1 hour 30 minutes before volatile collection. After the acclimation time, the collection chamber was connected to one charcoal filter at the air intake, and one Tenax filter (adsorbent material) was attached to the opening located at the top-center of the collection chamber (Cross 1980). Subsequently, the vacuum line was adjusted to pull 0.9–1 L air/min and was immediately attached to the Tenax filter in the collection chamber. Volatiles were collected for 20 minutes. After the collection time, the Tenax filter was disconnected from the collection chamber, sealed in a sterilized container, and kept at approximately −10°C for at least 18 hours before chemical analysis. Filters for volatile collections were prepared by packing 65 mg Tenax (Tenax TA, 60-80 mesh, Supelco, Pennsylvania, USA) into a splitless single-taper gas-chromatograph inlet liner with glass wool (5062-3587, Agilent, California, USA), and then packing a small plug of glass wool above the Tenax. At least 17 pheromone collections were made for each treatment using different males pooled together in groups of ten insects in each collection.

4.2.4.2 Pheromone analysis

After volatile collection, pheromone samples were analyzed using thermal desorption gas chromatography-mass spectrometry (TD-GC/MS) (Alborn 2018) using an HP 6890 gas chromatograph with a 30 m×0.25 mm ID, 0.25 μm film thickness DB-1 capillary column, and a mass selective detector in electron impact ionization mode (model 5973, Hewlett Packard, California, USA). The volatile samples were transferred directly from the Tenax filters into the column of the GC by thermal desorption injection using a conventional splitless inlet as the desorption oven with an attached cryofocusing device to achieve direct on-column injection and focusing of the sample without the use of additional flash heating (Alborn and Willms 2015). The GC inlet was set at 200°C to facilitate thermal desorption. The oven was held at 30°C for the first 3 minutes and then ramped at 10°C/min to a final temperature of 260°C and was held at the final temperature for 5 minutes. The helium-carrier gas flow rate was set at 1.4 mL/min for the first 2 minutes to aid in desorption and was then changed to 0.7 mL/min for the remainder of the run. Compounds were identified using a combination of standards (see below), comparison to the National Institute of Standards and Technology (NIST) mass spectral library, and the use of literature information for the expected components and elution order (Nation 1990; Rocca et al. 1992). Quantification of analytes was performed by comparison to serial dilutions of standards analyzed by the same chromatographic method but using a solvent injection into a regular splitless inlet without thermal desorption and cryofocusing. The β-ocimene was quantified using β-ocimene standard; (Z)-3-nonenol and (Z,Z)-3,6-nonadienol were quantified using 1-decanol standard; the three sesquiterpenes (α-trans-bergamotene, α-farnesene, and β-bisabolene) were quantified by comparison to β-caryophyllene standard; and the three lactones (suspensolide, anastrephin, and epianastrephin) were quantified using methyl tridecanoate standard (Nation 1990).

4.2.5 Statistical analyses

Fly line (line), radiation treatment (radiation), and their interaction (line×radiation) were included as fixed effects in all statistical analyses performed in this study. The response variables calling proportion (measured in percentage as calling propensity) and mating proportion (measured in percentage as mating propensity) for both focal and muted males were binomially distributed and thus analyzed using generalized linear models (GLMs) with a logit link function. The structural parameters of calling songs were also evaluated in GLMs with the total number of pulse trains, the number of short trains, and the number of long trains following a Poisson distribution with a log link function. The duration of all temporal and spectral parameters was analyzed using a two-way ANOVA. To improve the linear model fitting, the duration of calling songs, pulse trains, PTIs, and copulatory songs was log-transformed before data analysis. The total amount of pheromone produced by males was log-transformed and analyzed using a two-way ANOVA. The differences in pheromone composition among treatments were analyzed by a multivariate principal component analysis (PCA). Briefly, the relative peak areas from each compound identified by the TD-GC/MS were first converted to mass based on the standard for each volatile. The relative masses of the chemicals used in the PCA were log-transformed and scaled. A scatterplot of the rotated scores of the first two components from the PCA was used to obtain confidence ellipses (95%) representing group clusters based on male treatment. The first principal component (PC1) score, known to represent the greatest proportion of variance in the dataset (Abdi and Williams 2010), was then used as the dependent variable in a two-way ANOVA. All statistical analyses were conducted in R (R Core Team, version 3.4.3, 2018). A significance level of 0.05 was used for all analyses.

4.3 RESULTS

4.3.1 Acoustic communication

4.3.1.1 General insect performance

Two distinct sexual acoustic signals commonly identified in tephritid flies, calling and copulatory songs (Sivinski et al. 1984), were identified in the recordings evaluated in this study. Males across groups differed in both mating propensity and the

Table 4.1 Number of acoustic recordings performed, calling propensity, and proportion of mating of irradiated and nonirradiated WT and SOD2 5.2 *Anastrepha suspensa* males

Treatment	*Number of acoustic recordings*		*Calling propensity*[b]	*Mating propensity**	
	Total	*Analyzed*[a] *(%)*		*Focal males*[c]	*Muted males*[d]
WT 0 Gy	42	35 (83%)	48%	74%[a]	14%
SOD2 5.2 0 Gy	33	26 (78%)	30%	67%[a,b]	15%
WT 70 Gy	46	21 (45%)	35%	43%[b]	22%
SOD2 5.2 70 Gy	31	22 (71%)	39%	74%[a]	10%

* Different letters indicate significant differences between the treatments (Tukey contrasts, $p<0.05$)
[a] Number of processed acoustic recordings in which males exhibited calling and/or copulatory songs.
[b] Binary indicator of whether pheromone calling occurred during the recording period.
[c] Matings with focal males within 30 minutes maximum recording time.
[d] Matings with muted males without wings within 30 minutes maximum recording time.

temporal and spectral properties of their songs. The number of males recorded and analyzed is presented in Table 4.1. Irradiated WT-70 Gy (45%, 21 of 46) males engaged less often in sexual song production than WT-0 Gy (83%, 35 of 42), SOD2 5.2-0 Gy (78%, 26 of 33), and SOD2 5.2-70 Gy (71%, 22 of 31) males, as indicated by the number of acoustic recordings analyzed (Table 4.1). Although WT-0 Gy males exhibited calling songs more frequently than the other groups of males, their calling propensity did not differ (Table 4.1). In contrast, the mating propensity of focal males differed significantly among groups, with WT-0 Gy (74%) and SOD2 5.2-70 Gy (74%) being more likely to mate than WT-70 Gy (43%) males (Line×Radiation*: Table 4.1). That is, radiation reduced the mating propensity of irradiated WT males, but did not decrease the mating propensity of irradiated SOD2 5.2 males in nonchoice trials. A few muted males (wingless) did mate after competing with focal males (winged) from the same origin and radiation treatment, but there was no difference among groups of males in the number of muted males that mated (Table 4.1).

4.3.1.2 Signal structural parameters

Nonirradiated transgenic males sang less frequently and exhibited calling songs with lower numbers of pulse trains than nonirradiated WT males, indicating a potential side effect of transgenesis under nonstressful conditions (Line*: Table 4.2). Specifically, SOD2 5.2-0 Gy exhibited calling songs with a fewer number of pulse trains than WT-0 Gy, SOD2 5.2-70 Gy, and WT-70 Gy males (Figure 4.3a). Unlike the effect of fly lines on the structure of calling songs, the radiation treatment increased the total number of pulse trains in both irradiated WT and SOD2 5.2 males compared to nonirradiated insects (Radiation*: Table 4.2). However, although the total number of pulse trains increased in WT-70 Gy males relative to WT-0 Gy males, the difference was not statistically significant (Figure 4.3a). Neither fly line nor radiation treatment affected the number of short-pulse trains in calling songs, despite their marginal statistical significance

Table 4.2 Results of the generalized linear models (GLMs) testing for sexual performance aspects and structural parameters of songs produced by irradiated and nonirradiated WT and SOD2 5.2 *Anastrepha suspensa* males

	Sexual performance responses		
Predictors	*Calling propensity*[a]	*Mating propensity of focal males*[a]	*Mating propensity of muted males*[a]
Line	$P=0.41$ ($\chi^2=0.68$, $df=1$)	$P=0.12$ ($\chi^2=2.42$, $df=1$)	$P=0.34$ ($\chi^2=0.91$, $df=1$)
Radiation	$P=0.52$ ($\chi^2=0.41$, $df=1$)	$P=0.06$ ($\chi^2=3.54$, $df=1$)	$P=0.73$ ($\chi^2=0.10$, $df=1$)
Line×radiation	$P=0.18$ ($\chi^2=1.79$, $df=1$)	$P=0.02$* ($\chi^2=5.45$, $df=1$)	$P=0.28$ ($\chi^2=1.16$, $df=1$)
	Structural responses		
Predictors	*Total number of pulse trains*[b]	*Number of short-pulse trains*[b]	*Number of long-pulse trains*[b]
Line	$P<0.0001$* ($Z=4.04$, $df=1$)	$P=0.06$ ($Z=0.76$, $df=1$)	$P<0.0004$* ($Z=1.97$, $df=1$)
Radiation	$P<0.0003$* ($Z=2.91$, $df=1$)	$P=0.08$ ($Z=0.28$, $df=1$)	$P=0.13$ ($Z=0.35$, $df=1$)
Line×radiation	$P=0.29$ ($Z=-1.03$, $df=1$)	$P=0.35$ ($Z=0.93$, $df=1$)	$P=0.52$ ($Z=0.63$, $df=1$)

Fixed effects: line, radiation, and their interaction (×).

[a] GLM-Binomial distribution, *P* value (χ^2, *df*).
[b] GLM-Poisson distribution, *P* value (*Z*, *df*).
* Statistical significance ($P<0.05$).

(Table 4.2, Figure 4.3b). However, WT-70 Gy males exhibited a higher number of long pulses relative to SOD2 5.2 males (Line*: Table 4.2), regardless of radiation treatment (Figure 4.3b).

4.3.1.3 Signal temporal parameters

The duration of calling songs (Figure 4.3c) and pulse trains (Figure 4.3d) did not differ across treatments (Table 4.3). However, the lack of statistical support for the trend toward longer calling songs and pulse trains in WT-70 Gy males relative to WT-0 Gy and SOD2-70 Gy males may have been driven by outliers. Male line, radiation treatment, and their interaction affected the PTI of calling songs significantly (Table 4.3). Briefly, WT-0 Gy males exhibited calling songs with shorter intervals than WT-70 Gy, SOD2 5.2-70 Gy, and SOD2 5.2-0 Gy males (Figure 4.3e). Overall, SOD2 overexpression increased the interval between calling pulse trains produced by nonirradiated and irradiated transgenic males (Figure 4.3e), indicating that our transgenic manipulation may incur costs in critical parameters of acoustic signals used by *A. suspensa* males to attract females (see also Teets et al. 2019). Irradiated males displayed longer copulatory songs relative to their respective nonirradiated controls (Radiation*: Table 4.3), regardless of the line (Figure 4.3f).

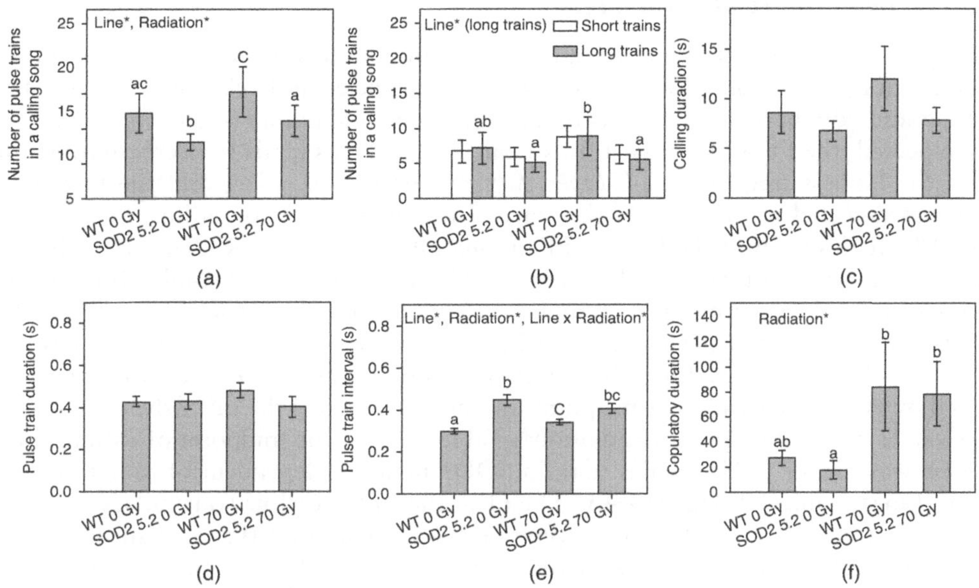

Figure 4.3 Structural and temporal parameters of the sexual songs emitted by SOD2 5.2 and WT *Anastrepha suspensa* males. (a) The total number of pulse trains in a calling song. (b) The number of pulse trains in a calling song is classified as short (>3 seconds) and long trains (≤3 seconds). (c) Duration of the calling song. (d) Duration of the pulse trains. (e) Duration of the pulse train intervals. (f) Duration of the copulatory song. In general, SOD2 overexpression improved the calling song produced by irradiated transgenic males, allowing SOD2 5.2-70 Gy males to call similarly to the WT-0 Gy males. Conversely, irradiation significantly affected the calling songs produced by WT-70 Gy males by inducing a higher number of bursts in a call and longer sexual songs (calling and copulatory). Transgenesis in SOD2 5.2-0 Gy males limited the production of calling songs and increased the duration of pulse train intervals. Bars represent means with standard errors of either the number of pulse trains or time (seconds). Means within each graph followed by different letters indicate statistical significance between groups ($P<0.05$, Tukey contrasts for GLM and Tukey HSD test for two-way ANOVA). (*) Results of the generalized linear models (GLMs).

Table 4.3 Results of the generalized linear models (GLMs) testing for temporal parameters of songs produced by irradiated and nonirradiated WT and SOD2 5.2 *Anastrepha suspensa* males

	Temporal responses			
Predictors	*Calling song duration*[a]	*Pulse train duration*[a]	*Pulse train interval*[a]	*Copulatory song duration*[a]
Line	$P=0.76$	$P=0.34$	$P<0.0001$*	$P=0.22$
	($F=0.09$, $df=1, 52$)	($F=0.92$, $df=1, 631$)	($F=34.67$, $df=1, 577$)	($F=1.54$, $df=1, 78$)
Radiation	$P=0.24$	$P=0.99$	$P=0.01$*	$P=0.0008$*
	($F=1.39$, $df=1, 51$)	($F=0.0002$, $df=1, 630$)	($F=5.81$, $df=1, 576$)	($F=12.29$, $df=1, 77$)
Line × radiation	$P=0.81$	$P=0.08$	$P=0.007$*	$P=0.45$
	($F=0.05$, $df=1, 50$)	($F=3.03$, $df=1, 629$)	($F=7.39$, $df=1, 575$)	($F=0.58$, $df=1, 76$)

Fixed effects: line, radiation, and their interaction (×).

[a] GLM-Gaussian distribution, *P* value (*F*, *df*).

* Statistical significance ($P<0.05$).

4.3.1.4 Signal spectral parameters

The spectral parameters evaluated in irradiated and nonirradiated males from WT and SOD2 lines were composed of fundamental frequencies and relative power levels of calling and copulatory songs measured at different time points (start, midpoint, end). As expected, the FF of the pulse trains from calling songs tends to decrease over time, with the highest frequencies observed at the start of the pulses and the lowest frequencies detected at the end. At the start of the pulse trains of calling songs, irradiated WT males showed the highest FF among all males (Line×Radiation*: Table 4.4, Figure 4.4a). Nonirradiated and irradiated SOD males produced similar fundamental frequencies at the start point of the pulse trains, but these frequencies were lower than the fundamental frequencies exhibited by WT males at the same time point (Line*: Table 4.4, Figure 4.4a). Although radiation did not affect the FF of calling songs produced by SOD2 males at any time point, radiation increased the FF of the calls sung by WT males at the start point (Radiation*: Table 4.4). At the midpoint of calling songs, the FF was higher in WT males than in SOD2 males, independent of radiation treatment (Line*: Table 4.4, Figure 4.4b). The FF at the end of calling song pulses was lower in SOD2 males than in WT males (Figure 4.4c), suggesting that our specific SOD2 overexpression construct results in the reduction of the FF of calling songs (Line*: Table 4.4). The reductive effect of radiation treatment on the FF at the end of calling songs was unclear and only marginally significant in the model (Radiation*: Table 4.4). Thus, SOD2 overexpression seems to decrease the FF of pulse trains within calls of transgenic males. At the same time, radiation increases the FF of calling songs of WT males, particularly at the start of these pulse trains.

Table 4.4 Results of the two-way ANOVA testing for spectral parameters of calling pulse trains produced by irradiated and nonirradiated wild-type and transgenic males

	Pulse train fundamental frequency		
Predictors	*Start*	*Midpoint*	*End*
Line	$P<0.0001$*	$P<0.0001$*	$P<0.0001$*
	(F=70.61, df=1, 650)	(F=45.06, df=1, 650)	(F=39.59, df=1, 650)
Radiation	$P=0.03$*	$P=0.38$	$P=0.04$*
	(F=4.46, df=1, 649)	(F=0.38, df=1, 649)	(F=3.86, df=1, 649)
Line×radiation	$P=0.03$*	$P=0.37$	$P=0.13$
	(F=4.43, df=1, 648)	(F=0.37, df=1, 648)	(F=2.28, df=1, 648)
	Pulse train relative spectrum level		
Predictors	*Start*	*Midpoint*	*End*
Line	$P<0.0001$*	$P=0.001$*	$P=0.001$*
	(F=14.72, df=1, 650)	(F=10.27, df=1, 650)	(F=10.84, df=1, 650)
Radiation	$P=0.11$	$P=0.06$	$P=0.32$
	(F=2.47, df=1, 649)	(F=3.39, df=1, 649)	(F=0.97, df=1, 649)
Line×radiation	$P<0.0001$*	$P<0.0001$*	$P<0.0001$*
	(F=108.41, df=1, 648)	(F=127.90, df=1, 648)	(F=117.30, df=1, 648)

Fixed effects: line, radiation, and their interaction (×).

* Statistical significance ($P<0.05$).

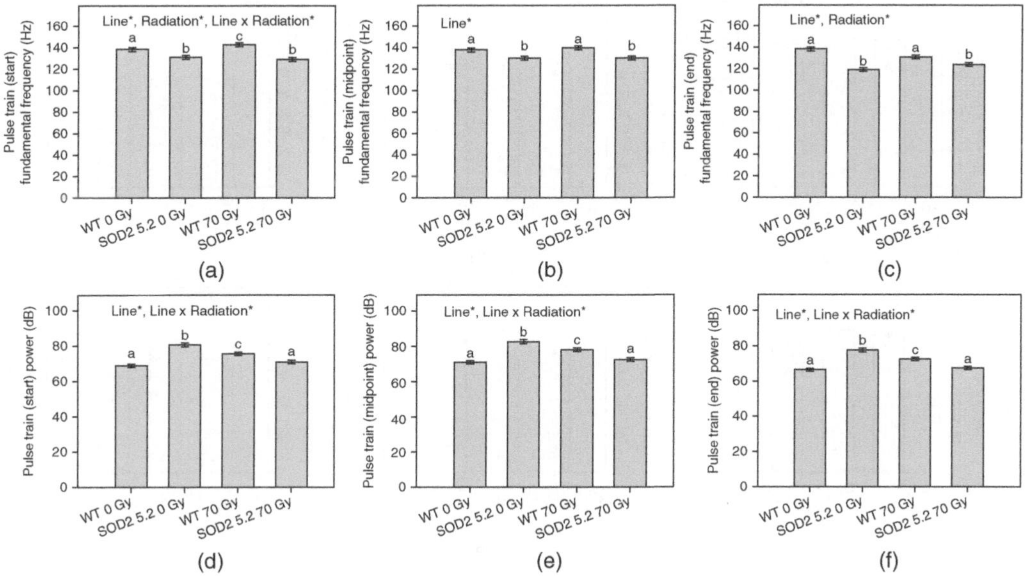

Figure 4.4 Spectral parameters of the calling songs produced by WT and SOD2 5.2 *Anastrepha suspensa* males. The fundamental frequency of the pulse trains from sexual calls at three different time points: (a) start, (b) midpoint, and (c) end. The relative power level of the pulse trains from calling songs at three different time points: (d) start, (e) midpoint, and (f) end. SOD2 overexpression reduced the pulse fundamental frequency of the SOD2 5.2-0 Gy and SOD2 5.2-70 Gy males at all time points while increasing the pulse power of SOD2 5.2-0 Gy compared to WT males. SOD2 5.2-70 Gy and WT-0 Gy males produced sexual calls with similar pulse power at all time points. Bars represent means with standard errors of either fundamental frequency (Hz) or relative power level (dB) of the calling songs. Means within each graph followed by different letters indicate statistical significance between groups ($P<0.05$, Tukey HSD test). (*) Results of the generalized linear models (GLMs).

The sound power remained constant throughout the pulse trains with no clear difference at the start, midpoint, and end of calling songs produced by all males (Figure 4.4d–f). However, the relative power level of male calling songs varied significantly across treatments. Nonirradiated SOD2 5.2 males showed the highest power level at all time points, indicating that SOD2 overexpression increased the power of calling songs (Line*: Table 4.4). The power levels of calling songs produced by WT and SOD2 were affected differently by radiation (Line×Radiation*: Table 4.4): WT-0 Gy and SOD2 5.2-70 Gy males exhibited calling songs with the same level of sound power at all time points, which differed significantly from SOD2 5.2-0 Gy and WT-70 Gy males (Figure 4.4d–f).

Overall, the FF of the copulatory song tends to decrease over time for all males (Figure 4.5a–c). Irradiated SOD2 5.2 displayed the highest FF, followed by WT 70 Gy males at the midpoint of the copulatory song (Radiation*: Table 4.5). However, WT-70 Gy males did not differ from nonirradiated males (Figure 4.5b). Irradiated WT males produced copulatory songs with the lowest relative power (Radiation*: Table 4.5), but did not differ statistically from SOD2 5.2-70 Gy and WT-0 Gy (Figure 4.5d).

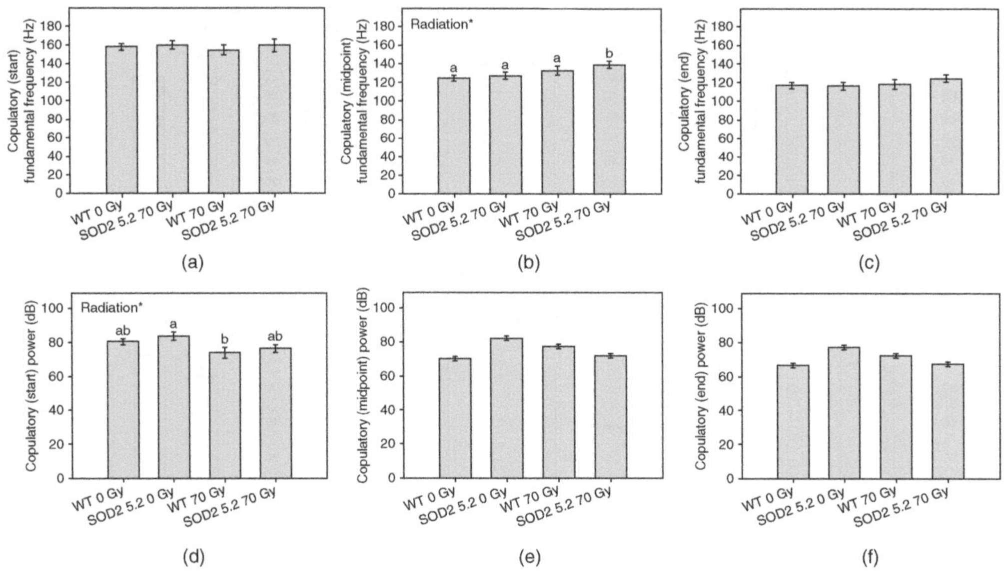

Figure 4.5 Spectral parameters of the copulatory songs produced by WT and SOD2 5.2 *Anastrepha suspensa* males. The fundamental frequency of the copulatory songs at three different time points: (a) start, (b) midpoint, and (c) end. The relative power level of the copulatory songs at the (d) start, (e) mid, and (f) end time points. Overall, the fundamental frequency of the copulatory songs decreased over time for all males. SOD2 overexpression increased the fundamental frequency of the copulatory song produced by SOD2 5.2-70 Gy at the midpoint relative to the other male groups. WT-70 Gy males produced copulatory songs with the lowest relative power level, particularly at the start point, while SOD2 5.2-0 Gy males exhibited copulatory songs with the highest power. Copulatory songs displayed by WT-0 Gy and SOD2 5.2-70 Gy males showed similar relative power levels. Bars represent means with standard errors of either fundamental frequency (Hz) or relative power level (dB) of the copulatory songs. Means within each graph followed by different letters indicate statistical significance between groups ($P<0.05$, Tukey HSD test). (*) Results of the generalized linear models (GLMs).

4.3.2 Volatile production

Nonirradiated and irradiated males from both lines produced and released the nine main pheromone compounds known for *A. suspensa* males (Figure 4.6; Nation 1983, 1990; Battiste et al. 1983; Chuman et al. 1988; Rocca et al. 1992). However, there was a significant quantitative difference in the total amount of pheromone released by the males (Figure 4.7a). Nonirradiated SOD2 5.2 males released the highest quantity of total pheromone volatiles among all males (Line*: Table 4.6). Irradiation significantly decreased the amount of volatile released by WT males, which was the lowest amount among all males (Radiation*: Table 4.6). Specifically, WT-70 Gy males released significantly less pheromone volatile than SOD2 5.2-70 Gy, SOD2 5.2-0 Gy, and, marginally significantly, WT-0 Gy males (Figure 4.7a).

Males also differed in the quantities of specific compounds from the volatile mixture collected (Figure 4.7b), particularly in the amount of anastrephin, epianastrephin, (Z)-3-nonenol, and (Z,Z)-3,6-nonadienol, which are the compounds known to elicit female antennal responses in *A. suspensa* (Robacker and Hart 1987). The variation

Table 4.5 Results of the two-way ANOVA testing for spectral parameters of copulatory songs produced by irradiated and nonirradiated wild-type and transgenic males

	Copulatory fundamental frequency (FF)		
Predictors	*Start*	*Midpoint*	*End*
Line	$P=0.55$	$P=0.14$	$P=0.49$
	($F=0.36$, $df=1, 77$)	($F=2.23$, $df=1, 77$)	($F=0.46$, $df=1, 77$)
Radiation	$P=0.67$	$P=0.01$*	$P=0.26$
	($F=0.18$, $df=1, 76$)	($F=6.27$, $df=1, 76$)	($F=1.24$, $df=1, 76$)
Line × radiation	$P=0.77$	$P=0.65$	$P=0.37$
	($F=0.08$, $df=1, 75$)	($F=0.20$, $df=1, 75$)	($F=0.78$, $df=1, 75$)
	Copulatory relative spectrum level		
Predictors	*Start*[a]	*Midpoint*[a]	*End*[a]
Line	$P=0.48$	$P=0.60$	$P=0.28$
	($F=0.51$, $df=1, 77$)	($F=0.28$, $df=1, 77$)	($F=1.17$, $df=1, 77$)
Radiation	$P=0.004$*	$P=0.58$	$P=0.51$
	($F=8.90$, $df=1, 76$)	($F=0.30$, $df=1, 76$)	($F=0.43$, $df=1, 76$)
Line × radiation	$P=0.87$	$P=0.19$	$P=0.09$
	($F=0.02$, $df=1, 75$)	($F=1.72$, $df=1, 75$)	($F=2.82$, $df=1, 75$)

Fixed effects: line, radiation, and their interaction (×).

* Statistical significance ($P<0.05$).

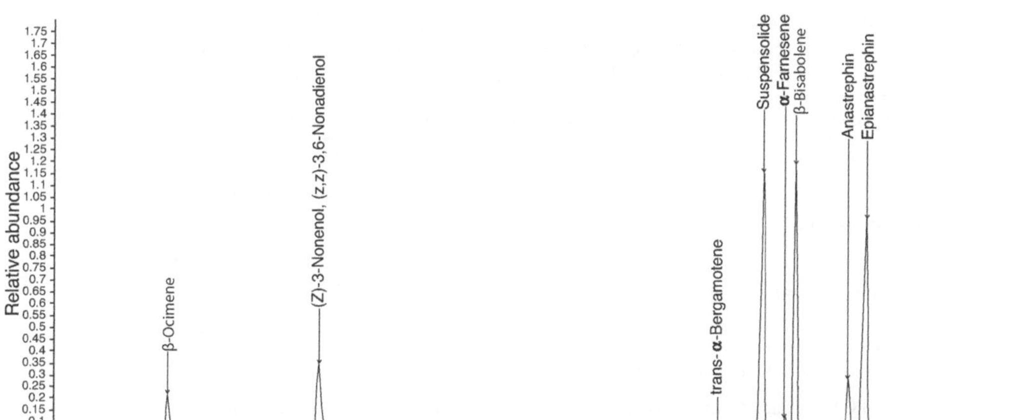

Figure 4.6 Typical thermal desorption gas chromatography-mass spectrometry (TD-GC/MS) chromatogram from analysis of the pheromone volatiles emitted by *Anastrepha suspensa* males. Eight peaks were identified as β-ocimene, (Z)-3-nonenol and (Z,Z)-3,6-nonadienol*, α-trans-bergamotene, suspensolide, α-farnesene, β-bisabolene, anastrephin*, and epianastrephin*. Volatile chemicals marked with an asterisk (*) are known for eliciting strong antennal responses in male and female *A. suspensa* (Robacker and Hart 1987). These chemicals were then used in the PCA followed by a two-way ANOVA using the principal component scores.

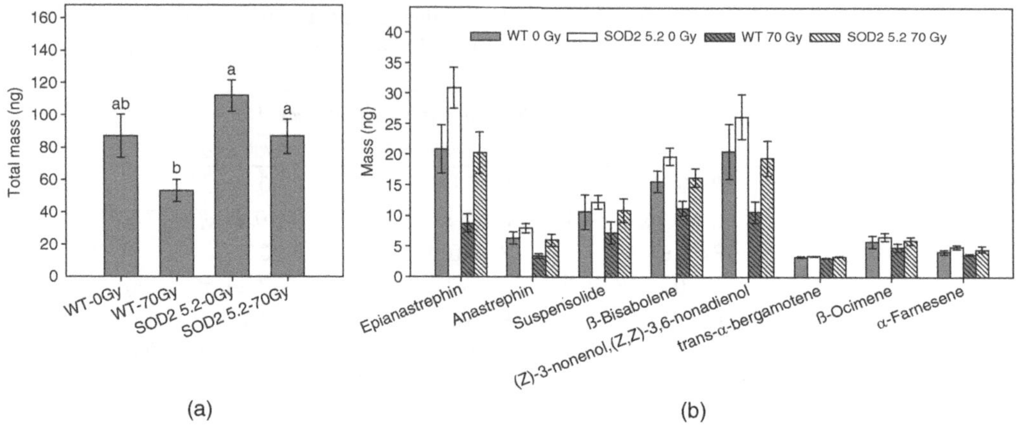

Figure 4.7 Quantities of the chemical compounds identified and evaluated in the pheromones produced by WT and SOD2 5.2 *Anastrepha suspensa* males. (a) The total amount of the eight pheromone volatiles combined and released by WT and SOD2 5.2 males, treated or untreated with radiation. (b) The relative amount of the nine main pheromone volatiles emitted by WT and SOD2 5.2 males treated or untreated with radiation. SOD2 5.2-0 Gy males released the greatest amount of pheromone, while WT-70 Gy males emitted the lowest amount of volatiles among all males. The total quantity of chemical volatiles released by WT-0 Gy and SOD2 5.2-70 Gy males did not differ significantly from each other. Bars represent means with standard errors of either volatile total mass (ng) or relative volatile mass (ng) based on the eight chemicals evaluated in this study. In graph a, means followed by different letters indicate statistical significance between groups ($P<0.05$, Tukey HSD test).

Table 4.6 Results of the two-way ANOVA testing for differences in chemical composition of the pheromone released by irradiated and nonirradiated wild-type and transgenic males

Predictors	*Total amount of pheromone*	*PC1 scores*[a]
Line	$P=0.0005$*	$P=0.0001$*
	($F=13.23$, $df=1, 76$)	($F=15.48$, $df=1, 76$)
Radiation	$P=0.002$*	$P=0.009$*
	($F=9.75$, $df=1, 75$)	($F=11.97$, $df=1, 75$)
Line × radiation	$P=0.65$	$P=0.57$
	($F=0.21$, $df=1, 74$)	($F=0.32$, $df=1, 74$)

Fixed effects: line, radiation, and their interaction (×).

[a] PCA extract from a matrix with three chemicals: anastrephin, epianastrephin, and (Z)-3-nonenol and (Z,Z)-3,6-nonadienol.

* Statistical significance ($P<0.05$).

in the amount of those four pheromone compounds released by irradiated and nonirradiated males from WT and SOD2 5.2 lines was then assessed by PCA. The first two principal components explained 99.1% (PC1: 93%, PC2: 6%) of the total variation in pheromone components (Figure 4.8a). Three clear groups were identified in the scores plot based on the dataset for anastrephin, epianastrephin, (Z)-3-nonenol, and (Z,Z)-3,6-nonadienol. Briefly, volatiles from nonirradiated SOD2 5.2 and irradiated WT clustered into distinct groups, while those from nonirradiated WT and irradiated

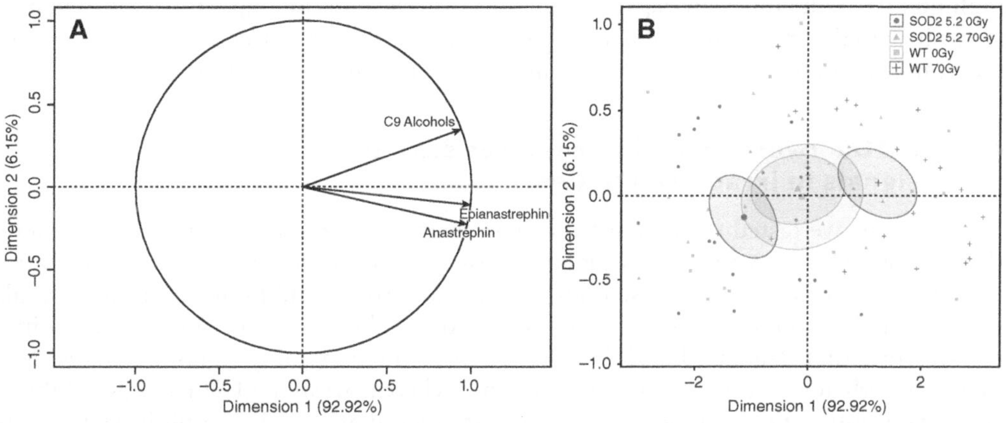

Figure 4.8 Principal component analysis (PCA) of a data matrix containing three chemical volatiles from the pheromone released by WT and SOD2 5.2 *Anastrepha suspensa* males. A total of 78 volatile collections were performed. At least 15 collections were done for each group of males. (a) Variables factor map with the circle of correlations of the variables (Z)-3-nonenol and (Z,Z)-3,6-nonadienol (C9 alcohols), epianastrephin, and anastrephin with components 1 and 2 extracted from the PCA. (b) Scatterplot of the PC_1 factor scores for individuals grouped by male line combined with radiation treatment. The limits of each male treatment were given by confidence ellipses (95%). There was a definite separation between the scores from SOD2 5.2-0 Gy (ellipse with circle symbol) and WT-70 Gy (ellipse with plus symbol) males. The volatiles emitted by WT-0 Gy and SOD2 5.2-70 Gy varied similarly in the PCA, as shown by the overlapping ellipses with the triangle and square symbols.

SOD2 5.2 overlapped, forming a single group around the origin (Figure 4.8b). The PC1 scores based on the three chemicals were used in the two-way ANOVA to test whether the variations in anastrephin, epianastrephin, (Z)-3-nonenol, and (Z,Z)-3,6-nonadienol differ across the groups of males tested. In summary, the PC1 scores from irradiated WT males were greater than those from irradiated SOD2 5.2 (Line*: Table 4.6), confirming a significant difference between the two lines. Likewise, the PC1 scores from irradiated WT males were higher than from nonirradiated males, regardless of the line (Radiation*: Table 4.6). Overall, irradiation reduced the emission of crucial pheromone volatiles in both WT and SOD2 males, but SOD2 males produced more of each of the four critical pheromone components than WT, thus blunting the effects of radiation stress on pheromone emission.

4.4 DISCUSSION

The ability of an organism to produce energy and manage oxidative stress is a fundamental factor in the expression of condition-dependent sexual signals, particularly index signals, such as calling songs and pheromones (Hill 2011; Hill and Johnson 2013; Koch and Hill 2018). Given that mitochondria are both the primary source of cellular energy and highly susceptible to oxidative damage, mitochondrial function likely plays a central role in the honesty of these sexual signals. Individuals with poor mitochondrial efficiency may struggle to produce energetically costly traits of high quality, reducing their reproductive success. Despite the importance of oxidative stress in signal honesty, to our knowledge no prior study has directly tested the role of manipulating mitochondrial antioxidants on maintaining sexual index signals.

4.4.1 SOD2 overexpression restores sexual signals in irradiated males

In this study, we tested the extent to which increasing enzymatic antioxidant capacity through SOD2 overexpression, a manipulation previously shown to mitigate oxidative damage from irradiation in these lines (Teets et al. 2019), could improve two key male sexual signals – songs and pheromone production – following irradiation. Our findings demonstrate that irradiated SOD2 males performed comparably to nonirradiated WT males in both acoustic and chemical signaling, whereas irradiated WT males exhibited significant impairments in these parameters after irradiation. This result is particularly notable given the extensive evidence documenting radiation-induced deterioration of fruit fly courtship signals (Sharp and Webb 1977; Webb et al. 1987; Kanmiya et al. 1987; Sivinski and Burk 1989; Ponce et al. 1993; Lance et al. 2000; Mankin et al. 2008; Briceño et al. 2009; Souza et al. 2015).

The observed benefits of SOD2 overexpression were most pronounced under severe stress conditions. While nonirradiated SOD2 males exhibited lower mating success compared to nonirradiated WT males, irradiated SOD2 males significantly outperformed irradiated WT males. This pattern aligns with previous findings (Teets et al. 2019; Dias et al. 2021) showing that SOD2 overexpression reduced male competitiveness in nonirradiated conditions but provided a significant advantage after radiation-induced oxidative stress. The mechanistic basis for this context-dependent response remains unclear but may involve interactions between SOD2 activity and other cellular stress response pathways. Radiation exposure triggers strong repair mechanisms in insects, including the upregulation of additional antioxidant defenses (Koval et al. 1978; Cheng et al. 2009; Cai et al. 2018; Li et al. 2018). Thus, the synergy between SOD2 overexpression, reduction of oxidative damage, and beneficial hormetic effects of radiation-induced repair responses may have enhanced pheromone production and calling song performance in irradiated transgenic males.

4.4.2 Pheromone production as an honest indicator of male condition

Although nonirradiated SOD2 males sang less frequently than their WT counterparts, they emitted ~30% more pheromone, suggesting a shift in resource allocation toward chemical signaling. Pheromone production in *A. suspensa* is believed to be energetically demanding because males release large quantities of volatiles (~2 μg/h for ~4 hours) relative to their limited pheromone storage capacity (~1 μg) (Nation 1989, 1990). If pheromone synthesis requires high metabolic efficiency, the increased pheromone output by SOD2 males may indicate enhanced mitochondrial function and respiratory efficiency.

Female mate choice is often influenced by pheromone quantity and composition, as pheromonal signals honestly reflect male condition (Johansson and Jones 2007; Peso et al. 2015; Steiger and Stökl 2014). Ionizing radiation exposure has been shown to reduce pheromone production in multiple insect species, including beetles (Haynes et al. 1977; Earle et al. 1978; Villavaso et al. 1983) and other fruit flies (Kanmiya et al. 1987; Moreno et al. 1991; López-Guillen et al. 2008). In *A. suspensa*, irradiation significantly decreases key pheromone compounds such as anastrephin and epianastrephin, both of which play a role in mate attraction (Ponce et al. 1993). Interestingly, SOD2 overexpression restored pheromone release to levels similar to those of nonirradiated WT males, suggesting that transgenesis could be a viable strategy to improve chemical signals in males used in SIT programs.

4.4.3 Potential costs of our SOD2 overexpression construct

We noted in previous publications that even though irradiated SOD2 males outperformed irradiated WT males, nonirradiated SOD males sometimes performed slightly more poorly than nonirradiated WT males, suggesting a potential cost of our transgenic approach (Teets et al. 2019; Dias et al. 2021). In this work, despite clear improvements in pheromone production, the acoustic signaling of SOD2 males showed limited enhancement. One possible explanation involves the by-products of increased SOD2 activity. SOD2 converts superoxide radicals into hydrogen peroxide (H_2O_2), a reactive molecule that can cause cellular damage if not adequately neutralized by additional antioxidant enzymes such as catalase (Imlay 2008). If SOD2 overexpression led to excessive H_2O_2 accumulation, it may have caused oxidative damage in flight muscle cells, limiting improvements in song performance. This hypothesis aligns with findings in *Drosophila*, where SOD2 overexpression increased H_2O_2 levels, disrupted locomotor activity, and altered behavioral rhythms (Grover et al. 2009). Considering that mitochondria can comprise up to 40% of insect wing muscle dry mass (Wigglesworth and Lee 1982) and that wing muscles power song production in fruit flies (Sharp et al. 1980), oxidative stress in these tissues could impair acoustic signaling.

In addition to the potential biochemical costs of increased H_2O_2 accumulation, there may be other costs associated with our transgenic construct, as have been found in other transgenic insect systems (Marrelli et al. 2006; Harvey-Samuel et al. 2014; Ahmed et al. 2019). Our transgenic construct overexpresses not only mitochondrial SOD but also two complementary fluorescent marker proteins (Teets et al. 2019). Thus, our SOD2 overexpression line may pay some costs to male performance of overexpressing SOD or either of the two marker proteins (Harvey-Samuel et al. 2014). Another nonmutually exclusive potential cost of transgenesis derives from the placement of the construct into the host insect line's genome. We used a *piggyBac* transposable element vector to randomly transpose our large construct into the *A. suspensa* genome (Teets et al. 2019). While we know from sex-linked segregation patterns that our construct has been integrated into the Y chromosome, we have not characterized the exact location of the insertion. Costs of random insertion of transposable elements have been well studied in the model fly *D. melanogaster* (Woodruff 1992) and may be due to elements disrupting gene structure directly or disrupting gene regulatory elements such as promoters or enhancers. Ultimately, application of this idea of transgenic improvement of male sexual competitiveness in SIT programs by enhancing antioxidants would require optimization of both the transgenic construct and placement of the construct in the genome to minimize such potential costs for male mating performance.

4.4.4 Implications for the role of mitochondria in sexual signaling

Our findings support the mitochondrial function hypothesis, which posits that variation in mitochondrial efficiency influences the expression of sexual traits, thereby providing honest information for female mate choice (Hill 2014; Koch et al. 2017; Weaver et al. 2017; Koch and Hill 2018). Given that oxidative phosphorylation simultaneously drives energy production and generates ROS, mitochondrial performance must be finely tuned to balance energetic demands and oxidative damage (Dowling and Simmons 2009; Hill and Johnson 2013; Hill 2014).

In species where sexual traits require substantial aerobic capacity, individuals with superior mitochondrial function may outperform competitors in reproductive displays. For instance, in birds, oxidative stress during neurogenesis impairs brain development, reducing the complexity of birdsongs in adulthood (Koch and Hill 2018). Similarly, in *A. suspensa*, energetically demanding lekking behaviors and pheromone signaling may serve as index signals of mitochondrial efficiency, helping females select males with high levels of cellular respiration and stress resistance. We encourage others studying the complex sexual courtship of other tephritid fruit fly species to test the extent to which oxidative stress and antioxidants may be important in their study systems of choice to test for the generality of our results.

4.5 CONCLUSION

Our study provides direct evidence that mitochondrial antioxidants play a key role in the maintenance of sexual index signals under oxidative stress. SOD2 overexpression in *A. suspensa* males effectively mitigated radiation-induced damage, restoring pheromone output and partially preserving calling song characteristics. These findings highlight the importance of mitochondrial function in honest sexual signaling and suggest that transgenic strategies could enhance the performance of sterile males in SIT programs. Future research should investigate potential trade-offs associated with SOD2 overexpression, including its effects on other antioxidant pathways, metabolic energy distribution, and long-term fitness costs. Additionally, exploring whether natural variation in mitochondrial function correlates with reproductive success could further elucidate the evolutionary role of mitochondria in sexual selection in general, as well as providing new tools and perspectives for improving the performance of males to enhance the efficacy and economy of SIT programs.

REFERENCES

Abdi H, Williams LJ. 2010. Principal component analysis. *Wiley Interdisciplinary Reviews: Computational Statistics* 2: 433–459.

Ahmed S, Hasan A, Kim Y. 2019. Overexpression of PGE2 synthase by in vivo transient expression enhances immunocompetency along with fitness cost in a lepidopteran insect. *Journal of Experimental Biology* 222: jeb207019.

Alborn HT. 2018. A technique for thermal desorption analyses suitable for thermally-labile, volatile compounds. *Journal of Chemical Ecology* 44: 103–110.

Alborn HT, Willms SD. 2015. Technique for thermal desorption analyses of thermo labile volatile compounds. *US Patent App* 14: 869–984.

Andersson M. 1994. *Sexual Selection*. Princeton, NJ: Princeton University Press.

Andersson M, Simmons LW. 2006. Sexual selection and mate choice. *Trends in Ecology & Evolution* 21: 296–302.

Battiste MA, Strekowski L, Vanderbilt DP, Visnick M, King RW, Nation JL. 1983. Anastrephin and epianastrephin, novel lactone components isolated from the sex-pheromone blend of male Caribbean and Mexican fruit-flies. Tetrahedron Letters 24: 2611–2614.

Biernaskie JM, Grafen A, Perry JC. 2014. The evolution of index signals to avoid the cost of dishonesty. *Proceedings of the Royal Society B-Biological Sciences* 281 (1790): 20140876.

Briceño RD, Hernández MR, Orozco D, Hanson P. 2009. Acoustic courtship songs in males of the fruit fly *Anastrepha ludens* (Diptera: Tephritidae) associated with geography, mass rearing and courtship success. *Revista de Biología Tropical* 57: 257–265.

Brieger K, Schiavone S, Miller FJ, Krause KH. 2012. Reactive oxygen species: From health to disease. *Swiss Medical Weekly* 142: w13659.

Brown WD, Wideman J, Andrade MCB, Mason AC, Gwynne DT. 1996. Female choice for an indicator of male size in the song of the black-horned tree cricket *Oecanthus nigricornis* (Orthoptera: Gryllidae: Oecanthinae). Evolution 50: 2400–2411.

Burk T. 1981. Signaling and sex in acalyptrate flies. *Florida* Entomologist 64: 30–43.

Burk T. 1983. Behavioral ecology of mating in the Caribbean fruit-fly, *Anastrepha suspensa* (Loew) (Diptera, Tephritidae). *Florida Entomologist* 66: 330–344.

Cai PM, Hong JF, Wang C, Yang YC, Yi CD, Chen JH, Ji QE. 2018. Effects of Co-60 radiation on the activities of three main antioxidant enzymes in *Bactrocera dorsalis* (Hendel) (Diptera: Tephritidae). *Journal of Asia-Pacific Entomology* 21: 345–351.

Calkins CO, Parker AG. 2005. Sterile insect quality. In: Dyck VA, Hendrichs J, Robinson AS, eds. *Sterile Insect Technique Principles and Practice in Area-Wide Integrated Pest Management*. Dordrecht, Netherlands: Springer, 269–296.

Candas D, Li JJ. 2014. MnSOD in oxidative stress response-potential regulation via mitochondrial protein influx. *Antioxidants & Redox Signaling* 20: 1599–1617.

Cheng IC, Lee HJ, Wang TC. 2009. Multiple factors conferring high radioresistance in insect Sf9 cells. Mutagenesis 24: 259–269.

Chuman T, Sivinski J, Heath RR, Calkins CO, Tumlinson JH, Battiste MA, Wydra RL, Strekowski L, Nation JL. 1988. Suspensolide, a new macrolide component of male Caribbean fruit-fly *Anastrepha suspensa* (Loew) volatiles. Tetrahedron Letters 29: 6561–6564.

Cross JH. 1980. A vapor collection and thermal-desorption method to measure semiochemical release rates from controlled release formulations. *Journal of Chemical Ecology* 6: 781–787.

Dias VS, Cáceres C, Parker AG, Pereira R, Demirbas-Uzel G, Abd-Alla AMM, Teets NM, Schetelig MF, Handler AM, Hahn DA. 2021. Mitochondrial superoxide dismutase overexpression and low oxygen conditioning hormesis improve the performance of irradiated sterile males. *Scientific Reports* 11: 20182.

Dowling DK, Simmons LW. 2009. Reactive oxygen species as universal constraints in life-history evolution. *Proceedings of the Royal Society of London* 276: 1737–1745.

Earle NW, Simmons LA, Nilakhe SS, Villavaso EJ, McKibben GH, Sikorowski P. 1978. Pheromone production and sterility in boll-weevils: Effect of acute and fractionated gamma-irradiation Coleoptera-Curculionidae. *Journal of Economic Entomology* 71: 591–595.

EPPO. 2025. *Anastrepha suspensa*. EPPO datasheets on pests recommended for regulation. https://gd.eppo.int (accessed 2025-02-28).

FAO/IAEA. 2017. *Guideline for Packing, Shipping, Holding and Release of Sterile Flies in Area-Wide Fruit Fly Control Programmes*. Rome, Italy: Food and Agriculture Organization.

FAO/IAEA/USDA. 2014. *Manual for Product Quality Control and Shipping Procedures for Sterile Mass-Reared Tephritid Fruit Flies*. Vienna, Austria: International Atomic Energy Agency.

Galvan I, Wakamatsu K, Camarero PR, Mateo R, Alonso-Alvarez C. 2015. Low-quality birds do not display high-quality signals: The cysteine-pheomelanin mechanism of honesty. Evo*lu*tion 69: 26–38.

Garratt M, Brooks RC. 2012. Oxidative stress and condition-dependent sexual signals: More than just seeing red. *Proceedings of the Royal Society Biological Sciences* 279: 3121–3130.

Grafen A. 1990. Biological signals as handicaps. *Journal of Theoretical Biology* 144: 517–546.

Grose J. 2011. Modelling and the fall and rise of the handicap principle. *Biology &* Philosophy 26: 677–696.

Grover D, Ford D, Brown C, Hoe N, Erdem A, Tavare S, Tower J. 2009. Hydrogen peroxide stimulates activity and alters behavior in *Drosophila melanogaster*. *Plos One* 4: e7580.

Hanlin MA, Liquido NJ, Nakamichi KAA, Santamaria J, Birnbaum AL. 2020. Host plant records of the Caribbean fruit fly, *Anastrepha suspensa* (Loew) (Diptera: Tephritidae), Version 1.0. Available online at: USDA Compendium of Fruit Fly Host Information (CoFFHI). https://coffhi.cphst.org/ (accessed 2025-02-28).

Harvey-Samuel T, Ant T, Gong H, Morrison NI, Alphey L. 2014. Population-level effects of fitness costs associated with repressible female-lethal transgene insertions in two pest insects. *Evolutionary Applications* 7: 597–606.

Haynes JW, Mitlin N, Davich TB, Dawson JR, McGovern WL, McKibben GH. 1977. Sterilization of boll-weevil pupae with fractionated doses of gamma-irradiation. *Entomologia Experimentalis Et Applicata* 21: 57–62.

Heath RR, Manukian A, Epsky ND, Sivinski J, Calkins CO, Landolt PJ. 1993. Bioassay system for collecting volatiles while simultaneously attracting tephritid fruit-flies. *Journal of Chemical Ecology* 19: 2395–2410.

Higham JP. 2014. How does honest costly signaling work? *Behavioral Ecology* 25: 8–11.

Higham JP, Hebets EA. 2013. An introduction to multimodal communication. *Behavioral Ecology and Sociobiology* 67: 1381–1388.

Hill GE. 2011. Condition-dependent traits as signals of the functionality of vital cellular processes. *Ecology Letters* 14: 625–634.

Hill GE. 2014. Cellular respiration: The nexus of stress, condition, and ornamentation. *Integrative and Comparative Biology* 54: 645–657.

Hill GE. 2015. Mitonuclear ecology. *Molecular Biology and Evolution* 32: 1917–1927.

Hill GE, Johnson JD. 2013. The mitonuclear compatibility hypothesis of sexual selection. *Proceedings of the Royal Society B-Biological Sciences* 280: 20131314.

Holman L. 2012. Costs and constraints conspire to produce honest signaling: Insights from an ant queen pheromone. *Evolution* 66: 2094–2105.

Imlay JA. 2008. Cellular defenses against superoxide and hydrogen peroxide. *Annual Review of Biochemistry* 77: 755–776.

Johansson BG, Jones TM. 2007. The role of chemical communication in mate choice. *Biological Reviews* 82: 265–289.

Jones AG, Ratterman NL. 2009. Mate choice and sexual selection: What have we learned since Darwin? *Proceedings of the National Academy of Sciences of the United States of America* 106: 10001–10008.

Kanmiya K, Tanaka A, Kamiwada H, Nakagawa K, Nishioka T. 1987. Time-domain analysis of the male courtship songs produced by wild, mass-reared, and by irradiated melon flies, *Dacus cucurbitae* Coquillett (Diptera, Tephritidae). *Applied Entomology and* Zoology 22: 181–194.

Kaspi R, Taylor PW, Yuval B. 2000. Diet and size influence sexual advertisement and copulatory success of males in Mediterranean fruit fly leks. *Ecological Entomology* 25: 279–284.

Knipling EF. 1955. Possibilities of insect control or eradication through the use of sexually sterile males. *Journal of Economic Entomology* 48: 459–462.

Koch RE, Hill GE. 2018. Behavioural mating displays depend on mitochondrial function: A potential mechanism for linking behaviour to individual condition. *Biological Reviews of the Cambridge Philosophical Society*. doi:10.1111/brv.12400.

Koch RE, Josefson CC, Hill GE. 2017. Mitochondrial function, ornamentation, and immunocompetence. *Biological* Reviews 92: 1459–1474.

Kotiaho JS. 2001. Costs of sexual traits: A mismatch between theoretical considerations and empirical evidence. *Biological* Reviews 76: 365–376.

Kotiaho JS, Puurtinen M. 2007. Mate choice for indirect genetic benefits: Scrutiny of the current paradigm. *Functional Ecology* 21: 638–644.

Koval TM, Myser WC, Hart RW, Hink WF. 1978. Comparison of survival and unscheduled DNA-synthesis between an insect and a mammalian-cell line following x-ray treatments. *Mutation* Research 49: 431–435.

Kuijper B, Pen I, Weissing FJ. 2012. A guide to sexual selection theory. *Annual Review of Ecology, Evolution, and Systematics* 43: 287–311.

Lachmann M, Szamado S, Bergstrom CT. 2001. Cost and conflict in animal signals and human language. *Proceedings of the National Academy of Sciences of the United States of America* 98: 13189–13194.

Landolt PJ, Sivinski P. 1992. Effects of time of day, adult food, and host fruit on incidence of calling by male caribbean fruit flies (Diptera: Tephritidae). *Environmental Entomology* 21(2): 382–387.

Laidre ME, Johnstone RA. 2013. Animal signals. Current Biology 23: R829–R833.

Lance DR, McInnis DO, Rendon P, Jackson CG. 2000. Courtship among sterile and wild *Ceratitis capitata* (Diptera: Tephritidae) in field cages in Hawaii and Guatemala. Annals of the Entomological Society of America 93: 1179–1185.

Lance DR, McInnis DO. 2005. Biological basis of the sterile insect technique. In: VA Dyck VA, Hendrichs J, Robinson AS, eds. *Sterile Insect Technique: Principles and Practice in Area-Wide Integrated Pest Management.* Dordrecht, Netherlands: Springer, 69–94.

Li X, Lingyan L, Karthi S, Zhang K, Jianjun L, Qiongbo H, Qunfang W. 2018. Effects of 200 Gy 60Co-γ radiation on the regulation of antioxidant enzymes, Hsp70 genes, and serum molecules of *Plutella xylostella* (Linnaeus). *Molecules* 23: 1–9.

López-Guillen G, Cruz-Lopez L, Malo EA, Gonzalez-Hernandez H, Cazares CL, Lopez-Collado J, Toledo J, Rojas JC. 2008. Factors influencing the release of volatiles in *Anastrepha obliqua* males (Diptera: Tephritidae). *Environmental Entomology* 37: 876–882.

López-Martínez G, Hahn DA. 2012. Short-term anoxic conditioning hormesis boosts antioxidant defenses, lowers oxidative damage following irradiation and enhances male sexual performance in the Caribbean fruit fly, *Anastrepha suspensa*. *Journal of Experimental Biology* 215: 2150–2161.

Mankin RW, Shuman D, Coffelt JA. 1996. Noise shielding of acoustic devices for insect detection. *Journal of Economic Entomology* 89: 1301–1308.

Mankin RW, Lemon M, Harmer AMT, Evans CS, Taylor PW. 2008. Time-pattern and frequency analyses of sounds produced by irradiated and untreated male *Bactrocera tryoni* (Diptera: Tephritidae) during mating behavior. *Annals of the Entomological Society of America* 101: 664–674.

Mankin RW, Petersson E, Epsky ND, Heath RR, Sivinski J. 2000. Exposure to male pheromones enhances *Anastrepha suspensa* (Diptera: Tephritidae) female response to male calling song. *Florida* Entomologist 83: 411–421.

Marrelli MT, Moreira CK, Kelly D, Alphey L, Jacobs-Lorena M. 2006. Mosquito transgenesis: What is the fitness cost? *Trends in Parasitology* 22: 197–202.

Maynard Smith J. 1956. Fertility, mating behaviour and sexual selection in Drosophila subobscura. *Journal of Genetics* 84: 17–35.

Maynard Smith J, Harper DGC. 1995. Animal signals: Models and terminology. *Journal of Theoretical Biology* 177: 305–311.

Maynard Smith J, Harper DGC. 2003. *Animal Signals*. Oxford: Oxford University Press.

Metcalfe NB, Alonso-Alvarez C. 2010. Oxidative stress as a life-history constraint: The role of reactive oxygen species in shaping phenotypes from conception to death. *Functional Ecology* 24: 984–996.

Monaghan P, Metcalfe NB, Torres R. 2009. Oxidative stress as a mediator of life history trade-offs: Mechanisms, measurements and interpretation. *Ecology Letters* 12: 75–92.

Moreno DS, Sanchez M, Robacker DC, Worley J. 1991. Mating competitiveness of irradiated Mexican fruit-fly (Diptera, Tephritidae). *Journal of Economic Entomology* 84: 1227–1234.

Murphy MP. 2009. How mitochondria produce reactive oxygen species. *Biochemical Journal* 417: 1–13.

Nation JL. 1983. Sex pheromone of the Caribbean fruit fly: Chemistry and field ecology. *Proceedings of the 5th International Congress of Pesticide Chemistry*, Kyoto, Japan, 109–110.

Nation JL. 1989. The role of pheromones in the mating system of *Anastrepha* fruit flies. In: Robinson AS, Hooper GHS, eds. *World Crop Pests: Fruit Flies, Their Biology, Natural Enemies and Control*. Amsterdam, the Netherlands: Elsevier, 189–205.

Nation JL. 1990. Biology of pheromone release by male Caribbean fruit flies, *Anastrepha suspensa* (Diptera: Tephritidae). *Journal of Chemical Ecology* 16: 553–572.

Norrbom A. 2022. *Anastrepha suspensa* (Caribbean fruit fly). *CABI Compendium* cabicompendium.5668.

Peso M, Elgar MA, Barron AB. 2015. Pheromonal control: Reconciling physiological mechanism with signalling theory. *Biological Reviews of the Cambridge Philosophical Society* 90: 542–559.

Ponce WP, Nation JL, Emmel TC, Smittle BJ, Teal PEA. 1993. Quantitative analysis of pheromone production in irradiated Caribbean fruit fly males, *Anastrepha suspensa* (Loew). *Journal of Chemical Ecology* 19: 3045–3056.

Robacker DC, Hart WG. 1987. Electroantennograms of male and female Caribbean fruit-flies (Diptera, Tephritidae) elicited by chemicals produced by males. *Annals of the Entomological Society of America* 80: 508–512.

Rocca JR, Nation JL, Strekowski L, Battiste MA. 1992. Comparison of volatiles emitted by male Caribbean and Mexican fruit-flies. *Journal of Chemical Ecology* 18: 223–244.

Ruxton GD, Schaefer HM. 2011. Resolving current disagreements and ambiguities in the terminology of animal communication. *Journal of Evolutionary Biology* 24: 2574–2585.

Santana GDS, Ronchi-Teles B, dos Santos CM, Soares MA, Souza PGC, Araújo FHV, Aguiar CVS, Silva RS. 2023. Climate suitability modeling for *Anastrepha suspensa* (Diptera: Tephritidae): Current and future invasion risk analysis. *International Journal of Biometeorology* 67: 1185–1197.

von Schantz T, Bensch S, Grahn M, Hasselquist D, Wittzel H. 1999. Good genes, oxidative stress and condition-dependent sexual signals. *Proceedings of the Royal Society of London* 266: 1–12.

Sharp JL, Carlysle TC, Little HF. 1980. Missing indirect flight muscles in Anastrepha suspensa (Diptera, Tephriidae) with droopy wing syndrome. Annals of the Entomological Society of America 73: 239–243.

Sharp JL, Webb JC. 1977. Flight performance and signaling sound of irradiated or unirradiated Anastrepha suspensa. *Proceedings of the Hawaiian Entomological* Society 22: 525–532.

Sivinski J. 1989. Lekking and the small-scale distribution of the sexes in the Caribbean fruit fly, *Anastrepha suspensa* (Loew). *Journal of Insect Behavior* 2: 3–13.

Sivinski J, Burk T. 1989. Reproductive and mating behavior. In: Robinson A, Hooper G, eds. *Fruit Flies: Their Biology, Natural Enemies and Control*. Amsterdam, Netherlands: Elsevier Publishers, 343–351.

Sivinski J, Burk T, Webb JC. 1984. Acoustic courtship signals in the Caribbean fruit-fly, *Anastrepha suspensa* (Loew). *Animal Behaviour* 32: 1011–1016.

Sivinski J, Webb JC. 1985. Sound production and reception in the Caribfly, *Anastrepha suspensa* (Diptera, Tephritidae). *Florida Entomologist* 68: 273–278.

South SH, House CM, Moore AJ, Simpson SJ, Hunt J. 2011. Male cockroaches prefer a high carbohydrate diet that makes them more attractive to females: Implications for the study of condition dependence. Evo*lu*tion 65: 1594–1606.

Souza JMGA, Lima-Filho PA, Molina WF, Almeida LM, Gouveia MB, Macêdo FP, Paranhos BAJ. 2015. Wing morphometry and acoustic signals in sterile and wild males: Implications for mating success in *Ceratitis capitata*. *The Scientific World Journal* 526969: 1–9.

Steiger S, Stökl J. 2014. The role of sexual selection in the evolution of chemical signals in insects. *Insects* 5: 423–438.

Számadó S. 2011. The cost of honesty and the fallacy of the handicap principle. *Animal* Behaviour 81: 3–10.

Számadó S, Penn DJ. 2015. Why does costly signalling evolve? Challenges with testing the handicap hypothesis. *Animal Behaviour* 110: 9–12.

Teets NM, Dias VS, Pierce BK, Schetelig MF, Handler AM, Hahn DA. 2019. Overexpression of an antioxidant enzyme improves male mating performance after stress in a lek-mating fruit fly. *Proceedings of the Royal Society B* 286: 20190531.

Velando A, Torres R, Alonso-Alvarez C. 2008. Avoiding bad genes: Oxidatively damaged DNA in germ line and mate choice. Bioessays 30: 1212–1219.

Villavaso EJ, Wiygul G, Thompson MJ. 1983. Pheromone production in boll-weevils (Coleoptera, Curculionidae) sterilized by 3 methods. *Journal of Economic Entomology* 76: 1038–1040.

Vreysen MJB, Hendrichs J, Enkerlin WR. 2006. The sterile insect technique as a component of sustainable area-wide integrated pest management of selected horticultural insect pests. *Journal of Fruit and Ornamental Plant Research* 14: 107–131.

Walder JMM, Calkins CO. 1993. Effects of gamma radiation on the sterility and behavioral quality of the Caribbean fruit fly, *Anastrepha suspensa* (Loew) (Diptera: Tephritidae). *Scientia Agricola* 50: 157–165.

Weaver RJ, Koch RE, Hill GE. 2017. What maintains signal honesty in animal colour displays used in mate choice? *Philosophical Transactions of the Royal Society B-Biological Sciences* 372: 1–8.

Webb JC, Sharp JL, Chambers DL, Benner JC. 1976. Acoustical properties of the flight activities of the Caribbean fruit fly. *Journal of Experimental Biology* 64: 761–772.

Webb JC, Sharp JL, Chambers DL, McDow JJ, Benner JC. 1975. The analysis and identification of sounds produced by the male Caribbean fruit fly, *Anastrepha suspensa* (Loew). *Annals of the Entomological Society of America* 69: 415–420.

Webb JC, Sivinski J, Litzkow C. 1984. Acoustical behavior and sexual success in the Caribbean fruit-fly, *Anastrepha suspensa* (Loew) (Diptera, Tephritidae). *Environmental Entomology* 13: 650–656.

Webb JC, Sivinski J, Smittle BJ. 1987. Acoustical courtship signals and sexual success in irradiated Caribbean fruit-flies (*Anastrepha suspensa*) (Diptera, Tephritidae). *Florida Entomologist* 70: 103–109.

Wigglesworth VB, Lee WM. 1982. The supply of oxygen to the flight muscles of insects - a theory of tracheole physiology. *Tissue & Cell* 14: 501–518.

Woodruff RC. 1992. Transposable DNA elements and life history traits. I. Transposition of P DNA elements in somatic cells reduces the lifespan of *Drosophila melanogaster*. *Genetica* 86: 143–154.

Wyatt TD. 2014. *Pheromones and Animal Behavior: Chemical Signals and Signatures*. New York, NY: Cambridge University Press.

Yuval B, Kaspi R, Shloush S, Warburg MS. 1998. Nutritional reserves regulate male participation in Mediterranean fruit fly leks. *Ecological Entomology* 23: 211–215.

Zahavi A. 1975. Mate selection: A selection for a handicap. *Journal of Theoretical Biology* 53: 205–214.

Zahavi A. 1977. The cost of honesty (further remarks on the handicap principle). *Journal of Theoretical Biology* 67: 603–605.

Zhou Y, Hu LF, Wu H, Jiang LW, Liu SQ. 2017. Genome-wide identification and transcriptional expression analysis of cucumber superoxide dismutase (SOD) family in response to various abiotic stresses. *International Journal of Genomics* 2017: 7243973.

Section II

Risk assessment, quarantine and post-harvest

Chapter 5

Host reproduction number for fruit fly management and trade

Bernard Dominiak

5.1 INTRODUCTION

We live in a changing world with a growing world population and a changing world climate (Bebber et al. 2014; Sultana et al. 2020). The world cannot continue to rely on single end-point disinfestation techniques such as methyl bromide and temperature-controlled treatments, which add to climate change impacts. How does the world feed this growing world population? New science has solved many problems (such as COVID) but sometimes old data can be used in new ways. One of these old ways is the metric of fruit fly adults emerging from a kilogram of fruit. This metric was recently termed the host reproduction number (HRN) (Dominiak 2022). HRN can be used to better inform activities in the fruit fly world such as management, surveillance and market access.

5.2 FRUIT FLY MANAGEMENT USING *BACTROCERA TRYONI* (QUEENSLAND FRUIT FLY (QFLY)) EXAMPLES

Management occurs traditionally in two situations, orchards and urban populations. Managers in both scenarios need to understand the risks posed, particularly by feral hosts. Previously, the New South Wales (NSW) government managed Queensland fruit fly, (Qfly), *Bactrocera tryoni* Froggatt in urban scenarios and growers managed orchard populations. However, both situations would benefit from knowledge that higher HRN hosts pose the greatest threat and, therefore, HRN rankings need to be better understood. In a recent review (Dominiak 2023), there were 277 published hosts for Qfly. But, there was HRN data on only 81 hosts (29.6%), so HRN data is not available for the other 196 hosts.

In urban settings, government officers would enter urban properties to spray the oranges and peaches in the backyard but walk past the roses in the front yard without treating the roses because they did not know the risk posed (Table 5.1). Growers were

Table 5.1 Host reproduction number data on several Qfly hosts in urban and orchard situations

Urban programmes	*Production or orchard sites*
Orange = 142	Orange = 142
Peaches = 179	Prickly pear = 6.1
Roses = 354	Olives = 33
Feijoa = 46	Grapes = 53

DOI: 10.1201/9781003644330-7

keen to have feijoa removed from towns based on a perceived threat, which appears to be largely unjustified now that the HRN is known (Table 5.1).

In citrus orchards, growers were concerned about prickly pears on their boundary but cared little for feral olives which were plentiful. For many years, grapes were considered a non-host (Dominiak 2011). But management could not understand why Qfly outbreaks were detected in citrus orchards adjacent to untreated vineyards. These errors in management were due to a lack of understanding of the risks posed by different HRN hosts (Table 5.1).

5.3 SURVEILLANCE

Australia and New Zealand do not have *Bactrocera dorsalis* (Hendel), and both countries are keen to prevent an incursion. Both countries conduct surveillance, but surveillance could be better informed by an improved understanding of HRN. New Zealand may choose different ideal surveillance hosts than Australia because New Zealand has no endemic fruit flies. *Bactrocera dorsalis* has entered Australia twice before and established but was eradicated (Hancock 2013). However, any *B. dorsalis* incursion must compete against the many entrenched tephritid species in northern Australia (Hancock 2013). The Australian choice of surveillance trees would be better informed by comparing the HRN of *B. dorsalis* and Qfly.

From Table 5.2, the "HRN for *B. dorsalis*" column may be the best surveillance hosts for New Zealand because of the HRN order. However, in Australia, the optimal surveillance host changes because of the competition effect from Qfly. Soursop might be a better surveillance host because Qfly does so poorly in this host and the competitive advantage favours *B. dorsalis*. Tropical almond is ideal for *B. dorsalis* alone but this host falls to the 15th position because Qfly also does so well in this host. Therefore, an understanding of HRN and the relative competitive advantage would better inform the choice of host trees for *B. dorsalis* surveillance in both Australia and New Zealand. Particularly for Australia, surveillance might be optimized by a better understanding of the reproductive advantages between *B. dorsals* and Qfly at least. There are six other tephritids in northern Australia (Hancock 2013), which would compete with *B. dorsalis* for fruit resources, but the HRN of these other species is largely unknown.

Table 5.2 HRN surveillance for *Bactrocera dorsalis* in *Bactrocera tryoni* territory

Host	*HRN for B. dorsalis*	*HRN for B. tryoni*	*Reproductive advantage for B. dorsalis (rank in brackets)*
Capsicum chilli pepper	750	24	31.2 (2)
Tropical almond	653	375	1.7 (15)
Mango	505	71	7.1 (4)
Soursop	454	10	45.4 (1)
Rambutan	370	60	6.2 (5)
Loquat	325	324	1
Paw paw, papaya	142	17	8.4 (3)

5.4 TRADE – CONSIDERING THE RISK BASED ON HRN

There are three traditional disinfestation trade options. The first option is fruit fly freedom for regions, production places and sites (new International Standards for Phytosanitary Measures (ISPM)). There is no risk from these freedom origins, and therefore, no disinfestation treatment is required. The second option is for endemic origins where usually a single endpoint treatment is used. Often, these treatments are usually methyl bromide, which is not consistent with the Montreal protocol of 1972 because it contributes to greenhouse gas (De Lima et al. 2007). Temperature control protocols for up to 21 days result in a large carbon footprint and now some regions such as Europe are less supportive of high carbon footprint activities. The third option is irradiation which has many advantages (Haynes and Dominiak 2018) but there are very few treatment facilities in Australia and road transport to these facilities adds to the carbon footprint.

Finally, there is the *Systems Approaches* that uses two independent measures (Dominiak 2019; van Klinken et al 2020). This is an ISPM and recognizes the risk in the overall package of the host to the importing country. The *Systems Approach* could be combined with the recognition of a short dispersal distance of most tephritids and linked to the current commodity traceability project in NSW. HRN is another component of the overall risk assessment. It has always intrigued me that a host is a host, no matter the risk it poses. Hosts such as peach (HRN=179), orange (142), apricot (63), pears (56), tomato (44), grapefruit and pomegranate (17), lemon (5) and eggplant (2.3) are subjected to the same disinfestation even though the risk is different in orders of magnitude.

Recently, the IAEA released "General guidelines to facilitate the opening of international markets for fruits and vegetables that are fruit fly hosts based on ISPM (FAO/IAEA 2022) as outlined by Griffin (2022) in this conference. The abstract contains the words "The key to opening markets ... is to focus on the pest risk, and especially on all the factors contributing to risk and mitigating risk ...". And further, "... by breaking from the dogma of historical designs to embrace the opportunities for innovation ...". And on page 5, "Curiously, the concept of host status has no general references standard ...". ISPM 37 is mentioned but it is a yes/no option with no ranking or risk assessment (IPPC 2018).

HRN could be part of these new trade opportunities. HRN could be linked to *Systems Approaches* (Dominiak 2019; van Klinken et al. 2020). Additionally, Australia is increasingly using the Bayes Net or Scenario Tree analysis to demonstrate statistical confidence of the absence or eradication of pests such as fire ants, yellow crazy ants, Fiji disease in sugarcane and horse flu (Dominiak et al. 2011; Sergeant and Wilson 2011; Magarey et al. 2019; Charlton et al. 2022). Scenario tree analysis could be used to generate confidence that fruit flies are absent from a fruit consignment to a standard equivalent to probit 9 or similar internationally accepted standards. HRN could be linked to *Systems Approaches*, statistical analysis and incorporate recognition of shorter natural dispersal distance (Dominiak and Fanson 2020). This would be enhanced using product traceability to develop a new market access option with no end-point treatment.

Towards this increased understanding of HRN, there are several recently published papers. Follett et al. (2021) reviewed the HRN in the Pacific. Similarly, the African species were reviewed and published (Dominiak 2022). The species-specific review of the Mediterranean fruit fly was published recently (Dominiak and Taylor-Hukins 2022). As yet unpublished, there is an HRN review of the New World tephritids with Peter Follett, the Qfly HRN and *B. dorsalis* HRN are in preparation.

5.5 IN CLOSING

HRN does not provide all the answers as demonstrated by our HRN knowledge for Qfly. But there are opportunities. Currently, NSW DPI is developing such a package for oranges from the Riverina (Kingham unpub.). There is another package for cherries produced from the Young district.

Generally, in plant biosecurity, HRN reports the worst-case data but there is variation in the data. For instance, the HRN for some tephritids declines with altitude or with increasing distance from the equator (Ekesi et al. 2009). For Qfly, the HRN for oranges is 142 in southern NSW but only <30 in Queensland and the Pacific islands. This is not a weakness but an opportunity to develop new protocols where some situations result in HRN values less than the worst-case scenario.

So, where to from here? First, return to your countries and consider how HRN can inform your tephritid surveillance, management and trade. Second, and particularly for those in the field, record the weight of hosts (along with frequency), calculate the HRN and publish your results so that the world tephritid community can benefit from your knowledge.

DISCLOSURE

I do not have any conflicts of interest. The review was funded by the NSW Department of Primary Industries.

REFERENCES

Bebber, P. D., T. Holmes, D. Smith, and S. J. Gurr. 2014. Economic and physical determinants of the global distributions of crop pests and pathogens. *New Phytologist* 202: 901–910.

Charlton, S., R. Henderson, E. S. G. Sergeant, P. Worsley, and B. C. Dominiak. 2022. Declaring Sydney free from red imported fire ants *Solenopsis invicta* Buren – a view from the branches of the scenario tree. *International Journal of Tropical Insect Science* 42: 1205–1214.

De Lima, C. P. F., A. J. Jessup, L. Cruickshank, C. J. Walsh, and E. R. Mansfield. 2007. Cold disinfestation of citrus (*Citrus* ssp.) for Mediterranean fruit fly (*Ceratitis capitata*) and Queensland fruit fly (*Bactrocera tryoni*) (Diptera: Tephritidae). *New Zealand Journal of Crop Horticultural Science* 35: 39–50.

Dominiak, B. C. 2011. Review of grapes *Vitis sp.* as an occasional host for Queensland fruit fly *Bactrocera tryoni* (Froggatt) (Diptera: Tephritidae). *Crop Protection* 30: 958–961.

Dominiak, B. C. 2019. Components of a systems approach for the management of Queensland fruit fly *Bactrocera tryoni* (Froggatt) in a post dimethoate fenthion era. *Crop Protection* 116: 56–67.

Dominiak, B. C. 2022. The use of Host Reproduction Number and Host Suitability Index to rank hosts of Fruit Flies in Africa. *International Journal of Tropical Insect Science* 42: 2717–2729.

Dominiak, B. C. 2023. Priority host plants of the Queensland fruit fly, *Bactrocera tryoni* (Froggatt), based on the host reproduction number, for tephritid management, surveillance and trade. *International Journal of Tropical Insect Science* 43: 1531–1538.

Dominiak, B. C., and B. G. Fanson. 2020. Current quarantine and suspension distances are excessive for incipient populations of Queensland fruit fly (*Bactrocera tryoni* (Froggatt)) (Diptera: Tephritidae) in southern New South Wales, Australia. *Crop Protection* 138: 105341.

Dominiak, B. C., K. Gott, D. McIver, T. Grant, P. S. Gillespie, P. Worsley, A. Clift, and E. S. G. Sergeant. 2011. Scenario tree analysis of zero detections and the eradication of yellow crazy ant (*Anoplepis gracilipes* (Smith)), in New South Wales, Australia. *Plant Protection Quarterly* 26: 124–129.

Dominiak, B. C., and R. Taylor-Hukins. 2022. Host Reproduction Number used to identify priority hosts for *Ceratitis capitata*, Mediterranean fruit fly, for surveillance, trade and eradication programs. *International Journal of Tropical Insect Science* 42: 3721–3727.

Ekesi, S., M. K. Billah, P. W. Nderitu, S. A. Lux, and I. Rwomushana. 2009. Evidence for competitive displacement of *Ceratitis cosyra* by the invasive fruit fly *Bactrocera invadens* (Diptera: Tephritidae) on mango and mechanisms contributing to the displacement. Journal of Economic Entomology 102: 981–991.

FAO/IAEA. 2022. *General Guidelines to Facilitate the Opening of International Markets for Fruit and Vegetables That Are Fruit Fly Hosts Based on International Standards for Phytosanitary Measures*. Vienna, Austria: FAO/IAEA.

Follett, P. A., F. E. M. Haynes, and B. C. Dominiak. 2021. Host Suitability Index for polyphagous tephritid fruit flies. Journal of Economic Entomology 114: 1021–1034.

Griffin, R. L. 2022. General guidelines to facilitate the opening of international markets for fruits and vegetables that are fruit fly host based on International Standards for Phytosanitary Measures. *11th International Symposium on fruit Flies of Economic Importance*, 13–18 November 2022, Sydney, Australia

Hancock, D. L. 2013. A revised checklist of Australia's fruit flies (Diptera: Tephritidae). *Australian Entomologist* 40: 219–236.

Haynes, F. E. M., and B. C. Dominiak. 2018. Irradiation for phytosanitary treatment of the Queensland fruit fly *Bactrocera tryoni* Froggatt benefits international trade. Crop Protection 112: 125–132.

IPPC. 2018. *Determination of Host Status of Fruit to Fruit Flies (Tephritidae). International Standards for Phytosanitary Standards* 37. International Plant Protection Convention, 37 pp.

van Klinken, R. D., K. Fielder, L. Kingham, K. Collins, and D. Barbour. 2020. A risk framework for using systems approaches to manage horticultural biosecurity risk for market access. *Crop Protection* 129: 104994.

Magarey, R. C., M. Reynolds, B. C. Dominiak, E. Sergeant, J. Agnew, A. Ward, and N. Thompson. 2019. Review of sugarcane Fiji leaf gall disease in Australia and the declaration of pest freedom in Central Queensland. *Crop Protection* 121: 113–120.

Sergeant, E. S. G., and G. Wilson. 2011. Demonstrating freedom from equine influenza in New South Wales, Australia, following the 2007 outbreak. *Australian Veterinary Journal* 89: 164–169.

Sultana, S., J. B. Baumgartner, B. C. Dominiak, J. E. Royer, and L. J. Beaumont. 2020. Impacts of climate change on high priority fruit fly species in Australia. *PLOS One* 15(2): e0213820; 1–19.

Chapter 6

General guidelines to facilitate the opening of international markets for fruits and vegetables that are fruit fly hosts based on International Standards for Phytosanitary Measures

Robert L. Griffin

6.1 REVIEW

The path to international trade in fruits and vegetables begins with opening new markets and continues with retaining and expanding those markets. The process can be complex, composed of diverse challenges and substantial uncertainty, but carefully navigating the early stages of market access is crucial to establishing the foundation for a successful and sustainable export programme.

A key factor in the market access process is addressing concerns about the potential for pests such as fruit flies. Many fruits and vegetables, especially from tropical and subtropical areas of the world, are subject to fruit fly infestation (species of Tephritidae). The geographic and host distribution of these pests and their potential for outbreaks and severe damage is well known. Likewise, there exists a substantial and growing reservoir of research information on their biology and control, making them one of the best-known pest groups. Their reputation easily lands them on lists for quarantine restrictions, where they nearly always become the primary barrier to market access for a wide range of hosts that would otherwise be eligible to enter trade.

The long history of fear regarding the potential for fruit fly introduction via commercial trade in fresh fruits and vegetables developed its own highly restrictive regulatory dogma, which has long resisted more objective attempts to inject research-based and field-tested alternatives to the more severe measures of prohibition or probit-9 treatments. The import requirements of virtually all countries have continued to strictly adhere to the doctrine of maximum restriction for decades, leaving little opening to explore possibilities for safe trade and thereby excluding enormous opportunities for trade in fruits and vegetables with countries whose economies are often strongly dependent on agricultural exports.

The winds of change came in 1994 with the establishment of the World Trade Organization (WTO), which evolved from the General Agreements on Tariffs and Trade. Three key aspects of this change were as follows:

1. the birth of the Agreement on the Application of Sanitary and Phytosanitary Measures (the SPS Agreement),
2. creation of a binding dispute settlement process in the WTO, and

DOI: 10.1201/9781003644330-8

3. formalization of a relationship between the WTO–SPS Agreement and the International Plant Protection Convention (IPPC) for international standard setting.

This new international regulatory framework challenged many historically held notions. It introduced the phytosanitary community to new terms, concepts, principles, and most importantly, harmonization via the elaboration of standards, which the IPPC quickly began to create.

The SPS Agreement requires all phytosanitary measures to be based on either a risk assessment or international standards, but fruit fly regulations around the world were, and many continue to be, glaringly inconsistent with this requirement. The IPPC was clear-eyed about this problem from the outset and devoted significant early effort to harmonization around risk assessment and fruit fly regulatory concepts. As a result, the first decade of standard setting was marked by significant advances in standards with general application to fruit fly regulation as well as specific standards on important fruit fly concepts and issues.

At the same time, the WTO was building a history of phytosanitary disputes that provided exceptional clarity on the proper interpretation of SPS obligations in practice as well as the role of standards. This jurisprudence became an important additional source of harmonization above what was offered by national interpretations of the obligations in the SPS, IPPC, and standards. As disputes were settled, it became increasingly clear to governments that the provisions of this new international regulatory framework were not open to being ignored or freely interpreted without facing the potential for a formal challenge from a trading partner and serious consequences from losing that challenge.

One important point that became immediately clear from disputes is that risk assessment is essential to establish the need for restrictions and to justify the strength of the restrictions. This means there must be a rational relationship between the risk and the phytosanitary measures put in place by the importing country. Measures could not continue to be precautionary simply because of fear or historical precedent. Rather, they must be based on policies designed to meet a demonstrated risk lest they be challenged. This realization shattered the historical paradigm for regulating fruit fly host material and put all countries on notice for their risk justification to maintain excessive restrictions. It also opened the door wide for a new look at phytosanitary policies toward fruit fly host material. The key to opening new markets was shifted from the simple question of the presence or absence of fruit flies of concern to the risk and measures available to mitigate the risk to acceptable levels. This makes it possible for trading partners to have new dialogues and collaborate with their industry on solutions such as systems approaches.

The simple shift toward a focus on negotiating risk management possibilities brought renewed interest in data and research (FAO/IAEA 2022). The treatment question was flipped from how many fruit flies are killed to how many survive, because risk is measured based on survivors. Likewise, scientific questions about the conditions for introduction replace the historical assumption of entry being equivalent to establishment. Now, a case can be made for risk attrition in the supply chain, and thresholds can be established for low prevalence, seasonal shipping windows, reduced treatment dosage, host status, and myriad other risk management possibilities to enrich market access negotiations.

The international regulatory framework and resulting harmonization have created a new playing field that strengthens the scientific underpinnings of regulatory decision-making and offers expanded opportunities for alliances with research and industry to create risk management solutions that promote safe trade in fruits and vegetables affected by fruit flies. Guidelines published by the International Atomic Energy Agency have been made available to support this transition (see link in references below).

REFERENCE

FAO/IAEA. 2022. *General Guidelines to Facilitate the Opening of International Markets for Fruits and Vegetables That Are Fruit Fly Hosts Based on International Standards for Phytosanitary Measures*. Vienna. https://doi.org/10.4060/cc0361en

Chapter 7

Towards a systems approach for mitigating fruit fly risk in export fruit commodities from Africa

Aruna Manrakhan

7.1 INTRODUCTION

Over the past 20 years, exports of unprocessed fresh fruits from Africa have steadily increased and reached a total value of US $ 9.4 billion in 2022 (Odjo et al. 2024). However, the export potential of commercially produced fruits from Africa is restricted by several factors, including challenges in meeting phytosanitary measures required by importing countries to mitigate the risk of pests of quarantine importance in fruit commodities (Odjo et al. 2024). These phytosanitary measures should follow international standards approved by the Commission on Phytosanitary Measures. This commission is the governing body of the International Plant Protection Convention, an intergovernmental treaty that aims to protect plants, agricultural products and natural resources from plant pests, across the world.

Several fruit fly species (Diptera: Tephritidae) are pests of quarantine importance on commercially produced fruits and can pose challenges to the export of fruits from a country where they are present (White and Elson-Harris 1994). The African continent is home to numerous pest species in the genera *Ceratitis*, *Dacus*, *Zeugodacus* and *Bactrocera*, which pose significant threats to the production and trade of fruits and vegetables that are hosts to them (Ekesi et al. 2016). Four International Standards for Phytosanitary Measures (ISPM) specific to fruit flies offer guidance on phytosanitary measures that importing countries may require and then exporting countries would need to implement to mitigate the risk of these pests in traded fruit commodities. These are: ISPM No. 26 on Establishment of pest-free areas for fruit flies (FAO 2006), ISPM No. 28 on Phytosanitary treatments for regulated pests, which include several specific treatments for fruit flies, as annexes (FAO 2016), ISPM No. 35 on Systems approach for pest risk management of fruit flies (FAO 2018) and ISPM No. 37 on Determination of host status of fruit to fruit flies (FAO 2024).

Due to the presence of multiple fruit fly pests within the same area in many African countries, the fragmented nature of agriculture across the continent and the high costs associated with the establishment of pest-free areas, it is not surprising that there are no fruit-producing regions in Africa that are free of all fruit fly pests. However, there are cases of pest-free areas for specific fruit fly pests, for example, the southern and western parts of South Africa are free of the invasive fruit fly pest – *Bactrocera dorsalis* (Hendel) (Manrakhan et al. 2015) and have been maintained as such to date.

Phytosanitary treatments, particularly heat (more specifically hot water immersion) and cold, are being either used or investigated in several African countries as stand-alone postharvest disinfestation treatments to mitigate risk of fruit flies and other pests in

DOI: 10.1201/9781003644330-9

export fruits (Self et al. 2012; Grout 2016; Mwando et al. 2021; Moore and Manrakhan 2022; Ndlela et al. 2022). When applied as a single measure, these treatments have to provide a high level of quarantine security and, depending on markets, this could be through the achievement of mortality of up to 99.9968% at the 95% confidence level (equivalent to zero survivors in 93,613 individuals tested), known as Probit 9 efficacy (Follett and Neven 2006). For tropical fruit commodities mostly produced across the African continent, heat may be more suitable as a postharvest treatment than cold, irradiation and controlled atmosphere, particularly regarding the preservation of fruit quality (Paull 1994). However, heat treatments could also be damaging to fruit quality if temperatures exceed specified thresholds and if conditions before and after treatment are not optimized (Paull 1994; Hansen et al. 2011). Additionally, disinfestation treatments are associated with significant costs (Aegerter and Folwell 2000). These treatments would not be required for fruits that are nonhosts and would possibly not be cost-effective for fruits that are either poor natural hosts or conditional hosts of quarantine fruit fly pests.

There have been several research studies carried out in Africa to classify the host status of several fruit commodities for selected fruit fly pests (De Graaf 2009; Cugala et al. 2014; Ware et al. 2016; Cugala et al. 2017; Manrakhan et al. 2018). Most of these were carried out after the adoption of ISPM No. 37 on determination of host status of fruit for fruit flies and after the introduction of the invasive fruit fly, *B. dorsalis*, in many parts of the continent. In many of these studies, fruit commodities previously listed as hosts for some of the fruit fly pests were categorized as either conditional hosts or nonhosts for the selected fruit fly pests. For fruits categorized as a nonhost, this would be a risk mitigation measure on its own, provided the importing country accepts the host status designation. This acceptance would largely be based on the associated confidence level as a function of the fruit sample size used in the study. For fruits categorized as either natural or conditional hosts, a host suitability index (HSI) as was introduced by Follett et al. (2021) could be used to further subcategorize these commodities from being very poor hosts to very good hosts, based on infestation rates. Such an index could be considered in the development of a systems approach for mitigation of fruit fly risk, especially if a single-point risk reduction measure, such as either a nonhost status of a fruit commodity or a phytosanitary treatment providing Probit-9 level efficacy, cannot be used. In a comprehensive review of fruit fly host information from Africa, Dominiak (2022) presented an HSI of more than 100 fruit species for nine fruit fly pest species in Africa. Commercially produced fruits, categorized as either moderately good or poor hosts for specific fruit fly species, would fit the requirements for the development of systems approach treatment for these pests.

In a fruit fly systems approach, at least two independent measures should be applied throughout various stages of the fruit production and trade process to reduce risk of fruit fly pests in fruit commodities (FAO 2018). For instance, measures could include the development of areas of low pest prevalence and the production of fruits that are less susceptible to fruit flies and postharvest treatments (FAO 2018). At least two of these independent measures should be combined to cumulatively reduce the risk of fruit fly pests in fruit commodities at a level acceptable to the importing country (FAO 2018). In an analysis of systems approach-based protocols implemented commercially for pathogens and various arthropods, including fruit flies, van Klinken et al. (2020) classified four categories of measures applied at various levels of the fruit production stage to reduce pest risk: pest exposure, host vulnerability, infestation rate and establishment in the import region. In the fruit fly systems approach protocols, three of these categories of measures were more commonly used: pest exposure, host vulnerability and infestation rate (van Klinken et al. 2020). Models based on Bayesian networks have been developed to quantify either the cumulative efficacy of independent risk reduction measures such as PRReSTO – Pest Risk Reduction Scenario Tool (Froese et al. 2024) or the combined efficacy of different risk reduction measures (Holt et al. 2018). These types of models can be used in the development of systems approach protocols and assessment of their efficacy. Ultimately, however, to prove equivalence to a single-point treatment with the desired level of efficacy – up to Probit 9, the efficacy of the specific systems approach treatment developed for the targeted fruit commodity and targeted fruit fly pest would need to be quantified. For example, this was done comprehensively for the false codling moth – *Thaumatotibia leucotreta* (Meyrick) (Lepidoptera: Tortricidae), in citrus fruit exported from South Africa (Moore et al. 2016b; Hattingh et al. 2020).

The aims of this review are to assess the status of a systems approach as a phytosanitary measure for mitigating fruit fly risks in export fruit commodities from Africa and to identify key steps required to advance the development and implementation of a fruit fly systems approach in Africa.

7.2 CURRENT STATUS OF FRUIT FLY SYSTEMS APPROACH IN AFRICA

To determine the status of the use of systems approaches for mitigation of fruit fly risks in fruit commodities exported from Africa, protocols that were publicly available online were first sourced from the Google and Google Scholar search engines on the internet. The key words used in the search were "systems approach", "fruit flies", "Tephritidae" and "Africa". In addition, the publicly available annual edition of the Code of Federal Regulation (https://www.govinfo.gov/app/collection/CFR) was searched, using the key terms "systems approach" and all available final rules on the use of a systems approach for mitigation of fruit fly risks in fruit commodities from Africa published in the Federal Register by the Department of Agriculture Animal and Plant Health Inspection Service were selected for further review. Only protocols where systems approach was mentioned as a fruit fly treatment and where at least two independent risk reduction measures could be found were considered for examination of target fruit commodities, target fruit fly pest species and risk reduction measures implemented.

Six publicly available fruit fly systems approach treatment protocols approved for trade of fruits from Africa were considered (Table 7.1). Most of the protocols, more specifically those for fruits destined to continental USA, did not explicitly identify certain measures as risk reduction measures. Instead, these measures were interpreted as such and subsequently categorized under the risk reduction objective defined by van Klinken et al. (2020)

An examination of current fruit fly systems approach protocols for fruits produced and traded from Africa revealed that minimizing pest exposure and infestation rates were the foci of risk reduction measures (Table 7.1). Pest exclusion was the predominant method for minimizing pest exposure, while the reduction of fruit infestation rates

Table 7.1 Fruit fly systems approach protocols used for trade of fruits from African countries categorized by fresh fruit commodity, fruit fly pest species targeted, country/countries of origin, destination countries and risk reduction measures

				Risk reduction objectives		
Fruit commodity	*Fruit fly species*	*Country of origin*	*Destination country*	*Pest exposure*	*Host vulnerability*	*Infestation rate*
Baby Squash[a]	*Dacus ciliatus*, *Dacus bivittatus*, *Dacus frontalis*, *Dacus lounsburyii*, *Dacus punctatifrons*, *Dacus vertebratus*	Zambia	USA	1. Pest exclusion 2. Low prevalence of *Dacus* spp. (<0.7 flies per protein baited McPhail trap per week) outside of pest exclusionary structure		Protein bait application on shade trees and host plants within 200 m from pest exclusionary structure
Tomato[b]	*Ceratitis capitata*	Souss-Massa-Draa, Morocco	Continental USA	Pest exclusion	Pink at the time of packing	Exclusion of shade trees and fruit fly host trees within 50 m of insect proof production and packing structure. Pest exclusionary packing house procedures
Tomato[c]	*Zeugodacus cucurbitae*, *Bactrocera dorsalis*, *Ceratitis capitata*, *Ceratitis rosa*	West African States	Continental USA	Pest exclusion	Packing within 24 hours of harvest	Exclusion of shade trees and fruit fly host trees within 50 m of insect proof production and packing structure. Pest exclusionary packing house procedures

(Continued)

Table 7.1 (Continued) Fruit fly systems approach protocols used for trade of fruits from African countries categorized by fresh fruit commodity, fruit fly pest species targeted, country/countries of origin, destination countries and risk reduction measures

				Risk reduction objectives		
Fruit commodity	*Fruit fly species*	*Country of origin*	*Destination country*	*Pest exposure*	*Host vulnerability*	*Infestation rate*
Citrus (lemon and limes)[d]	Tephritidae (non-European)	South Africa	Europe		Commercial export grade lemons and limes are nonhost	Inspection of a 2% sample packed for export
Citrus (other than lemon and limes)[d]	Tephritidae (non-European)	South Africa	Europe	Fruit fly good agricultural practices		Packhouse grading and inspection Cold shipping conditions
Citrus[e]	*Bactrocera dorsalis*	Botswana	South Africa	Pest management		Postharvest sorting Inspection

The risk reduction measures are categorized according to three risk reduction objectives used in the risk framework described in van Klinken et al. (2020). References for these protocols are provided as superscripts for each specific fruit commodity as follows:

[a] 7CFR319 (2008).
[b] 7CFR319 (2009).
[c] 7CFR319 (2012).
[d] DALRRD (2024a).
[e] DALRRD (2024b).

after harvest was largely achieved through the implementation of fruit fly exclusionary handling procedures (Table 7.1).

Besides the existing fruit fly systems approach protocols, there are a few ongoing studies on the efficacy of fruit fly systems approaches to mitigate the risk of quarantine fruit fly pests on fresh fruit commodities produced in Africa for trade. One of them was the evaluation of a systems approach for fruit flies and false codling moth on Hass avocado produced in South Africa (Grove 2025). The systems approach for fruit flies on Hass avocados produced in South Africa, much like current protocols, emphasized minimizing pest exposure and infestation rates (Grove 2025).

7.3 IDENTIFICATION OF KEY STEPS TO ADVANCE THE USE OF FRUIT FLY SYSTEMS APPROACHES IN AFRICA

Alternative methods could be explored to broaden and advance the use of fruit fly systems approach treatment for fruits produced in and traded from Africa, specifically addressing the risk reduction objectives of pest exposure and infestation rate in existing protocols.

As an alternative to pest exclusion for minimizing pest exposure within a fruit fly systems approach, the establishment and maintenance of areas of low fruit fly prevalence (FF-ALPP), as outlined in ISPM 35 (FAO 2018), could be explored. Annex 1 of ISPM 35 provides specific guidelines for establishment of FF-ALPP (FAO 2018). Fruit fly prevalence levels as required for trade are mostly determined by trapping (Jang et al. 2014). In the establishment of FF-ALPP, the determination of a specified level of low prevalence is key and should be established in relation to fruit infestation risk (aiming for a fruit fly-free fruit commodity for export) (FAO 2018). In a few studies conducted in Africa, significant positive correlations were found between fruit fly infestation and fruit fly trap levels (Vayssières et al. 2009; Theron et al. 2017; Cugala et al. 2020). Some of these studies have utilized infestation rates and fruit fly trap levels to estimate indices such as economic thresholds and economic injury levels that aim to balance control costs with yield losses and as such aid in decision-making for implementing control (Vayssières et al. 2009; Cugala et al. 2020). However, a predetermined level of fruit fly infestation is tolerated with these indices. To date, no published studies have specifically addressed the level of fruit fly prevalence that can be tolerated for production of fruit fly-free fruits. A recent model examining the impact of pest surveillance data and consignment sampling on fruit infestation levels demonstrated that surveillance systems with adequate trap density and moderate sensitivity can effectively predict low fruit infestation rates (Caley et al. 2024). Future field studies are needed to validate these findings.

In a systems approach protocol, where risk reduction measures are implemented at multiple stages before and after harvest, the infestation rates of prepared fruit commodities are typically low. For fruits that have low infestations, postharvest treatments to achieve mortality at the Probit 9 level may not be necessary (Follett and McQuate 2001). For those commodities, postharvest treatments can be modified (Follett and McQuate 2001) to provide additional risk reduction within the systems approach protocol. On another fruit infesting pest in citrus produced from South Africa, modified cold treatments were developed for inclusion in a systems approach

(Moore et al. 2016a, 2022). These modified cold treatments, also referred to as partial cold treatments, did not lead to complete mortality of the targeted insects (Moore et al. 2016a) but when integrated as a component of a systems approach, they contributed to an acceptable level of overall risk reduction (Moore et al. 2016b). Similar strategies on the development of modified postharvest treatment methods could be explored for fruit fly pests in Africa.

In the context of host vulnerability as a third risk reduction objective in a fruit fly systems approach, general information is available on the suitability of fruits produced in Africa for several African fruit fly pests (Follett et al. 2021; Dominiak 2022). Additionally, a few studies were conducted on the continent to determine the suitability of varieties of selected commercially produced fruits for fruit fly pests. For example, the suitability of different cultivars of mango for selected fruit fly pests was quantified in separate studies conducted in West Africa (Felicitas et al. 2012; Mokam et al. 2024). The susceptibility and suitability of three avocado cultivars to *B. dorsalis* were quantified in laboratory and field studies in East Africa (Ware et al. 2016). There is significant potential to expand this body of work to include a wider range of fruits. However, tests to assess the vulnerability of commercially grown export fruits should focus on fruits that remain attached to the tree since these would better represent the vulnerability of fruits produced commercially for trade. These fruits can either be sampled from the tree to determine natural infestation or be exposed to a selected number of fruit fly individuals of a targeted pest species in enclosed bags, as outlined in ISPM 37 (FAO 2024). In a study done in Mexico, Aluja et al. (2014) used the approach of bagged fruit bearing branches with a predetermined number of fruit fly individuals to quantify the susceptibility of 15 mango cultivars to two *Anastrepha* pest species.

There is opportunity in the development of fruit fly systems approaches for fruits from Africa to explore the fourth risk reduction objective, which focuses on minimizing establishment risk. Several climate-based models were developed to predict the potential geographical distribution of important African fruit fly pests, such as *B. dorsalis* (De Villiers et al. 2016), *Bactrocera latifrons* (Hendel) (Wei et al. 2024), *C. capitata* (Wiedemann) (Rao et al. 2024), *Ceratitis rosa* Karsch (Tanga et al. 2018), *Ceratitis quilicii* De Meyer, Mwatawala & Virgilio (Tanga et al. 2018) and *Ceratitis cosyra* (Li et al. 2009). The existing information on fruit fly establishment risk can be included as a risk reduction objective, when relevant, in the development of a systems approach for targeted fruit fly pests and their host fruit. Climate-based models on the potential distribution of other important fruit fly pest species from Africa, such as *Dacus* species, other important *Ceratitis* species and *Zeugodacus cucurbitae* (Coquillett), should be developed in the future. Existing climate-based models on potential distribution have primarily quantified general (annual) establishment risk of the African fruit fly pests in countries that are still free of them, but have not accounted for seasonal variations in establishment risk in those areas. Studies by Szyniszewska and Tatem (2014) and Szyniszewska et al. (2016) showed that there is variation across the year in the potential establishment of *C. capitata*. The identification of establishment risk across different time periods, as demonstrated in the above-mentioned studies, could provide valuable insights for assessing establishment risk specific to seasonal fruit trade from Africa. In this process, safe "trade windows" in terms of fruit fly risk could be identified.

Once a fruit fly systems approach is developed for fruit fly pests on fruit commodities in Africa, it may be necessary to demonstrate the equivalence of these systems to single-point disinfestation treatments, where applicable, in providing quarantine security. These validations would require fruit sampling to determine infestations at different points along the fruit production chain, so that the cumulative mitigation of risk can be calculated at the appropriate confidence levels, as determined by sample sizes.

A potential driver for increased interest in, and use of, a fruit fly systems approach for managing the phytosanitary risk of fruit fly host commodities from Africa would be a cost-and-benefit analysis of the treatment. In the analysis, the systems approach treatment could be compared to other scenarios, such as no treatment (no market access) or a single-point disinfestation treatment.

Finally, as outlined in the guidelines of ISPM 35 (FAO 2018), the National Plant Protection Organization (NPPO) of the exporting country is responsible for the development, implementation and verification of the fruit fly systems approach. In this context, capacity building within the NPPOs of African countries is essential, particularly in areas related to fruit fly identification, fruit fly detection and control. Strengthening these capacities will provide a robust foundation for the effective application of this treatment for fruits traded from Africa. Previous capacity-building activities on fruit fly identification across Africa have included participation from NPPOs (Jordaens et al. 2024). To ensure the sustainability of this knowledge, such activities should be consistently maintained and expanded over time.

7.4 CONCLUSION

The use of a fruit fly systems approach to mitigate the phytosanitary risk of fruits that are hosts to fruit fly pests in Africa remains limited. However, there is potential for expansion, provided that key gaps are addressed through further research. These gaps include the development of areas with low fruit fly prevalence, the integration of phytosanitary treatments within a systems approach and the determination of susceptibility of commercially produced fruits to fruit fly pests. The research on these topics could be carried out in collaborative projects across countries in different regions in Africa so that the objectives and methodologies are aligned, and the resulting information would be applicable to broader regions across Africa.

ACKNOWLEDGEMENTS

I would like to thank Sean Moore for a critical review of the manuscript. This review was funded by Citrus Research International (Pty) Ltd.

REFERENCES

7CFR319.2008. Federal Register Volume 73, Issue 244 (December 18, 2008). In *APHIS, United States Department of Agriculture*, ed. Office of the Federal Register, 76863–76868. USA: National Archives and Records Administration.

7CFR319.2009. Federal Register Volume 74, Issue 210 (November 2, 2009). In *APHIS, United States Department of Agriculture*, ed. Office of the Federal Register, 56523–56526. USA: National Archives and Records Administration.

7CFR319.2012. Federal Register Volume 77, Issue 113 (June 12, 2012). In *APHIS, United States Department of Agriculture*, ed. Office of the Federal Register, 34781–34783. USA: National Archives and Records Administration.

Aegerter, A. F., and R. J. Folwell. 2000. Economic aspects of alternatives to methyl bromide in the postharvest and quarantine treatment of selected fresh fruits. *Crop Protection* 19:161–168.

Aluja, M., J. Arredondo, F. Díaz-Fleischer, et al. 2014. Susceptibility of 15 mango (Sapindales: Anacardiaceae) cultivars to the attack by *Anastrepha ludens* and *Anastrepha obliqua* (Diptera: Tephritidae) and the role of underdeveloped fruit as pest reservoirs: Management implications. *Journal of Economic Entomology* 107:375–388.

Caley, P., D. W. Gladish, L. Kingham, et al. 2024. Inferring fruit infestation prevalence from a combination of pre-harvest monitoring and consignment sampling data. *Scientific Reports* 14:13022.

Cugala, D., S. Ekesi, D. Ambasse, et al. 2014. Assessment of ripening stages of Cavendish dwarf bananas as host or non-host to *Bactrocera invadens*. *Journal of Applied Entomology* 138:449–457.

Cugala, D., J. Jordane, and S. Ekesi. 2017. Non-host status of papaya cultivars to the oriental fruit fly, *Bactrocera dorsalis* (Diptera: Tephritidae), in relation to the degree of fruit ripeness. *International Journal of Tropical Insect Science* 37:19–29.

Cugala, D., V. Massimiliano, M. Maulid, et al. 2020. Economic injury level of the Oriental fruit fly, *Bactrocera dorsalis* (Diptera : Tephritidae), on commercial mango farms in Manica Province, Mozambique. *African Entomology* 28:278–289.

DALRRD. 2024a. *Fruit Fly Management System for Export of Fresh Citrus Fruit Produced in South Africa (Citrus FF-MS) March 2024*. Department of Agriculture, Land Reform and Rural Development, South Africa.

DALRRD. 2024b. *Phytosanitary Import Requirements for Citrus (Citrus spp.) Fresh Fruit from Botswana to South Africa*. Department of Agriculture, Land Reform and Rural Development, South Africa.

De Graaf, J. 2009. Host status of Avocado ('Hass') to *Ceratitis capitata,* Ceratitis rosa, *and* Ceratitis cosyra (Diptera: Tephritidae) in South Africa. *Journal of Economic Entomology* 102:1448–1459.

De Villiers, M., V. Hattingh, D. J. Kriticos, et al. 2016. The potential distribution of *Bactrocera dorsalis*: Considering phenology and irrigation patterns. *Bulletin Entomological Research* 106:19–33.

Dominiak, B. C. 2022. The use of host reproduction number and host suitability index to rank hosts of fruit flies in Africa. *International Journal of Tropical Insect Science* 42:2717–2729.

Ekesi, S., M. De Meyer, S. A. Mohamed, et al. 2016. Taxonomy, ecology, and management of native and exotic fruit fly species in Africa. *Annual Review of Entomology* 61:219–238.

FAO. 2006. *ISPM 26: Establishment of Pest Free Areas for Fruit Flies (Tephritidae). International Standards for Phytosanitary Measures*. Rome, Italy: International Plant Protection Convention.

FAO. 2016. *ISPM 28: Phytosanitary Treatments for Regulated Pests. International Standards for Phytosanitary Measures*. Rome, Italy: International Plant Protection Convention.

FAO. 2018. *ISPM 35: Systems Approach for Pest Risk Management of Fruit Flies (Tephritidae). International Standards for Phytosanitary Measures*. Rome, Italy: International Plant Protection Convention.

FAO. 2024. *Determination of Host Status of Fruit to Fruit Flies (Tephritidae). International Standards for phytosanitary measures* No. 37. Rome, Italy: International Plant Protection Convention.

Felicitas, A. C., M. K. Billah, K. Afreh-Nuamah, et al. 2012. Susceptibility of four mango varieties to the Africa Invader fly, *Bactrocera invadens* Drew, Tsuruta and White. *Journal of Applied Biosciences* 49:3425–3434.

Follett, P. A., F. E. M. Haynes, and B. C. Dominiak. 2021. Host suitability index for polyphagous tephritid fruit flies. *Journal of Economic Entomology* 114:1021–1034.

Follett, P. A., and G. T. McQuate. 2001. Accelerated development of quarantine treatments for insects on poor hosts. *Journal of Economic Entomology* 94:1005–1011.

Follett, P. A., and L. G. Neven. 2006. Current trends in quarantine entomology. *Annual Review of Entomology* 51:359–385.

Froese, J. G., J. V. Murray, N. J. Beeton, et al. 2024. The Pest Risk Reduction Scenario Tool (PRReSTo) for quantifying trade-related plant pest risks and benefits of risk-reducing measures. *Crop Protection* 176:106484.

Grout, T. G. 2016. Cold and heat treatment technologies for post-harvest control of fruit flies in Africa. In Fruit Fly Research and Development in Africa: Towards a Sustainable Management to Improve Horticulture, ed. S. Ekesi, M. A. Mohamed, and M. De Meyer, 465–473. Switzerland: Springer Nature.

Grove, T. 2025. Investigating the potential of a systems approach for false codling moth, Thaumatotibia leucotreta (Lepidoptera: Tortricidae) and fruit flies (Diptera: Tephritidae) on 'Hass' Avocado. *African Journal of Agricultural Research* 21(1):40–55.

Hansen, J., J. Johnson, and D. Winter. 2011. History and use of heat in pest control: A review. *International Journal of Pest Management* 57:267–289.

Hattingh, V., S. Moore, W. Kirkman, et al. 2020. An improved systems approach as a phytosanitary measure for *Thaumatotibia leucotreta* (Lepidoptera: Tortricidae) in export citrus fruit from South Africa. *Journal of Economic Entomology* 113:700–711.

Holt, J., A. W. Leach, S. Johnson, et al. 2018. Bayesian networks to compare pest control interventions on commodities along agricultural production chains. *Risk Analysis* 38:297–310.

Jang, E. B., W. Enkerlin, C. "Ed" Miller, et al. 2014. Trapping related to phytosanitary status and trade. In Trapping and the Detection, Control, and Regulation of Tephritid Fruit Flies: Lures, Area-Wide Programs, and Trade Implications, ed. T. Shelly, N. Epsky, E. B. Jang, J. Reyes-Flores, and R. Vargas, 589–608. Netherlands, Dordrecht: Springer.

Jordaens, K., M. De Meyer, M. Van Nuffel, et al. 2024. A best way forward to the organisation of entomological training courses in sub-Saharan Africa. *International Journal of Educational Development* 107:103026.

van Klinken, R. D., K. Fiedler, L. Kingham, et al. 2020. A risk framework for using systems approaches to manage horticultural biosecurity risks for market access. *Crop Protection* 129:104994.

Li, B., J. Ma, X. Hu, et al. 2009. Potential geographical distributions of the fruit flies *Ceratitis capitata*, *Ceratitis cosyra*, and *Ceratitis rosa* in China. *Journal of Economic Entomology* 102:1781–1790.

Manrakhan, A., J. H. Daneel, R. Beck, et al. 2018. Non-host status of commercial export grade lemon fruit, *Citrus limon* (L.) Burman f. cv. Eureka for *Ceratitis capitata*, *Ceratitis rosa*, *Ceratitis quilicii* and *Bactrocera dorsalis* (Diptera: Tephritidae) in South Africa. *African Entomology* 26:202–214, 213.

Manrakhan, A., J. Venter, and V. Hattingh. 2015. The progressive invasion of *Bactrocera dorsalis* (Diptera: Tephritidae) in South Africa. *Biological Invasions* 17:2803–2809.

Mokam, D. G., N. Atougour, Z. Tadu, et al. 2024. Susceptibility of *Mangifera indica* (Sapindales: Anacardiaceae) cultivars to fruit flies (Diptera: Tephritidae) in 2 agroecological zones of Cameroon. *Journal of Insect Science* 24:9.

Moore, S. D., W. Kirkman, S. Albertyn, et al. 2016a. Partial cold treatment of citrus fruit for export risk mitigation of *Thaumatotibia leucotreta* (Lepidoptera: Tortricidae) as part of a systems approach. *Journal of Economic Entomology* 109(4):1578–1585.

Moore, S. D., W. Kirkman, and V. Hattingh. 2016b. Verification of inspection standards and efficacy of a systems approach for *Thaumatotibia leucotreta* (Lepidoptera: Tortricidae) for export citrus from South Africa. *Journal of Economic Entomology* 109:1564–1570.

Moore, S., and A. Manrakhan. 2022. Postharvest disinfestation treatments for false codling moth and fruit flies in citrus from South Africa. *Horticulturae* 8:221.

Moore, S. D., M. Peyper, M., Kirkman, et al. 2022. Efficacy of various low temperature and exposure time combinations for *Thaumatotibia leucotreta* (Meyrick) (Lepidoptera: Tortricidae) Larvae. *Journal of Economic Entomology* 115(4):1115–1128.

Mwando, N. L., S. Ndlela, R. Meyhöfer, et al. 2021. Hot water treatment for post-harvest disinfestation of *Bactrocera dorsalis* (Diptera: Tephritidae) and its effect on cv. tommy atkins mango. *Insects* 12:1070.

Ndlela, S., F. Obala, N. L. Mwando, et al. 2022. Hot water disinfestation treatment does not affect physical and biochemical properties of export quality mango fruit [*Mangifera indica* L.]. *Agriculture* 12:570.

Odjo, S., F. Traore, and C. Zaki. 2024. *Africa Agriculture Trade Monitor 2024*. Kigali and Washington, DC: International Food Policy Research Institute.

Paull, R. E. 1994. Response of tropical horticultural commodities to insect disinfestation treatments. *HortScience* 29:988–996.

Rao, J., Y. Zhang, H. Zhao, et al. 2024. Projecting the global potential geographical distribution of *Ceratitis capitata* (Diptera: Tephritidae) under current and future climates. *Biology* 13:177.

Self, G., M. N. Ducamp, P. Thaunay, et al. 2012. The effects of phytosanitary hot water treatments on West African mangoes infested with *Bactrocera invadens* (Diptera: Tephritidae). *Fruits* 67:439–449.

Szyniszewska, A. M., N. C. Leppla, Z. Huang, et al. 2016. Analysis of seasonal risk for importation of the Mediterranean fruit fly, *Ceratitis capitata* (Diptera: Tephritidae), via air passenger traffic arriving in Florida and California. *Journal of Economic Entomology* 109:2317–2328.

Szyniszewska, A. M., and A. J. Tatem. 2014. Global assessment of seasonal potential distribution of Mediterranean fruit fly, *Ceratitis capitata* (Diptera: Tephritidae). *PLoS One* 9:e111582.

Tanga, C. M., F. M. Khamis, H. E. Z. Tonnang, et al. 2018. Risk assessment and spread of the potentially invasive *Ceratitis rosa* Karsch and *Ceratitis quilicii* De Meyer, Mwatawala & Virgilio sp. Nov. using life-cycle simulation models: Implications for phytosanitary measures and management. *PLoS One* 13:e0189138.

Theron, C., A. Manrakhan, and C. W. Weldon. 2017. Host use of the oriental fruit fly, *Bactrocera dorsalis* (Hendel) (Diptera: Tephritidae), in South Africa. *Journal of Applied Entomology* 141:810–816.

Vayssières, J. F., S. Korie, O. Coulibaly, et al. 2009. The mango tree in central and northern Benin: Damage caused by fruit flies (Diptera Tephritidae) and computation of economic injury level. *Fruits* 64:207–220.

Ware, A. B., C. L. N. du Toit, E. du Toit, et al. 2016. Host suitability of three avocado cultivars (*Persea americana* Miller: Lauraceae) to oriental fruit fly *Bactrocera (invadens) dorsalis* (Hendel) (Diptera: Tephritidae)). *Crop Protection* 90:84–89.

Wei, Y., X. Xian, H. Zhao, et al. 2024. Projection of the potential global geographic distribution of the solanum fruit fly *Bactrocera latifrons* (Hendel, 1912) (Diptera: Tephritidae) based on CLIMEX models. *Horticulturae* 10:977.

White, I. M., and M. M. Elson-Harris. 1994. *Fruit Flies of Economic Significance: Their Identification and Bionomics*. UK: CAB International.

Section III

Sterile insect technique

Chapter 8

Status of the tephritid fruit fly production worldwide for the sterile insect technique application

Abdeljelil Bakri, Rui Pereira, Walther Enkerlin, Daguang Lu, Carlos Cáceres, Mayren Sanchez, Pablo Liedo, and Jorge Toledo

8.1 INTRODUCTION

Several tephritid fruit fly species are economically important, as they cause damage to a large range of fruit and vegetable species, hindering trade, directly as those are quarantine pests, and indirectly when large amounts of insecticide residues are detected in the commodities or in the environment. Thus, the application of efficient, clean, and species-specific control techniques of these agricultural pests is essential.

One of the proven environment-friendly and species-specific techniques that can be applied at an area-wide scale is the sterile insect technique (SIT) (Hendrichs 2000; Dyck et al. 2021). This technique involves the mass production, irradiation, and release of sterile insects to overflood the local wild population. Suppression, and ultimately eradication, of the target species might be achieved over time after continuous inundative release. Furthermore, SIT can be combined with other integrated pest management programmes. This review presents the data related to the worldwide production of sterile fruit flies of economic importance for the SIT application.

8.2 PRODUCTION OF THE STERILE TEPHRITID FRUIT FLIES IN THE WORLD

Currently, there are 39 tephritid fruit fly species that have been subject to research on irradiation (IDIDAS 2025). These include 16 *Bactrocera species* (data of synonymized species within *Bactrocera* complex were pooled together (Schutze et al. 2014)), nine *Anastrepha* species, seven *Rhagoletis* species, three *Ceratitis* species, two *Zeugodacus* species, one *Dacus* species, and one *Myiopardalis* species.

Out of the 39 tephritid species, only 24 species were subject to radiation for sterilization and 10 were mass-produced for SIT programmes. There is a slight increase in the number of flies subject to sterilization compared to the past review where 21 tephritid species were recorded (Bakri and Hendrichs 2002). The remaining 15 tephritid species were mostly subject to phytosanitary irradiation where the goal was essentially achieving sterility in immature stages (IDIDAS 2025).

Various strains of these fruit fly species were mass-produced (Table 8.1), and it is worth mentioning that most SIT programmes aiming to control the Mediterranean fruit fly, *Ceratitis capitata*, use genetic sexing strains (GSSs) and temperature-sensitive

DOI: 10.1201/9781003644330-11

Table 8.1 Main tephritid fruit fly species produced in the SIT operational programmes

Species	*Country*	*Strain*	*Current production (million/week)*	*Rearing capacity (million/week)*	*Sterilization dose (Gy)*
Anastrepha fraterculus	Peru	Bisexual (original strain from la Molina facility)	5	60	80
Anastrepha ludens	Guatemala	Gua-10 Black Pupae GSS (BPS)	100	200	80
	Mexico	Bisexual strain (original strain with many refreshments)	20	290	80
	Mexico	Tapachula 7 (GSS colour pupa)	30	150	80
	USA	Bisexual Willacy County, 2010	150	200	70 (minimum dosage)
Anastrepha obliqua	Mexico	Bisexual	60	65	80
Bactrocera correcta	Thailand	Thailand	10	40	80
Bactrocera dorsalis	Mauritius	GSS Hawaii	1	15	70
	Thailand	Pakchong/July 30, 2000	10	100	90
	Thailand	Genetic marking strain (GMS)	10	100	85
Bactrocera latifrons	Japan	Yonaguni	0.3	1	70
Bactrocera oleae	Israel	Argov (2008) Yael (2010)	0.12	0.25	100
Bactrocera tryoni	Australia	Local *B. tryoni*	50-60	60	70

(*Continued*)

Table 8.1 (Continued) Main tephritid fruit fly species produced in the SIT operational programmes

Species	*Country*	*Strain*	*Current production (million/week)*	*Rearing capacity (million/week)*	*Sterilization dose (Gy)*
Ceratitis capitata	Argentina	*tsl* VIENNA 8	250	600	120
	Argentina	*tsl* 2006	40	50	
	Brazil	*tsl* Vienna 8	10	100	100
	Chile	*tsl* VIENNA 8 Mix/2006	22	50	120
	Costa Rica	Bisexual	2.5	5	140
	Guatemala	VIENNA 8 - INV D53/ Tolimán	1200	3000	100 (suppression); 145 (eradication)
	Israel	*tsl* VIENNA 8	65	90	100
	Mexico	*tsl* VIENNA 8-Tol/October 2002	500	1000	125
	Morocco	*tsl* Vienna-8 FDF	5	130	100
	Peru	Vienna-8 "*tsl*" "wp"	13	327	120
	South Africa	*tsl* VIENNA 8	68	75	85
	Spain	*tsl* Vienna 8/refreshed 2015	250	500	95
	USA	Vienna7/Mixed-99	112	200	130
Zeugodacus (Bactrocera) cucurbitae	Japan	Taiwan 6/2011	64	200	70
	Mauritius	GSS Hawaii	1	15	70

Presently, there are 16 countries around the world with such programmes: Argentina, Australia, Brazil, Chile, Costa Rica, Guatemala, Israel, Japan, Mauritius, Mexico, Morocco, Peru, South Africa, Spain, Thailand, and USA. *tsl* strain: temperature-sensitive lethal strain for male-only release. GSS: genetic sexing strain (DIR-SIT 2025).

lethal (*tsl*) strains. GSSs allow the release of sterile males-only and thus increase the efficiency and the cost-effectiveness of the programmes. *tsl* strains allow male-only release, as high temperatures can be lethal to eggs containing *tsl* females (Hendrichs et al. 1995; Caceres 2002; Caceres et al. 2023).

Presently, there are 23 facilities in 16 countries producing sterile tephritid fruit flies with a total production volume of over 3 billion/week (Table 8.1). The majority of the fruit fly production sites are in the American continent. However, further fruit fly production facilities were either planned or are at an advanced stage of development elsewhere. Some large facilities produce more than one species or strain at the same time, and others are capable of exporting sterile fruit flies abroad [(e.g., shipment of sterile flies to Croatia, Dominican Republic, Morocco, and the USA (Florida and California) (FAO/IAEA 2022a)]. The production size in a facility may vary in time according to the need of the SIT programmes, but is limited by the production capacity (Table 8.1).

8.3 IRRADIATION DOSES FOR TEPHRITID FRUIT FLY STERILIZATION

According to the reported data in the scientific publications (IDIDAS 2025), all *Anastrepha* species, except *Anastrepha suspensa,* required less than 60 Gy for sterilization (Figure 8.1). Similarly, the genus *Zeugodacus*, with only two species tested (*Z. tau* and *Z. cucurbitae*), required less than 60 Gy. The genus *Bactrocera* is less homogeneous: the dose varies from 52 Gy for *B. tryoni* to 90 Gy for *B. minax* and *B. tsuneonis*. The average dose for *Ceratitis* species is between 60 Gy for *C. rosa* and 90 Gy for *C. capitata*. The most striking and less sensitive tephritid species is *Dacus ciliatus,* requiring a very high dose of up to 140 Gy for full sterilization, while at 120 Gy, residual fertility was reported (Rempoulakis et al. 2015). However, previous authors (Huque and Ahmad 1969) reported that 85–100 Gy induces full sterility in *D. ciliatus*.

The average doses (Figure 8.1) were estimated from IDIDAS data (IDIDAS 2025). While the data presented here might have some limitations, they provide an indication of an upper absorbed dose limit for effective sterilization of major tephritid species. With the exception of the Mediterranean fruit fly outbreak eradication programme or preventive released programmes in the USA, which uses a dose of up to 145 Gy (Table 8.1). On average, about or less than 100 Gy is sufficient to ensure no residual fertility in the released flies.

In general, the wide range of sterilization doses (Table 8.1) observed in some species (e.g., *C. capitata*) can be attributed to different parameters involved during the irradiation process conducted by different independent studies and also due to the level of sterility required by the SIT programme's objective (eradication, suppression, containment, or prevention). This could also explain the slight difference, in some species, between the doses reported in the literature (IDIDAS 2025) (Figure 8.1) and the doses applied by the operational programmes (DIR-SIT 2025) (Table 8.1).

To standardize the sterile insect production protocols and minimize the discrepancies, quality control international standards for the sterile insect production (FAO/IAEA/USDA 2019) and the related standard operating procedures (SOPs) were set and updated regularly.

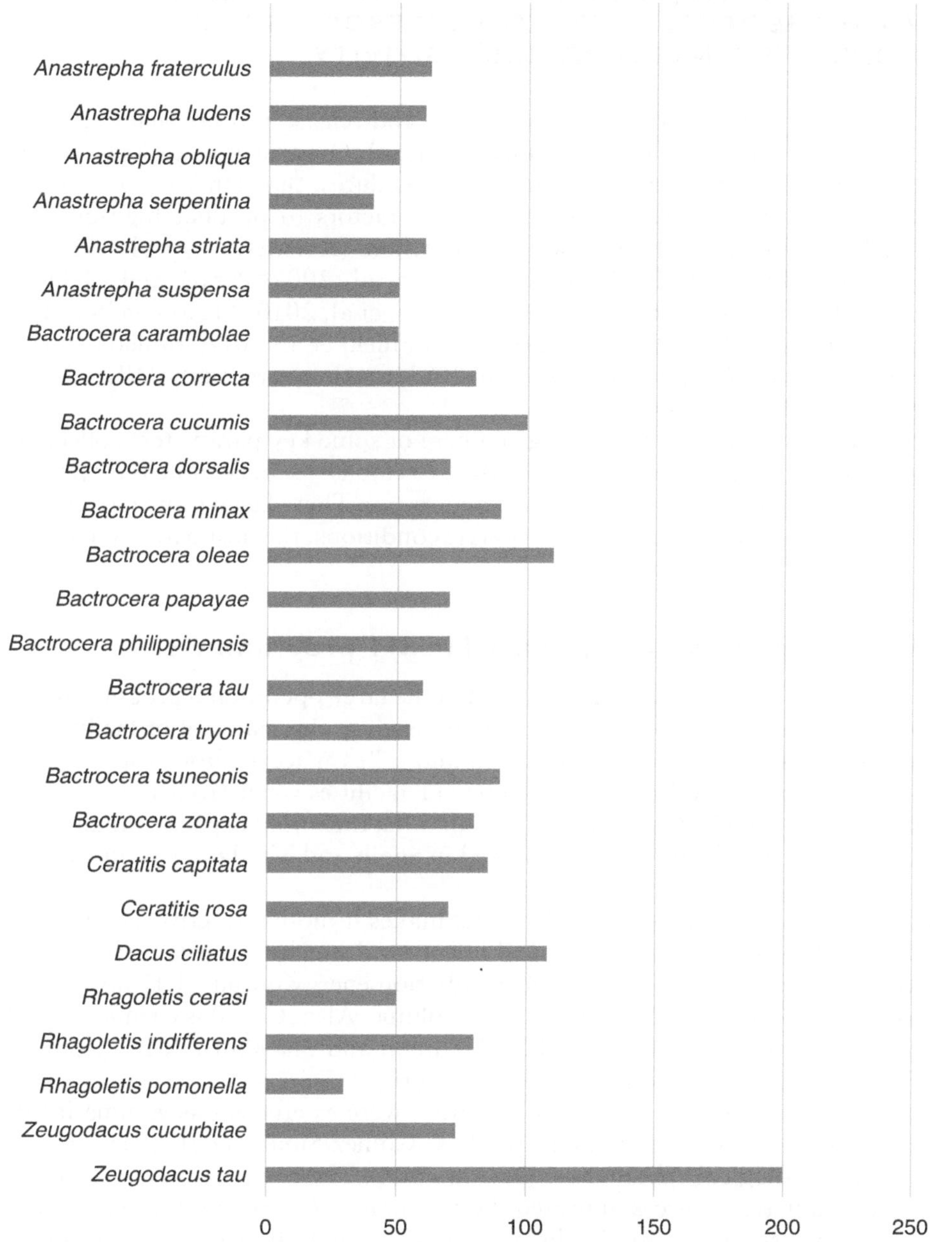

Figure 8.1 Radio-sterilization doses (Gy) per species of tephritid fruit fly males treated at the pupal stage under air atmosphere. **Rhagoletis* species treated as adults, as diapausing pupae are not practical for age synchronization (Brown 1978; IDIDAS 2025).

8.4 SOME BASIC PARAMETERS THAT MIGHT INTERFERE WITH THE RADIO-STERILIZATION PROCESS AND MODIFY THE INSECT RADIOSENSITIVITY

Irradiation is only one part of the production and release process that can affect the quality of sterile flies (e.g., mating competitiveness). Overviews of endogenous (biotic) and exogenous (abiotic) factors, and human procedures, that can affect ionizing radiation treatment, and the way to optimize these factors to produce high-quality sterile flies, have thoroughly been described elsewhere (Rananavare et al. 1991; Toledo et al. 2004; Bakri et al. 2005, 2021; Caceres et al. 2007; Parker and Mehta 2007; Collins et al. 2009; Helinski et al. 2009; Hallman et al. 2010; Mastrangelo et al. 2010; Dominiak et al. 2014; Arthur et al. 2015; Bloomfield et al. 2017; Yamada et al. 2019, 2022; Kruger et al. 2021; Panduranga et al. 2021) and in the recent FAO/IAEA manuals (FAO/IAEA 2022b, 2022c).

Hence, only a short description is given here of some key parameters collected from SIT operational programmes (DIR-SIT 2025) to emphasize once more their importance in the optimization of the radio-sterilization process. These key parameters are the irradiation source, the dose rate, the atmospheric conditions, the dosimetry system, and the age and stage of the insects.

8.4.1 Ionizing radiation types used in SIT programmes

Four types of irradiation machines are used in the large operational programmes releasing sterile insects: cobalt-60 (Co-60) source, used by the majority of facilities (73%), caesium-137 (Cs-137) (12%), X-ray (12%), and e-Beam accelerator (3%). The energies of the photons produced by X-rays used by SIT facilities range from 125 keV/18 mA to 150 keV/45 mA. For Co-60 and Cs-137 irradiators, the initial capacity ranges from 846 Ci to 100 kCi. The e-beam accelerator used in Spain and Thailand has an energy of 10 MeV 40 kW.

Although its longer half-life (30.07 years) makes it more attractive as an irradiation source than Co-60 (5.27 years), Cs-137 is not used much in sterile insect production facilities because cobalt-60 offers higher radiation energy output (1.17 and 1.33 MeV) against caesium-137 (0.66 MeV) for a given volume. Also, Co-60 is chemically stable in the metallic form in contrast to Cs-137 (Hallman and Blackburn 2016). Additionally, Cs-137 presents some concerns in terms of security.

Several studies reported that X-ray machines were as efficient as gamma irradiators (Mastrangelo et al. 2010; Yamada et al. 2014; Gomez-Simuta et al. 2021) and had the same relative biological effectiveness (Mehta 2004). The comparison of certain parameters of the radiation sources indicated the advantage of gamma rays, self-shielded and panoramic, in terms of their reliability (FAO/IAEA 2004). In selecting an irradiator, the throughput and the availability of accessories such as gas feed systems, turntables, and field flatters should be considered.

In terms of depth-dose penetration, X-rays have similar penetration as gamma rays, while electron beams have low penetration but deliver very high dose rates. However, the dose rate is hypothesized to directly affect the irradiation efficacy. That is, a faster dose rate may lead to increased efficacy because it overwhelms radiation damage repair mechanisms (Hallman et al. 2010; Hallman and Blackburn 2016).

As sterility is linked to dominant lethal mutations, radiations of higher ion densities were not recommended for insect sterilization, since a greater proportion of the cells would degenerate and not continue through spermatogenesis (Alexander et al. 1959). In one of the rare studies on radio-sterilization of tephritids using fast neutron, higher efficiency than gamma rays was found in producing sterile Mediterranean fruit fly and olive flies with good mating competitiveness (Cavalloro and Delrio 1974). In addition, the authors reported no damage to the mesenteron and the mating competitiveness in both species was about four times greater for males exposed to fast neutrons than for those exposed to gamma-radiation. In any case, the use of a reactor to produce fast neutron seems impractical at the present state of the technology.

The aim of the sterilization process is to deliver the prescribed irradiation dose to every single insect in the canister. However, the nature of the radiation is such that it is inevitable that there is an uneven distribution of the dose in the canister loaded with insects. Therefore, the goal is to select a configuration of the radiation source and the irradiated insect sample's volume size and geometry that would result in an acceptable dose variation within the canister (Mehta 2004). It is recommended to maintain this inevitable range of absorbed doses at a maximum dose uniformity ratio of <1.3, and preferably at 1.05 between the maximum and minimum absorbed doses across the target volume (ICRU 1979). A dosimetry system is essential to be able to measure the dose range within the canister.

8.4.2 Dosimetry and frequency

Dosimetry can be useful for both canister dose mapping and in the routine dosimetry for the quality control to confirm that the insects received the appropriate sterilizing dose (besides the sterility testing). Various dosimetry systems are used by the SIT programmes (DIR-SIT 2025). Unfortunately, in the scientific literature not all programmes indicate the type of dosimetry system nor the frequency of use. The most common dosimetries are the Gafchromic® system that covers the dose range from 50 to 600 Gy (FAO/IAEA 2020) and the Fricke dosimeter (Fricke and Hart 1966) used in a dose range of 5–400 Gy and for dose rates of up to 106 Gy/s. However, certain cautions should be taken as it has been reported that Fricke dosimetry has some disadvantages related to its high sensitivity to impurities and the presence of oxygen in the solution (Prieto Miranda et al. 1999; de Almeida et al. 2014).

The dosimetry system should be calibrated in accordance with international standards or appropriate national standards. Comprehensive information is included in the SOPs for the Gafchromic™ Film Dosimetry System for Gamma Radiation and X-ray manuals (FAO/IAEA 2022b, 2022c). Also, other technical standards (e.g., ISO/ASTM 2004, 2013) provide details on the standard guide for the selection and calibration of dosimetry systems for radiation processing. Adherence to the present dosimetry SOPs will result in improved data consistency and meeting the highest standards for research quality.

8.4.3 Dose rate

In some scientific publications and SIT programmes, the dose rate information is not reported and when reported the range can be large (DIR-SIT 2025). In the past, there were conflicting data about the impact of the dose rate on the quality of sterile insects

(Hooper 1975; Hallman 2000; Bakri et al. 2005, 2021; Collins et al. 2008; Hallman et al. 2010; Ernawan et al. 2017, 2022). Recently, it has been shown in mosquitoes that below full sterility doses, there is a U-shape correlation when pairing dose and dose rate for radiation sterilization (Yamada et al. 2022). That is, not only does the dose rate alter irradiation-induced effects, but the interaction is not linear and may change with the dose. Further studies might be needed to confirm this relationship in tephritid fruit flies. This might help explain some divergences in the dose of radiation for sterilization. Therefore, it is highly recommended to record this parameter during the irradiation process.

8.4.4 Air atmosphere, anoxia, and hypoxia

In SIT programmes, irradiation is being conducted under different atmospheric conditions (DIR-SIT 2025): Air, 1h, 2h, 12h, 20h hypoxia, or nitrogen, but most programmes use an ambient air condition termed normoxia. Under well-oxygenated conditions, the radiosensitivity is about three times higher (Steel 1993). However, under very low oxygen tension, anoxia, or more frequently under purged nitrogen, higher doses are required to achieve the same level of sterility but with less damage to the insect (Hooper 1976; Ohinata et al. 1977; Fisher 1997). Therefore, the time under hypoxia needs to be accurately determined to avoid data discrepancies and also controlled to avoid insect stress (Rull et al. 2011; Benelli et al. 2021).

8.4.5 Age/stage

Irradiation of homogeneous batches of insects having approximately the same age is essential for consistent sterility results. The timing of pupal or adult irradiation exposures may impact the resulting sterility, longevity, and mating competitiveness (Fezza et al. 2021). Generally, differentiated cells that have lost their ability to divide are less sensitive to radiation than stem cells (Bakri et al. 2021). Thus, insects at an advanced development stage (with more differentiated cells) are less radiosensitive than younger insects. Furthermore, a greater decreasing trend in adult flight ability and an increasing trend in adult mortality were observed for pupae irradiated at a young age compared to those irradiated 24 hours before emergence (Toledo et al. 2004).

In most SIT programmes, mature fruit fly pupae are preferred for irradiation (DIR-SIT 2025). However, the current technique used for age determination based on pupal eye colour (Ruhm and Calkins 1981; Resilva and Pereira 2014; Resilva et al. 2019a, 2019b) may not provide accurate and rapid results as it is based on human judgment. At the moment, some work has been done on the machine reading of the eye colour with the support of artificial intelligence (González-López et al. 2022). Nestel et al. (2007) proposed the age selection based on the pupal respiratory rate. These methods may enable greater precision of age selection, but still their practicability under large operational production conditions needs to be explored.

8.4.6 Strain

To increase the efficiency of the SIT technique, different fruit fly GSSs were developed and mass-produced (Table 8.1): for example, *C. capitata* male-only Vienna-*tsl* (temperature sensitivity lethal) (Caceres et al. 2002), *A. ludens* and *A. fraterculus* GSS colour pupa, *B. dorsalis* genetic marking strain, and *Z. cucurbitae* GSS.

According to data in the literature, strains with some special genotype characters did not have any perceptible advantage or disadvantage vis-a-vis the irradiation effect. Allinghi et al. (2007) reported that within the range of tested irradiation doses, *A. fraterculus* males from laboratory strain and wild population did not differ in mating competitiveness evaluated under semi-natural field cage conditions. Also, the copulation start-time and the copulation duration did not differ significantly between the two strains. Giustina et al. (2021) reported that males of two *A. fraterculus* strains, a bisexual strain and a novel GSS, required the same dose (74 Gy), administered under hypoxia, to induce 99% sterility in females, while the same dose between 80 and 90 Gy can induce complete sterility.

Moreover, it is very important to consider keeping voucher specimens deposited in a national collection of insects just in case a future taxonomic verification is needed. Furthermore, this may seem obvious, but it is necessary to emphasize that species taxonomy should be confirmed with experts before launching any large-scale mass-rearing and the taxonomic name of the species should be updated.

8.5 CONCLUSION

Extensive experience in SIT for tephritid fruit fly control is now available in several countries and routinely very large mass-rearing and release operations are conducted in several sites. Optimizing the irradiation dose, one of the production process components, requires a trade-off between the objectives of producing competitive adult sterile males and the reduction in residual fertility of flies being released into the environment. There is a positive correlation between increased irradiation and induced sterility of males. Conversely, a negative correlation is observed with emergence rate, survival, and mating competitiveness. Several pre- and post-irradiation factors can also be considered to improve the fitness of the sterile male flies. There is still room for optimizing the key parameters cited above and one of the very efficient mechanisms, available to IAEA Member States, is through coordinated research projects (IAEA 2024) under the patron of the Insect Pest Control Section of the Joint FAO/IAEA Centre of Nuclear Techniques in Food and Agriculture.

REFERENCES

Alexander, M.L., J. Bergendahl, and M. Brittain. 1959. Biological damage in mature and immature germ cells of *Drosophila virilis* with ionizing radiations. *Genetics* 44(6):979–999. doi:10.1093/genetics/44.6.979.

Allinghi, A., C. Gramajo, E. Willink, et al. 2007. Induction of sterility in *Anastrepha fraterculus* (Diptera: Tephritidae) by gamma radiation. *Florida Entomologist* 90:96–102. doi:10.1653/0015-4040(2007)90[96:IOSIAF]2.0.CO;2.

de Almeida, C. E., R. Ochoa, M. C. Lima, et al. 2014. A feasibility study of Fricke dosimetry as an absorbed dose to water standard for 192Ir HDR sources. *PloS One* 9(12):e115155. doi:10.1371/journal.pone.0115155

Arthur, V., A. Machi, and T. Mastrangelo. 2015. *Ionizing Radiations in Entomology*, Evolution of Ionizing Radiation Research, Mitsuru Nenoi, IntechOpen. doi:10.5772/60409. Available from: https://www.intechopen.com/chapters/48737

Bakri, A., N. Heather, J. Hendrichs, et al. 2005. Fifty years of radiation biology in entomology: Lessons learned from IDIDAS. *Annals of the Entomological Society of America* 98:1–12. doi :10.1603/0013-8746(2005)098[0001:FYORBI]2.0.CO;2

Bakri, A. and J. Hendrichs. 2002. Radiation doses for sterilization of tephritid fruit flies. *Proceedings of 6th International Fruit Fly Symposium*, 6–10 May, 2002, Stellenbosch, South Africa, pp. 475–479.

Bakri, A., K. Mehta, and D. R. Lance. 2021. Sterilizing insects with ionizing radiation. In: *Sterile Insect Technique, Principles and Practice in Area-Wide Integrated Pest Management* (2nd ed.). Dyck, V.A., Hendrichs, J., and Robinson, A. S. (Eds.), CRC Press, New York. pp. 355–398. doi:10.1201/9781003035572

Benelli, M., B. Mainali, P. W. Taylor, et al. 2021. Reduced quality of sterile Queensland fruit fly following post-production stress from hypoxia, irradiation and vibration. *Journal of Pest Science* 94:473–485. doi:10.1007/s10340-020-01269-9

Bloomfield, C. I. A., B. G. Fanson, R. Mirrington, et al. 2017. Optimising irradiation dose in mass-produced Queensland fruit fly, *Bactrocera tryoni*, for sterile insect release: The incorporation of residual effects on F1 progeny. *Entomologia Experimentalis et Applicata* 162(2):168–177. doi:10.1111/eea.12538

Brown, R. D. 1978. *Gamma irradiation of the western cherry fruit fly for the sterile male control technique*, Thesis (Ph. D.). Oregon State Univ., Corvallis (USA), 218 p.

Caceres, C. (2002). Mass rearing of temperature sensitive genetic sexing strains in the Mediterranean fruit fly (*Ceratitis capitata*). *Genetica* 116(1):107–116.

Caceres, C., K. Bourtzis, G. Gouvi, et al. 2023. Development of a novel genetic sexing strain of *Ceratitis capitata* based on an X-autosome translocation. *Scientific Reports* 13:16167. doi:10.1038/s41598-023-43164-0

Caceres, C., J. P. Cayol, W. Enkerlin, et al. 2002. Comparison of Mediterranean fruit fly (*Ceratitis capitata*) (Tephritidae) bisexual and genetic sexing strains: Development, evaluation and economics. *Proceedings of 6th International Fruit Fly Symposium*, 6–10 May 2002, Stellenbosch, South Africa, pp. 367–381.

Caceres, C., D. McInnis, T. Shelly, et al. 2007. Quality management systems for fruit fly (Diptera: Tephritidae) sterile insect technique. *Florida Entomologist* 90:1–9. doi:10.1653/0015-4040(2007)90[1:QMSFFF]2.0.CO;2

Cavalloro, R., and G. Delrio. 1974. Sterilization of *Dacus oleae* Gmel. and *Ceratitis capitata* Wied. with gamma radiation and fast neutrons. *Redia* 55:373–392.

Collins, S. R., C. W. Weldon, C. Banos, et al. 2008. Effects of irradiation dose rate on quality and sterility of Queensland fruit flies, *Bactrocera tryoni* (Froggatt). *Journal of Applied Entomology* 132(5):398–405. doi:10.1111/j.1439-0418.2008.01284.x

Collins, S. R., C. W. Weldon, C. Banos, et al. 2009. Optimizing irradiation dose for sterility induction and quality of *Bactrocera tryoni*. *Journal of Economic Entomology* 102(5):1791–1800. doi:10.1603/029.102.0509

DIR-SIT. 2025. The world-wide directory of SIT facilities. Available online: https://nucleus.iaea.org/sites/naipc/dirsit/SitePages/All%20Facilities.aspx (accessed on 31 March 2025).

Dominiak, B. C., S. Sundaralingam, L. Jiang, et al. 2014. Evaluating irradiation dose for sterility induction and quality control of mass-produced fruit fly *Bactrocera tryoni* (Diptera: Tephritidae). *Journal of Economic Entomology* 107(3):1172–1178. doi:10.1603/ec13421

Dyck, V. A., J. Hendrichs, and A. S. Robinson (Eds.). 2021. *Sterile Insect Technique: Principles and Practice in Area-Wide Integrated Pest Management* (2nd ed.). CRC Press, New York. doi:10.1201/9781003035572

Ernawan, B., U. S. F. Tambunan, I. Sugoro, et al. 2017. Effects of gamma irradiation dose-rate on sterile male *Aedes aegypti*. *AIP Conference Proceedings* 1854:020010. doi:10.1063/1.4985401

Ernawan, B., A. Tjandra, Y. Sri, et al. 2022. Investigation of developmental stage/age, gamma irradiation dose, and temperature in sterilization of male *Aedes aegypti* (Diptera: Culicidae) in a sterile insect technique programme. *Journal of Medical Entomology* 59(1):320–327. doi:10.1093/jme/tjab166

FAO/IAEA. 2004. *Dosimetry System for SIT: Manual for Gafchromic® Film*. Joint FAO/IAEA Centre of Nuclear Techniques in Food and Agriculture, IAEA, Vienna, Austria.

FAO/IAEA. 2020. *Dose mapping by scanning Gafchromic film to measure the absorbed dose of insects during their sterilization*. Parker, A., Gomez-Simuta, Y., and Yamada, H. (Eds.), Food and Agriculture Organization of the United Nations/International Atomic Energy Agency, Vienna, Austria. 17 pp.

FAO/IAEA. 2022a. *International Guideline for Transboundary Shipments of Irradiated Sterile Insects*. Food and Agriculture Organization of the United Nations/International Atomic Energy Agency, Vienna, Austria. 38 pp.

FAO/IAEA. 2022b. *Dosimetry for SIT: Standard operating procedures for Gafchromic™ film dosimetry system for low energy X radiation* v. 1.0. Parker, A., Mehta, K., and Gómez-Simuta, Y. (Eds.), Food and Agriculture Organization of the United Nations/International Atomic Energy Agency, Vienna, Austria. 42 pp. https://www.iaea.org/sites/default/files/x-ray-sop-en.pdf

FAO/IAEA. 2022c. *Dosimetry for SIT: Standard operating procedures for Gafchromic™ film dosimetry system for gamma radiation* v. 1.0. Parker, A., Mehta, K., and Gómez-Simuta, Y. (Eds.), Food and Agriculture Organization of the United Nations/International Atomic Energy Agency, Vienna, Austria. 43 pp. https://www.iaea.org/sites/default/files/gamma-sop-en.pdf

FAO/IAEA/USDA. 2019. *Product Quality Control for Sterile Mass-Reared and Released Tephritid Fruit Flies, Version 7.0*. International Atomic Energy Agency, Vienna, Austria. 148 pp.

Fezza, T. J., P. A. Follett, and T. E. Shelly. 2021. Effect of the timing of pupal irradiation on the quality and sterility of oriental fruit flies (Diptera: Tephritidae) for use in sterile insect technique. *Applied Entomology and Zoology* 56(4):443–450. doi:10.1007/s13355-021-00751-9

Fisher, K. 1997. Irradiation effects in air and in nitrogen on Mediterranean fruit fly (Diptera: Tephritidae) Pupae in Western Australia. *Journal of Economic Entomology* 90(6):1609–1614. doi:10.1093/jee/90.6.1609

Fricke, H., and E. J. Hart. 1966. *Chemical Dosimetry. Radiation Dosimetry. Vol II, Ed. Attix, Roesch, Tocholin*. Academic Press, New York and London.

Giustina, P. D., T. Mastrangelo, S. Ahmad, et al. 2021. Determining the sterilization doses under hypoxia for the novel black pupae genetic sexing strain of *Anastrepha fraterculus* (Diptera, Tephritidae). *Insects* 12:308. doi:10.3390/insects12040308

Gomez-Simuta, Y., A. Parker, C. Caceres, et al. 2021. Characterization and dose-mapping of an X-ray blood irradiator to assess application potential for the sterile insect technique (SIT). *Applied Radiation and Isotopes* 176:109859. doi:10.1016/j.apradiso.2021.10985

González-López, G. I., G. Valenzuela-Carrasco, E. Toledo-Mesa, et al. (2022). Determination of the physiological age in two tephritid fruit fly species using artificial intelligence. *Journal of Economic Entomology* 115: 1513–1520. doi:10.1093/jee/toac133

Hallman, G. J. 2000. Expanding radiation quarantine treatments beyond fruit flies. *Agricultural and Forest Entomology* 2(2):85–95. doi:10.1046/j.1461-9563.2000.00057.x

Hallman, G., and C. Blackburn. 2016. Phytosanitary irradiation. *Foods* 5(4):8. doi:10.3390/foods5010008.

Hallman, G. J., N. M. Levang-Brilz, J. L. Zettler, et al. 2010. Factors affecting ionizing radiation phytosanitary treatments, and implications for research and generic treatments. *Journal of Economic Entomology* 103(6):1950–1963. doi:10.1603/ec10228

Helinski, M. E., A. G. Parker, and B. G. Knols. 2009. Radiation biology of mosquitoes. *Malaria Journal* 8(Suppl 2):S6. doi:10.1186/1475-2875-8-s2-s6.

Hendrichs, J. 2000. Use of the sterile insect technique against key insect pests. *Sustainable Development International* 2:75–79.

Hendrichs, J., G. Franz, and P. Rendon. 1995. Increased effectiveness and applicability of the sterile insect technique through male-only releases for control of Mediterranean fruit flies during fruiting seasons. *Journal of Applied Entomology* 119:371–377. doi:10.1111/j.1439-0418.1995.tb01303.x

Hooper, G. H. S. 1975. Sterilization of *Dacus cucumis* French (Diptera: Tephritidae) by gamma radiation. II. Effect of dose rate on sterility and competitiveness of adult males. *Australian Journal of Entomology* 14(2):175–177. doi:10.1111/J.1440-6055.1975.TB02021.X

Hooper, G. H. S. 1976. Sterilization of *Dacus cucumis* French (Diptera: Tephritidae) by gamma radiation. III. Effect of irradiation in nitrogen on sterility, competitiveness and mating propensity. *Journal of the Australian Entomological Society* 15(1): 13–18. doi:10.1111/j.1440-6055.1976.tb02067.x

Huque, H. and C. R. Ahmad 1969. Studies on the control of *Dacus ciliatus* Loew (Tephritidae: Diptera) by sterile male release technique. *International Journal of Applied Radiation and Isotopes* 20: 791–795. doi:10.1016/0020-708X(69)90042-8

IAEA. 2024. *Coordinated Research Activities*. International Atomic Energy Agency, Vienna, Austria. Available online: https://www.iaea.org/services/coordinated-research-activities (accessed on 08 June 2024).

ICRU. 1979. Quantitative concepts and dosimetry in radiobiology. International Commission on Radiation Units and Measurement. Report no. 30. https://www.icru.org

IDIDAS. 2025. International database on insect disinfestation and sterilization. Available online: https://nucleus.iaea.org/sites/naipc/ididas/Pages/Browse-IDIDAS.aspx (accessed on 31 March 2025).

ISO/ASTM. 2004. *Standard Guide for Dosimetry for Irradiation of Insects for Sterile Release Programmes, ISO/ASTM 51940, Annual Book of ASTM Standards 12.02*. American Society for Testing and Materials International, Philadelphia, PA.

ISO/ASTM. 2013. *Practice for Calibration of Routine Dosimetry Systems for Radiation Processing. International Standard ISO/ASTM 51261:2013*. American Society for Testing and Materials International, Philadelphia, PA.

Kruger, A. P., M. S. Ferreira, L. N. Martins, et al. 2021. Impact of gamma radiation dose on sterility and quality parameters of *Anastrepha fraterculus* (Diptera: Tephritidae). *Anais da Academia Brasileira de Ciências* 93:2. doi:10.1590/0001-3765202120190249

Mastrangelo, T., A. G. Parker, A. Jessup, et al. 2010. A new generation of X ray irradiators for insect sterilization. *Journal of Economic Entomology* 103: 85–94. doi:10.1603/ec09139

Mehta, K. 2004. Absorbed dose and sterilization process. In: *Optimizing Sterilization Procedures in SIT Programmes*. Report of the FAO/IAEA Consultants Meeting held in Vienna, Austria from 19 to 23 April 2004. IAEA-04T02362. IAEA, Vienna, Austria.

Nestel, D., E. Nemny-Lavy, and V. Alchanatis. 2007. Gas-exchange patterns of Mediterranean fruit fly pupae (Diptera: Tephritidae): A tool to forecast developmental stage. *Florida Entomologist* 90(1):71–79. doi:10.1653/0015-4040(2007)90[71:gpomff]2.0.co;2

Ohinata, K., M. Ashraf, and E. J. Harris. 1977. Mediterranean fruit flies: Sterility and sexual competitiveness in the laboratory after treatment with gamma irradiation in air, carbon dioxide, helium, nitrogen or partial vacuum. *Journal of Economic Entomology* 70:165–68. doi:10.1093/jee/70.2.165

Panduranga, G. S., K. Sharma, B. Singh, et al., 2021. Effect of gamma irradiation on quality parameters, sterility and mating competitiveness of melon fly, *Bactrocera cucurbitae* (Coquillett). *International Journal of Tropical Insect Science* 42:875–883. doi:10.1007/s42690-021-00613-5

Parker, A., and K. Mehta. 2007. Sterile insect technique: A model for dose optimization for improved sterile insect quality. *Florida Entomologist* 90(1):88–95. doi:10.1653/0015-4040(2007)90[88:SITAMF]2.0.CO;2

Prieto Miranda, E.F., G. Cuesta Fuente, and A. Chavez Ardanza. 1999. *Development and Current State of Dosimetry in Cuba (IAEA-TECDOC-1070)*. International Atomic Energy Agency (IAEA).

Rananavare, R. D., M. R. Harwalkar, and G. W. Rahalkar. 1991. Influence of modifying factors on induction of sterility and mating ability of potato tuberworm, *Phthorimaea operculella* (Zeller). *Journal of Nuclear Agriculture and Biology* 20:199–205.

Rempoulakis, P., R. Castro, E. Nemny-Lavy, et al. 2015. Effects of radiation on the fertility of the Ethiopian fruit fly, *Dacus ciliatus*. *Entomologia Experimentalis et Applicata* 155(2):117–122. doi:10.1111/eea.12289

Resilva, S. R., N. B. Barnes, and B.G. Obra. 2019a. Documented pupal eye colour of Mediterranean fruit fly as a tool for radiation sterilization. *Philippine Journal of Science* 148(1): 87–93.

Resilva, S. R., E. Hernandez, and B. G. Obra. 2019b. Radiation sterilization of Mexican fruit fly *Anastrepha ludens* (Leow) based on pupal eye colour. *Philippine Journal of Science* 148(1):45–50.

Resilva, S. R., and R. Pereira. 2014. Age and temperature-related pupal eye colour changes in various tephritid fruit fly species with a view to optimizing irradiation timing. *International Journal of Tropical Insect Science* 34:S59–S65. doi:10.1017/S1742758414000095

Ruhm, M. E., and C. O. Calkins. 1981. Eye-colour changes in *Ceratitis capitata* pupae, a technique to determine pupal development. *Entomologia Experimentalis et Applicata* 29:237–240. doi:10.1111/j.1570-7458.1981.tb03064.x

Rull, J., A. Birke, R. Ortega, et al. 2011. Quantity and safety vs. quality and performance: Conflicting interests during mass rearing and transport affect the efficiency of sterile insect technique programmes. *Entomologia Experimentalis et Applicata* 142(1):78–86. doi:10.1111/j.1570-7458.2011.01196.x

Schutze, M. K., N. Aketarawong, W. Amornsak, et al. 2014. Synonymization of key pest species within the *Bactrocera dorsalis* species complex (Diptera: Tephritidae): Taxonomic changes based on a review of 20 years of integrative morphological, molecular, cytogenetic, behavioural and chemoecological data. *Systematic Entomology* 40(2):456–471. doi:10.1111/syen.12113

Steel, G. G. (Ed.) 1993. *Basic Clinical Radiobiology for Radiation Oncologists*. Edward Arnold Publications, London.

Toledo, J., J. Rull, A. Oropeza, et al. 2004. Irradiation of *Anastrepha obliqua* (Diptera: Tephritidae) revisited: Optimizing sterility induction. *Journal of Economic Entomology* 97(2):383–389. doi:10.1093/jee/97.2.383

Yamada, H., H. Maiga, J. Juarez, et al. 2019. Identification of critical factors that significantly affect the dose-response in mosquitoes irradiated as pupae. *Parasites and Vectors* 12(1):435. doi:10.1186/s13071-019-3698-y

Yamada, H., H. Maiga, C. Kraupa, et al. 2022. Effects of chilling and anoxia on the irradiation dose-response in adult *Aedes* mosquitoes. *Frontiers in Bioengineering and Biotechnology* 10:856780. doi:10.3389/fbioe.2022.856780

Yamada, H., A. G. Parker, C. F. Oliva, et al. 2014. X-Ray-induced sterility in *Aedes albopictus* (Diptera: Culicidae) and male longevity following irradiation. *Journal of Medical Entomology* 51(4):811–816. doi:10.1603/me13223

Chapter 9

A focus on Morocco's sterile insect technique initiative

Jamaa Zim, Yassine Aoutil, Amine Leghlid, Yahya Machal, Hassan Zouhry, Zhor Dehbi, Siham Boujdig, and Carlos Cáceres

9.1 INTRODUCTION

In Morocco, the citrus industry is an important component of the national agricultural sector. It contributes to the value added of the agricultural sector with a value of 489 million US dollars. Citrus production occupies a total area of 105,000 ha, with the Sous-Massa region being the most important production area accounting for 32% of the total area, followed by Rabat-Sale-Kenitra accounting for 20% (MMAFRDWF 2023).

In addition to water scarcity, one of the main challenges in Morocco, the citrus industry must face pest management requirements, due to the high-quality standards demanded by the market, and in some cases quarantine restriction. Many key citrus pests are present in Morocco such as *Ceratitis capitata* (Wiedemann), *Aonidiella aurantii* Maskell, *Panonychus citri* McGregor, and *Phyllocnistis citrella* Stainton. However, *C. capitata* is the most damaging pest with both direct and indirect economic impacts and is regarded as a critical threat to fruit production, diversification, and export (Mazih 2015; Elaini and Mazih 2018; Assouguem et al. 2021).

9.2 MEDFLY STATUS IN MOROCCO

The Mediterranean fruit fly (Medfly) (*C. capitata*) is the primary tephritid pest in Morocco, remaining present year-round. Its spatiotemporal occurrence and phenology are determined by suitable environmental conditions, host plants, and fruit availability (Elaini and Mazih 2018). Medflies are active nearly all year, completing up to ten generations annually if host fruits are available. With increasing global temperatures, the species may soon expand its host range and infestation period to spring, early summer, and even winter (Boulahia-Kheder 2021). Medfly populations in Morocco display substantial genetic diversity, particularly at higher altitudes, with marked differentiation between populations (Alaoui et al. 2010). Besides citrus, other fruit crops such as pears, apples, peaches, nectarines, plums, and pomegranates are notably impacted by Medflies, especially in Northern Morocco (Elaini and Mazih 2018). In the Souss valley within the Sous-Massa region of Central Morocco, a major citrus production area, citrus orchards are adjacent to Argan forests (*Argania spinosa*), which serve as significant reservoirs for Medfly individuals (Sacantanis 1957). Infestation rates in Argan fruits vary seasonally, remaining low from February to May but rising sharply in June and July, with ripe fruits being most vulnerable (Mazih and Debouzie 1996). Additionally,

DOI: 10.1201/9781003644330-12

while *Opuntia ficus-indica* was once a potential reservoir, its density in Morocco has drastically declined due to infestations by *Dactylopius opuntiae* (El Aalaoui et al. 2019).

In Morocco, citrus is one of the crops most severely affected by Medfly-related losses. Despite extensive efforts to control the Medfly population, this pest remains a significant threat to the national citrus industry, impacting both exports and production (Boulahia-Kheder 2021). Farmers must closely monitor Medfly prevalence during fruit maturation phases and may need to apply multiple pesticide treatments to manage the pest. However, even with these control measures, fruit infestation levels can still average between 10% and 20%, sometimes reaching even higher levels (Elaini and Mazih 2018). In the Berkane region (northeast Morocco), notably high infestation rates have been recorded resulting in a temporary ban on the commercialization of certain products, including clementines, tangerines, mandarins, and sweet oranges produced or shipped from this area (APHIS 2016).

9.3 CONVENTIONAL PEST CONTROL METHODS

Chemical control is the most widely employed method against Medflies and other pests (Assouguem et al. 2021). Typically, treatments involve full-cover spraying with various broad-spectrum pesticides, which are highly detrimental to natural pest enemies, including beneficial insects, birds, and bats (Berni et al. 2021). Alternatively, foliar spot spray using hydrolyzed protein baits, applied to one row out of every three or four, constitute the other primary chemical application method currently in use (Elaini and Mazih 2018).

Adult Medfly prevalence and fruit infestation levels determine the timing and frequency of chemical sprays. The trigger threshold for initiating spraying is 3–5 adult males per trap per day from a trapping grid network of 50–100 traps/km^2. However, in smaller farms, sprays are typically applied every 8–10 days during late summer and autumn for early citrus varieties and from the end of winter to spring for late varieties, depending on the region (Mazih 2015). To meet the requirements of certain international markets, such as the Russian market, some farmers lower the chemical spray threshold to 1 male or 0.5 female per trap per day (Assouguem et al. 2021).

As farmers prioritize economic benefits over health and environmental considerations, chemical pesticide treatments remain the most common control method. However, there are growing concerns about pesticide residues and their impact on human health and the environment (Aberkani et al. 2022; Khfif et al. 2022). Furthermore, exceeding the maximum residue limits would impede adherence to international standards for Moroccan citrus exports (Addi et al. 2013; Choubbane et al. 2022).

In this context, the demand for biopesticides has increased as a safer alternative to hazardous chemical pesticides. Consequently, biopesticides such as *Beauveria bassiana*, Azadirachtin, and Spinosad have been approved for use (Mokrini et al. 2020). In particular, Spinosad-based products have become more preferred for chemical treatments, especially during the harvest period, due to their short environmental persistence and a preharvest interval of just one day (Elaini and Mazih 2018).

Increasingly, more environmentally friendly approaches are being adopted within the framework of integrated pest management, such as mass trapping. This method relies on a highly selective and potent combination of three chemical compounds (putrescine, trimethylamine, and ammonium acetate) designed to attract female Medflies (Bakri et al. 1998; Epsky et al. 1999; Assouguem et al. 2021).

9.4 STERILE INSECT TECHNIQUE

Based on the previously cited problems with conventional control methods of the Medfly, it becomes imperative to employ a control strategy that not only effectively manages the target pest population but also minimizes its environmental impact. The sterile insect technique (SIT) emerges as a highly promising method capable of satisfying these criteria adeptly. SIT is based on mass production, reproductive sterilization using ionizing radiation, and the release of the target sterile insect into infested areas. Within these environments, sterile males mate with wild females, resulting in the inability to produce viable offspring and consequently lowering pest populations (Knipling 1955; Dyck et al. 2021).

Globally, this technique is employed for containing, excluding, suppressing, and eradicating a significant variety of pests, including various species of flies, mosquitoes, moths, and beetles (Vreysen et al. 2006; Bakri et al. 2021; Bourtzis and Vreysen 2021). Its extensive use is due to several advantages it offers. It is species-specific, targeting only the intended pest species without affecting the population of nontarget insects. In addition, it preserves the genetic integrity of the target species by not introducing foreign genetic material into the population. Moreover, the method minimizes the reliance on hazardous chemicals, thereby mitigating potential environmental and health risks associated with their use. Furthermore, this technique is compatible with other control measures, such as the bait application technique and augmentative biological control (Knipling 1955; Nagel and Peveling 2005; Toledo et al. 2017; Montoya et al. 2020).

SIT programs have been implemented in several countries (Figure 9.1) in North, Central, and South America, Europe, the Middle East, Asia, Africa, and Australia (Enkerlin 2005) and have become an essential measure to control the Medfly population worldwide. Currently, the different fruit fly factories are weekly producing several

Figure 9.1 Medfly SIT units in the world based on Bakri et al. (2022). Countries having units mass-rearing sterile Medflies are shown on the map (dark grey).

billion sterile Medfly males that are released in the field to mate with wild females (Llopis-Giménez et al. 2017; Bakri et al. 2020, 2022).

SIT, incorporated in an area-wide integrated pest management approach, has been a successful Medfly control strategy in many countries around the world. For example, in Mexico, this strategy has helped maintain a pest-free status for over 30 years and proved to be highly cost-effective (Enkerlin et al. 2015). Also, Spain's Valencian Community implemented an area-wide integrated pest management program in 2007, with SIT as the main control method, resulting in a 90% reduction in aerial insecticide use (Plá et al. 2021). Also, Argentina successfully applied an integrated area-wide SIT program against Medflies in the Patagonia and Cuyo regions. This achievement redefined the phytosanitary status of these regions, enabling exports to previously closed markets without the need for postharvest treatments (Guillén and Sánchez 2007). Additionally in Peru, the released Medflies in the Santa Anita district and along the coastal areas led to the eradication of Medflies in the Tacna and Moquegua regions (Pérez-Staples et al. 2021). Similarly, in Chile, the continued releases of sterile Medflies prevented outbreaks in northern Chile (Arica province) and southern Peru (Tacna Valley) (Gonzalez and Troncoso 2007).

9.5 SIT IMPLEMENTATION IN MOROCCO

The first implementation of SIT in Morocco began in 2006, with imported sterile males released in the pilot area of 4,864 ha of citrus orchards with a 1 km buffer zone around this area within the Ouled-Berhil commune in the Souss-Massa region. The releases were at a density of 2,000 sterile males per hectare and managed by ALBRA (Association de Lutte Biologique Contre les Ravageurs des Agrumes et Primeurs). After three years of releases, despite favorable environmental conditions for Medflies between 2010 and 2011, the abundance of female Medflies in the examined zone declined compared to previously recorded levels. Consequently, chemical sprays were reduced by 25%–75% in conjunction with this approach (ALBRA 2015).

Another pilot area in the Moulouya perimeter in northeastern Morocco underwent sterile male releases in 2017 to test the effectiveness of SIT combined with trapping in an area of 1,200 ha of citrus orchards. The results confirmed its effectiveness, showing a reduction in citrus infestation rates from 5% in areas with no sterile male release to 2.4% and 1.6% in fields where sterile males were released at densities of 1,000 and 3,000 males per hectare, respectively (Yazid et al. 2020).

Recognizing the many advantages of SIT as an effective method of Medfly control, the Moroccan Ministry of Agriculture, Fisheries, Rural Development, Water and Forests (MMAFRDWF) decided to establish a Medfly mass-rearing facility to meet national needs of sterile males. The application of this technique on a national scale will benefit the citrus sector by improving fruit quality, decreasing pesticide use, protecting the environment and the consumers, and preserving the current export markets and exploration of other potential international markets. In collaboration with the International Atomic Energy Agency (IAEA), the unit was established in 2020 with a budget exceeding six million dollars. Partial equipment was sourced from the previous unit in the Madeira Islands, while the remaining materials were supplied by the National Office of Food Safety (ONSSA) with support from the IAEA.

The facility, located on an 8,340 m² site in the municipality of Tikiouine, prefecture of Agadir Ida-Outanane, includes several buildings for administration, mass rearing, filter and quality control, sterile adult emergence, and an area for future production capacity extension (Figure 9.2a and b). The unit was built to have an estimated production capacity of 130 million pupae per week, making it the largest SIT Medfly production facility in Africa.

Currently, the facility is utilizing the Vienna-8 FDF strain (Porras et al. 2020) and has recently been equipped with a Cobalt-60 irradiator model Foss-812 (Figure 9.2c). The planned radio-sterilization dose is set at 100 Gy. The facility's infrastructure is sufficient to provide sterile males to cover the entire Souss area, encompassing approximately 40,000 hectares of citrus-producing zones and neighboring regions. The ultimate goal is to implement the SIT as a nationwide program, covering all citrus-producing areas in the country.

Currently, the infestation rate of the Medfly in citrus orchards is being assessed in the Souss Valley, particularly in the Ouled-Aissa area, which spans approximately 6,000 hectares of citrus orchards. Using GIS software, the area was delineated and divided into equal sections of roughly 100 hectares each (Figure 9.3). Then, in each section

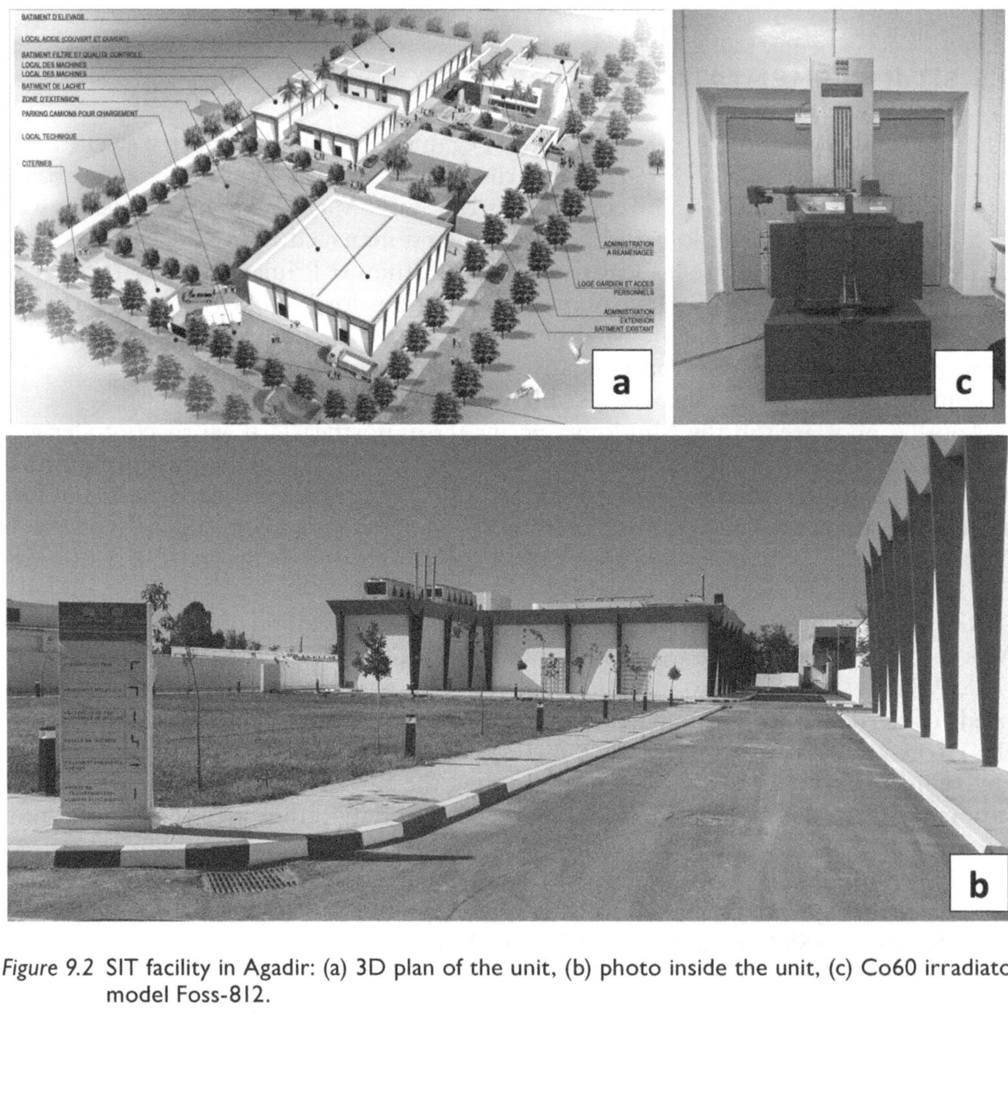

Figure 9.2 SIT facility in Agadir: (a) 3D plan of the unit, (b) photo inside the unit, (c) Co60 irradiator model Foss-812.

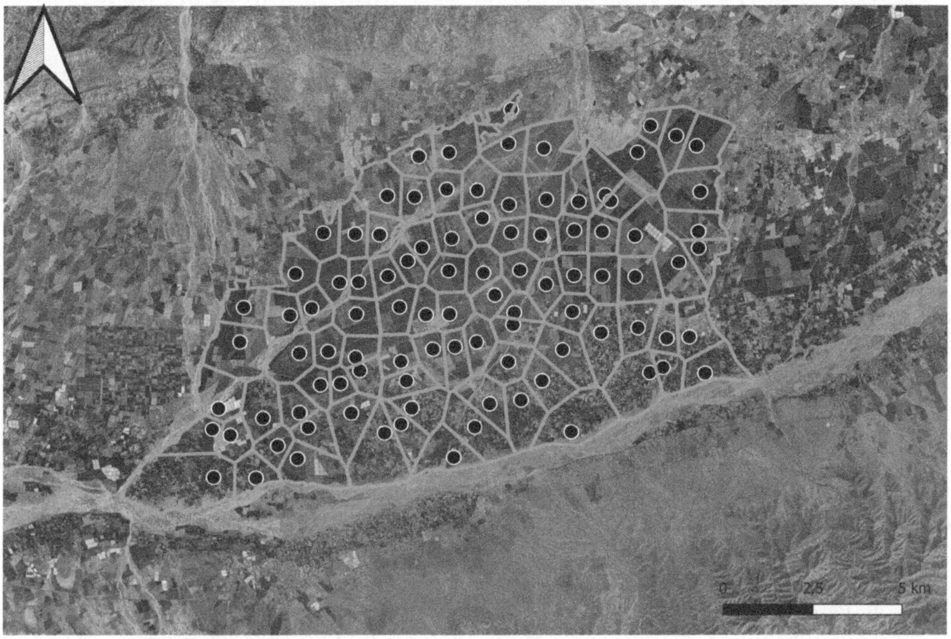

Figure 9.3 Trap positions in the monitoring area.

were placed Net-traps baited with a food attractant dispenser composed of a mixture of three components: ammonium acetate, trimethylamine, and diaminopentane. The trap's transparent lid was impregnated with deltamethrin insecticide on the inner surface. Data were recorded every 14 days since December 2023. This monitoring program aims to understand the population dynamics within the area, to help with a better optimization of sterile male releases in the future.

9.6 CONCLUSION

It is anticipated that the implementation of SIT will significantly reduce economic losses caused by the Medfly, thereby mitigating one of the major threats to the citrus industry in Morocco. By lowering the pest population through environmentally friendly and species-specific methods, SIT will reduce reliance on chemical pesticides, leading to healthier ecosystems and safer agricultural practices.

REFERENCES

Aberkani, K., F. Z. Briache, H. Ghazal, et al. 2022. Knowledge about pesticides use and consumption of fruit and legumes in growers and consumers in the Province of Nador, Morocco. Results from a survey study. *Revista Española de Nutrición Humana y Dietética* 26(4):268–276. DOI:10.14306/renhyd.26.4.1708

Addi, E. H., A. A. Addi, and F. Achemchem. 2013. Evaluation of four pesticides residues contents in the citrus fruits from Souss-Massa area (Morocco). *Environmental Engineering and Management Journal* 13:663–667. DOI: 10.30638/eemj.2014.070

Alaoui, A., A. Imoulan, Z. El-Alaoui-Talibi, et al. 2010. Genetic structure of Mediterranean fruit fly (*Ceratitis capitata*) populations from Moroccan endemic forest of *Argania spinosa*. *International Journal of Agriculture and Biology* 12:291–298.

ALBRA. 2015. Projet pilote d'introduction de la technique des insectes steriles (tis) contre la ceratite dans la plaine du souss, 78. Agadir: ALBRA.

APHIS. 2016. APHIS prohibits the importation of certain citrus fruits from Morocco due to Mediterranean fruit fly (*Ceratitis capitata*). https://www.aphis.usda.gov/import_export/plants/plant_imports/federal_order/downloads/2016/2016-10.pdf.

Assouguem, A., M. Kara, S. Benmessaoud, et al. 2021. *Ceratitis capitata* Wiedemann (Diptera: Tephritidae) in Moroccan Citrus orchards: Chemical and sustainable control methods and priorities for future research. *Tropical Journal of Natural Product Research* 5:1703–1708.

Bakri, A., W. Enkerlin, R. Pereira, et al. 2020. Tephritid-related databases: TWD, IDIDAS, IDCT, DIR-SIT. In Area-Wide Management of Fruit Fly Pests, ed. D. Pérez-Staples, F. Díaz-Fleischer, P. Montoya, and M. T. Vera, 369–385. Boca Raton: CRC Press.

Bakri, A., H. Hadis, N. D. Epsky, R. R. Heath, and J. Hendrichs. 1998. Female *Ceratitis capitata* (Diptera: Tephritidae) capture in a dry trap baited with food-based synthetic attractant in an argan forest in Morocco. Part I: Low population field test. *The Canadian Entomologist* 130(3):349–356. DOI:10.4039/Ent130349-3

Bakri, A., K. Mehta, and D. R. Lance. 2021. Sterilizing insects with ionizing radiation. In *Sterile Insect Technique: Principles and Practice in Area-Wide Integrated Pest Management*, ed. V. A. Dyck, J. Hendrichs, and A. S. Robinson, 355–398. Boca Raton: CRC Press.

Bakri, A., R. Pereira, W. Enkerlin, et al. 2022. Status of the tephritid fruit fly production worldwide for the sterile insect technique application. The 11th ISFFEI, 13–18 November 2022, Sydney, Australia.

Berni, I., A. Menouni, I. G. El, et al. 2021. Understanding farmers' safety behavior regarding pesticide use in Morocco. *Sustainable Production and Consumption* 25(1):471–483. DOI:10.1016/j.spc.2020.11.019

Boulahia-Kheder, S. 2021. Review on major fruit flies (Diptera: Tephritidae) in North Africa: Bio-ecological traits and future trends. *Crop Protection* 140:105416. DOI:10.1016/j.cropro.2020.105416

Bourtzis, K., and M. J. Vreysen. 2021. Sterile insect technique (SIT) and its applications. *Insects* 12(7):638. DOI:10.3390/insects12070638

Choubbane, H., A. Ouakhssase, A. Chahid, et al. 2022. Pesticides in fruits and vegetables from the Souss Massa region, Morocco. *Food Additives & Contaminants* 15:79–88. DOI:10.1080/19393210.2022.2028196

Dyck, V. A., J. Hendrichs, and A. S. Robinson. 2021. *Sterile Insect Technique: Principles and Practice in Area-Wide Integrated Pest Management*. Boca Raton: CRC Press/Taylor & Francis. DOI:10.1201/9781003035572

El Aalaoui, M., R. Bouharroud, M. Sbaghi, et al. 2019. Predatory potential of eleven native Moroccan adult ladybird species on different stages of *Dactylopius opuntiae* (Cockerell) (Hemiptera: Dactylopiidae). *EPPO Bulletin* 49:374–379. DOI:10.1111/epp.12565

Elaini, R., and A. Mazih. 2018. Current status and future prospects of *Ceratitis capitata* Wiedemann (Diptera: Tephritidae) control in Morocco. *Journal of Entomology* 15:47–55. DOI:10.3923/je.2018.47.55

Enkerlin, W. R. 2005. Impact of fruit fly control programmes using the sterile insect technique. In *Sterile Insect Technique: Principles and Practice in Area-Wide Integrated Pest Management*, ed. V. A. Dyck, J. Hendrichs, and A. S. Robinson, 651–676. Dordrecht: Springer Netherlands. DOI:0.1201/9781003035572-30

Enkerlin, W., J. M. Gutiérrez-Ruelas, A. V. Cortes, et al. 2015. Area freedom in Mexico from Mediterranean fruit fly (Diptera: Tephritidae): A review of over 30 years of a successful containment program using an integrated area-wide SIT approach. *Florida Entomologist* 98(2):665–681. DOI:10.1653/024.098.0242

Epsky, N. D., J. Hendrichs, B. I. Katsoyannos, et al. 1999. Field evaluation of female-targeted trapping systems for *Ceratitis capitata* (Diptera: Tephritidae) in seven countries. *Journal of Economic Entomology* 92(1): 156–164. DOI:10.1093/jee/92.1.156

Gonzalez, J., and P. Troncoso. 2007. The fruit fly exclusion programme in Chile. In *Area-Wide Control of Insect Pests: From Research to Field Implementation*, ed. M. J. B. Vreysen, A. S. Robinson, and J. Hendrichs, 641–651. Dordrecht: Springer Netherlands. DOI:10.1007/978-1-4020-6059-5_59

Guillén, D., and R. Sánchez. 2007. Expansion of the national fruit fly control programme in Argentina. In Area-Wide Control of Insect Pests: From Research to Field Implementation, ed. M. J. B. Vreysen, A. S. Robinson, and J. Hendrichs, 653–660. Dordrecht: Springer Netherlands. DOI:10.1007/978-1-4020-6059-5_60

Khfif, K., J. Benyazid, R. Bouharroud, et al. 2022. Study of the control methods practiced against the Mediterranean fruit fly (*Ceratitis capitata* Wiedemann, 1824) in citrus orchards of the Moulouya region. *AFRIMED AJ –Al Awamia* 135:192–207. DOI: 10.34874/IMIST.PRSM/afrimed-i135.33456

Knipling, E. F. 1955. Possibilities of insect control or eradication through the use of sexually sterile males. *Journal of Economic Entomology* 48:459–462. DOI:10.1093/jee/48.4.459

Llopis-Giménez, A., R. M. González, A. Millán-Leiva, et al. 2017. Novel RNA viruses producing simultaneous covert infections in *Ceratitis capitata*. Correlations between viral titers and host fitness, and implications for SIT programs. *Journal of Invertebrate Pathology* 143:50–60. DOI:10.1016/j.jip.2016.11.014

Mazih, A. 2015. Status of citrus IPM in the Southern Mediterranean basin Morocco, North Africa. *Acta Horticulturae* 1065:1097–1103. DOI:10.17660/ActaHortic.2015.1065.138

Mazih, A., and D. Debouzie. 1996. Infestation rate of argan fruit (*Argania spinosa*) by the Mediterranean fruit fly (*Ceratitis capitata*) in relation to phenology and maturation of the fruit. *Entomologia Experimentalis et Applicata* 81:31–38. DOI:10.1111/j.1570-7458.1996.tb02011.x

MMAFRDWF. 2023. Filière agrumicole | Ministère d'agriculture. https://www.agriculture.gov.ma/fr/filiere/agrumicole#:~:text=Une%20gamme%20diversifi%C3%A9e%20de%20vari%C3%A9t%C3%A9s,32%20%25%20(40%20344%20Ha) (accessed January 31, 2025).

Mokrini, F., S. E. Laasli, Y. Benseddik, et al. 2020. Potential of Moroccan entomopathogenic nematodes for the control of the Mediterranean fruit fly *Ceratitis capitata* Wiedemann (Diptera: Tephritidae). *Scientific Reports* 10:1–11. DOI:10.1038/s41598-020-76170-7

Montoya, P., S. Flores, S. Campos, et al. 2020. Simultaneous use of SIT plus disseminator devices of *Beauveria bassiana* enhance horizontal transmission in *Anastrepha ludens*. *Journal of Applied Entomology* 144:509–518. DOI:10.1111/jen.12766

Nagel, P., and R. Peveling. 2005. Environment and the sterile insect technique. In *Sterile Insect Technique: Principles and Practice in Area-Wide Integrated Pest Management*, ed. V. A. Dyck, J. Hendrichs, and A. S. Robinson, 499–524. Dordrecht: Springer Netherlands. DOI:10.1201/9781003035572-23

Pérez-Staples, D., F. Díaz-Fleischer, and P. Montoya. 2021. The sterile insect technique: success and perspectives in the Neotropics. *Neotropical Entomology* 50(2):172–185. DOI:10.1007/s13744-020-00817-3

Plá, I., J. García de Oteyza, C. Tur, et al. 2021. Sterile insect technique programme against Mediterranean fruit fly in the Valencian Community (Spain). *Insects* 12(5):415. DOI:10.3390/insects12050415

Porras, M. F., J. S. Meza, E. G. Rajotte, et al. 2020. Improving the phenotypic properties of the *Ceratitis capitata* (Diptera: Tephritidae) temperature-sensitive lethal genetic sexing strain in support of sterile insect technique applications. *Journal of Economic Entomology* 113(6):2688–2694. DOI:10.1093/jee/toaa220

Sacantanis, K. B. 1957. La forêt d'arganier. Le plus grand foyer de *Ceratitis capitata* (Wied.) connu dans le monde. *Bollettino del Laboratorio di Entomologia. Agraria* 15:1–53.

Toledo, J., S. Flores, and S. Campos, et al. 2017. Pathogenicity of three formulations of *Beauveria bassiana* and efficacy of autoinoculation devices and sterile fruit fly males for dissemination of conidia for the control of *Ceratitis capitata*. *Entomologia Experimentalis et Applicata* 164:340–349.

Vreysen, M. J., J. Hendrichs, and W. Enkerlin. 2006. The sterile insect technique as a component of sustainable area-wide integrated pest management of selected horticultural insect pests. *Journal of Fruit and Ornamental Plant Research* 14:107–131.

Yazid, J. B., Z. Chafik, I. Bibi, et al. 2020. Effectiveness of sterile insect technique for medfly (*Ceratitis capitata*, Wiedemann, 1824) control in citrus orchards of Moulouya perimeter North East of Morocco. *ISPEC Journal of Agricultural Sciences* 4(3):405–421. DOI:10.46291/ispecjasvol4iss3pp405-421

Chapter 10

Packing and release methods for sterile fruit flies

Jose Luis Zavala-Lopez, Rubén Luis Leal-de la Garza, and Rene Cano-Pérez

10.1 INTRODUCTION

The sterile insect technique (SIT) includes two main components: one is the mass rearing and sterilization to obtain high-quality insects and the other is the management of the sterile insects, where they are prepared, fed, and matured before their release in the field (FAO/IAEA 2017). The methods of packing fruit flies not only include different equipment such as paper bags, boxes, or towers but also imply the design of the place or packing center that will be used for the management of the flies, as well as defining, based on the packing method, how to carry out the release.

The purpose of this document is to emphasize the needs in infrastructure and equipment for the packing, emergence, holding, and release of sterile insects. We based this document on the experience gained and the methods developed during the eradication of the Mediterranean fruit fly, *Ceratitis capitata* (Wiedemann), in Mexico, as well as with the Mexican fruit fly, *Anastrepha ludens* (Loew), and *Anastrepha obliqua* (Macquart) in the National Campaign against Fruit Flies in Mexico. These flies were eradicated from 50% of the Mexican territory from the northern border of Mexico with the United States of America to the State of Sinaloa. Also, the same methodology with some variants is being used in Central and South America.

10.2 PACKING CENTERS

The final step of the SIT is the release of the sterile insects in the field. Traditionally, three packing and release systems or methodologies have been used: chilled adult (plastic adult rearing container (PARC) boxes, towers), adults in paper bags, and adults in cardboard boxes (Zavala et al. 2020). These methods are suitable for aerial release. However, sometimes fly releases are conducted from the ground due to weather conditions or other reasons such as the extent of the program, costs, and releases in protected areas. For ground releases, plastic containers can also be used to hold and release the flies.

The selection of the location for the packing center is of utmost importance. It should be close to an airport or airplane runway if the release is going to be aerial, preferably within the release area to avoid damage to the fly due to long trips (ferry), either in the plane or in the ground release vehicle. Aspects related to logistics transportation from production centers, flight optimization to avoid long ferries, services, etc. must also be considered. It is suggested that the location of packaging centers be defined using

DOI: 10.1201/9781003644330-13

an analysis matrix, which considers all the factors involved in an optimal placement. Working with packing centers close to infested areas results in savings in ferry hours and improves the quality of the insects to be released. Instead of sending emerged adult fruit flies long distances, pupae are sent and given optimal conditions for emergence in centers built for this purpose near infested areas (Dowell et al. 2020).

Once the location has been defined, it will be necessary to determine the type of installation, either fixed or mobile, based on the urgency of starting the use of the SIT, the time the operations will take place in this location, and other factors.

10.2.1 Fixed packing centers

The fixed packing centers are constructions designed exclusively for the process of packing and storing sterile fruit flies until they are placed in the cold rooms and prepared for release. These centers provide the appropriate conditions at each step: packing, emergence, chilling, and quality control rooms. Also, they have a food preparation room, machine house, maintenance, and storage warehouses for materials and equipment, according to the needs of the project. Another important area for the fixed packing centers is the wastewater treatment plant because cleaning is a crucial activity and the management of the wastewater is regulated (Zavala et al. 2014). Examples are presented in Figure 10.1.

10.2.2 Mobile packing centers

The mobile packing center consists of a single autonomous room equipped with an air conditioning system and dehumidifying equipment. This allows the center to be used for all the technical activities of the sterile flies' prerelease process, modifying the temperature and humidity according to the needs of the flies. The mobile room is a modified Thermo King box designed to meet the specific conditions required by the flies. It facilitates the process, from packing to chilling knocking down to preparing the flies for placement into the release box, which then is loaded onto the plane for field release. As in the fixed system, it should have the same technical areas: reception of pupae, packing, emergence, cold rooms, quality control, and support areas such as food preparation, washing and drying of materials, storage, equipment repair area, machine house, wastewater treatment, offices, roads, and green areas. The only difference is that in the mobile packing units, the technical activities are carried out in a single room, while in a fixed packing unit, there is a room equipped for each technical activity; that is, there is a reception area, packing area, emergence area, chilling area, and a separate quality control area. Mobile packaging units are ideal for attending to or implementing emergence plans. The autonomous rooms, with controlled environmental conditions, enable a packaging, emergence, and collection process with established quality standards. These rooms have the same working areas as the fixed packing center; however, the logistics of the operations change because each autonomous room is used individually as a packing room and an emergence room and serves as its own chilling and collection room. With this work scheme, the handling of the flies is reduced from packing through emergence until collection. The towers that contain the biological material remain in their assigned room and do not move, avoiding stress and changes in temperature and

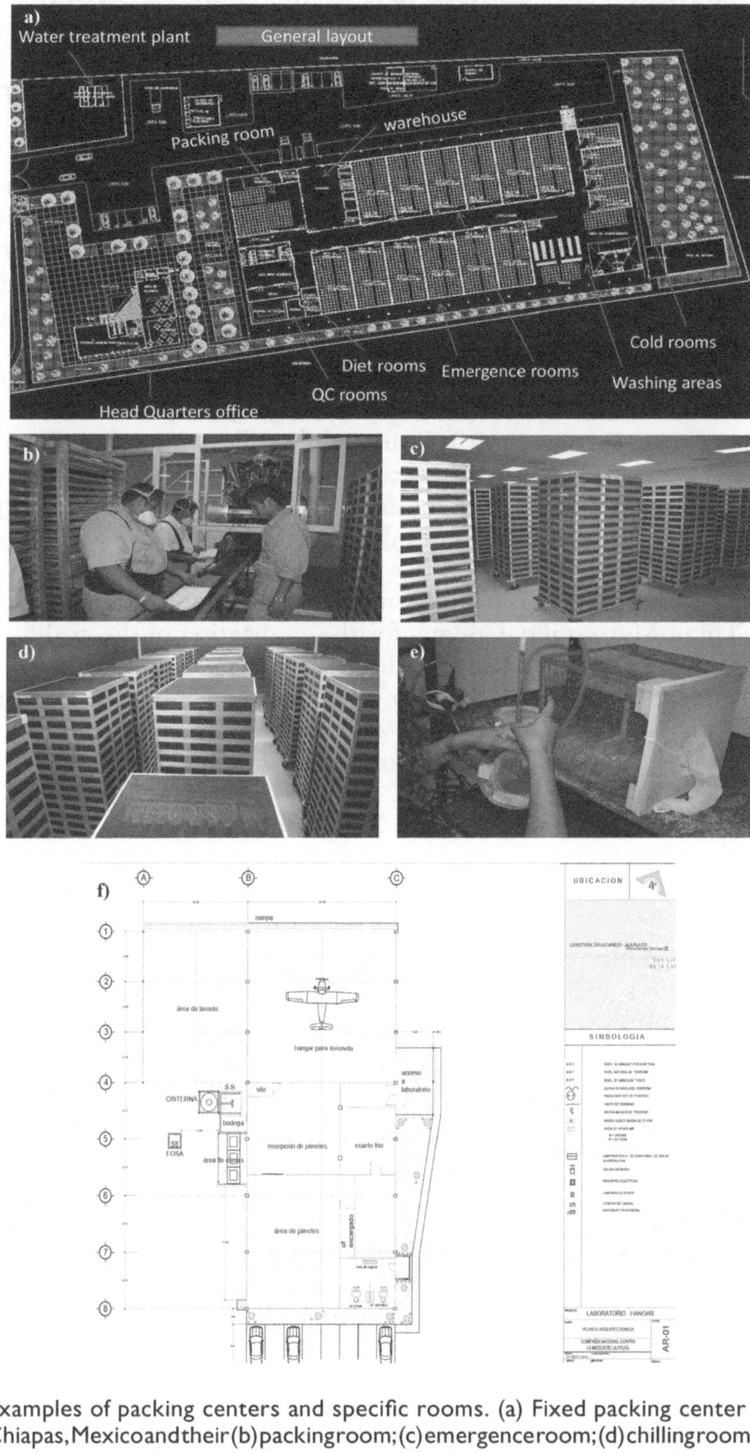

Figure 10.1 Examples of packing centers and specific rooms. (a) Fixed packing center in Tapachula, Chiapas, Mexico and their (b) packing room; (c) emergence room; (d) chilling room; and (e) quality control room. (f) Fixed packing center in El Tablero, Tamaulipas, Mexico.

humidity conditions that are generated due to the transfer between rooms. Another benefit of this new methodology is that each room represents a release flight. This means that from the packing stage, a certain number of towers are designated to each room, which at the time of collection is equivalent to a full flight. This provides the advantage of not needing to chill all the biological material, but only the flies required for the specific release flight.

By organizing the biological material among the rooms, it reduces the pupae density per level, providing better conditions for the adults inside the cage. This includes more space in the resting area, greater availability of space and food in the food tray, and better access to the hydration pad. Additionally, reducing the density results in improved aeration inside the cage, significantly lowering the stress levels in the adults (Zavala Lopez et al. 2020).

This new concept allows for immediate preparation and response for contingency plans and permanent programs according to emergency circumstances. Also, it avoids spending money on expensive infrastructure while still maintaining the quality of the sterile insects with the chilled adult release system. However, the investment in air conditioning equipment and the need for greater investment in backup equipment compared to fixed installations result in a disadvantage for the processing of larger continuous production.

The Moscamed Program successfully managed this mobile packing concept in the Emergency Plan of Manzanillo, Colima, for the eradication of the medfly; they pack a total of 739 million pupae and release sterile flies by air in the designated polygons of the Emergency Plan (SENASICA 2020). Also, the National Campaign against Fruit Flies in Mexico (Figure 10.2) used the Mobile Packing Center concept, created and designed by the company Servicios Mubarqui, for ground or aerial release of *Anastrepha* spp.

Figure 10.2 Mobile packing center used in Colima, Mexico.

10.2.3 Hybrid packing center

This is a fusion that combines the best aspects of the two previous modalities and consists of an infrastructure that can be adapted to handle packing and emergence in modular rooms, as well as collection in mobile cold rooms. This results in faster project implementation and savings in construction costs (Figure 10.3). The National Fruit Fly Program in Mexico and the Fruit Growers Association of Soconusco, Chiapas have a program that uses SIT in an area-wide management program for the suppression of *A. ludens* (Liedo et al. 2024). An average of 20 million sterile males is released per week in the municipalities of Tapachula, Tuxtla Chico, Mazatán, Huehuetán, and Huixtla in the State of Chiapas, Mexico (covering approximately 18,000 ha).

10.3 PACKAGING METHODS

At the beginning of the large-scale operational programs, many materials for the packing and emergence of fruit flies have been used, ranging from cardboard and plastic boxes, plastic buckets, and paper bags. Later, mesh cages and towers were introduced. The original concept of using towers was John Worley's idea, aiming to maximize space in emergence rooms and reduce the workers required for packaging and handling processes. At that time, PARCs were used to pack large quantities of sterile flies, requiring very large packing centers.

10.3.1 Worley towers

Worley Towers (Figure 10.4) were the first to revolutionize fruit fly packing systems, facilitating the packing of large quantities of flies in a small space: one million flies per square meter. These towers considered the space required, food, and environmental

Figure 10.3 Hybrid Packing Center used in Tapachula, Chiapas, Mexico for the *Anastrepha* Fruit Fly Program.

Figure 10.4 Worley tower.

conditions during the emergence period until chilling. They worked so well that they began to be used in other countries. However, problems arose later with the electrical systems, related to the fans used to force air flow inside the towers, leading to high fly mortality due to the lack of air circulation when the fans installed at the top of the towers failed. Nevertheless, these towers are still successfully used in the United States of America for the medfly and Mexican fruit fly, and in Panama for the screw worm, with excellent results in terms of fly quality.

10.3.2 Mubarqui towers

These towers were designed considering the advantages of the Worley towers while addressing the problem of a lack of air circulation. Each of the 18 levels was designed to provide enough space for the flies to live, feed, drink water, and rest, with ventilation on all four sides (Figure 10.5). This incurred an increase in the number of workers involved in the process compared to the Worley towers system.

Automated packing facilitates loading of the towers and provides greater precision in the number of pupae deposited per level, as well as a cleaner and safer environment for the operators. Recyclable materials, such as pupae containers, food trays, resting areas, and pads to provide water, result in cost savings and reduced waste generation. Currently, the Mubarqui Tower is used in Mexico and in other countries for different fruit fly species. Servicios Mubarqui, part of the group that designed, developed, and registered the tower, has implemented the tower system in Mexico, Argentina, Australia, Croatia, El Salvador, and Ecuador. It is used to pack Mediterranean fruit

Figure 10.5 Mubarqui tower.

flies and certain *Anastrepha* and *Bactrocera* fruit flies. Also, it is used for the cattle screwworm fly in the eradication program in Honduras, as well as in Mexico with the hybrid packing system, in Tuxtla Gutierrez, Chiapas, and in Tapachula, Chiapas, with the fixed packing system.

10.3.3 Guatemala tower

This tower was also part of efforts to solve the problem of fruit fly mortality due to a lack of air circulation in the Worley Tower. It uses a pair of opposite structures that allow the pupae placed inside to emerge, feed, hydrate, and rest on both the lower and upper mesh. Originally, food was placed on the upper part of the mesh, but nowadays, a "pelum" fabric is used to provide protein, sugar, and water. These structures are placed on shelves, giving them the shape of a tower and providing air circulation (Figure 10.6). These towers are in use for the Mediterranean Fruit Fly Program in Guatemala.

All the materials of the packing systems used for the emergence of flies have a useful life expectancy according to their components. In the case of the towers, the most fragile component is the mesh, which must be changed due to use. Also, they constantly tear or move out of place, allowing newly emerged flies to escape. A maintenance and cleaning program is required not only for the meshes but also for the metal structures and of course for the plastic parts and fabric that is used to avoid leaks or proliferation of microorganisms due to food residues or dirt, caused by the dye used to mark the sterile flies, which can affect the quality of the flies during their emergence and handling in the packing centers.

Figure 10.6 Guatemala tower.

10.4 ADULT DIETS PROVIDED

Fruit flies need different types of food after emergence to reach sexual maturity (Hendrichs and Prokopy 1994; Papanastasiou et al. 2019; Pereira et al. 2021), and this applies also in the packing center before the release. The type of diet provided to the adults varies across the facilities with those that use only sugar and those that use sugar plus a protein source.

10.4.1 Sugar

Sugar (as the only food) is provided in a gel matrix. Generally, the preparation involves adding sugar and agar to water, cooking it until it boils, and then placing it in trays to cool (Figure 10.7). Once cooled, it is placed on the mesh of the tower or PARC box, making it available for the flies when they emerge. This method of feeding flies was first used in some fruit fly action programs such as those for *C. capitata* and certain *Anastrepha* species in Mexico, Guatemala, and the United States of America and also *Cochliomyia hominivorax* in the United States of America, Mexico, and Panama. The food for *C. hominivorax* agar was replaced by carrageenan to give excellent support for feeding.

10.4.2 Mubarqui diet

This diet consists of a formulated mixture with extra fine flour of amaranth, peanuts, egg, and sugar, with a content confirmed to be at least 85% carbohydrates and 4%–10% natural protein. There should be no scale waste and nondigestible solids for flies. This type of food

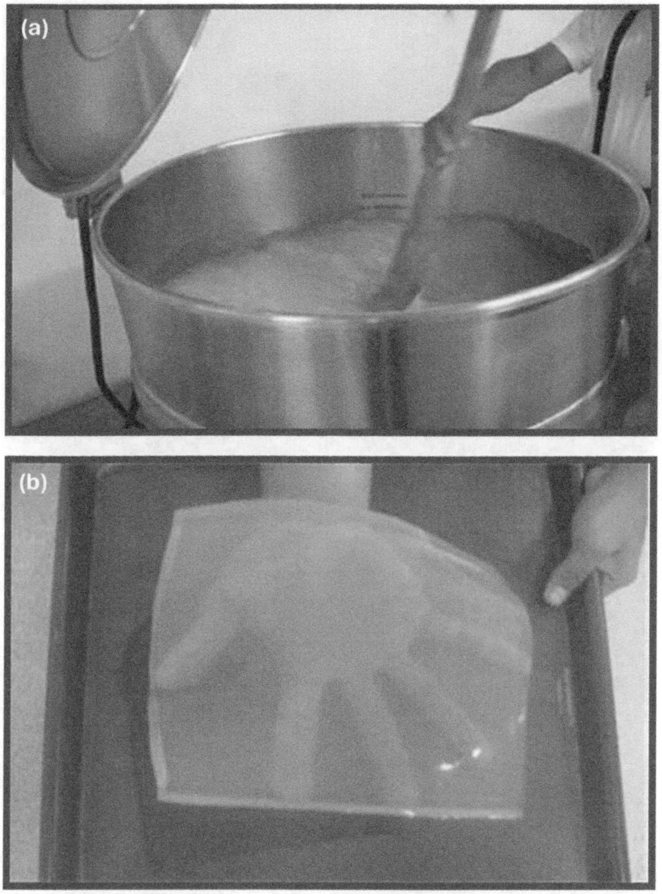

Figure 10.7 (a) Food preparation with agar and sugar and (b) final product.

was formulated by the Mubarqui company in a dry presentation. The powder is mixed with cold water to obtain a paste, which is then distributed in the food tray and placed on the mesh of the tower's level where the flies emerge and feed (Figure 10.8). Sugar and protein, along with nutritional supplements, enhance sexual maturity and thus sexual behavior (Pereira et al. 2021). The Mubarqui diet has been in use for *C. capitata* and *A. ludens* and *A. obliqua* in Mexico, Croatia, Ecuador, and Bolivia.

10.4.3 Other combinations of sugar and protein

In addition to the use of sugar and natural protein, different proportions of sugar with hydrolyzed yeast protein have been used to feed and promote the fertility of females in the mass rearing of fruit flies. Based on this experience, different proportions of sugar with hydrolyzed yeast protein have been tested. For example, Liedo et al. (2013) reported that 4%–10% of protein in the proportion with sugar is enough to increase the sexual behavior of flies of the genus *Anastrepha*. Some packaging centers such as Mendoza, Argentina, for *C. capitata*, provide a mix of sugar plus hydrolyzed protein to the flies to reach sexual maturity.

Figure 10.8 Mubarqui diet. (a) preparation of the diet and (b) final appearance.

10.5 PREPARATION OF ADULTS FOR RELEASE

After their emergence, the adults remain in the management or emergence rooms for 5–7 days to achieve sexual maturity or at least get close to it, so that when they are released, they are nearly ready to mate with the wild females. For this purpose, the towers or the equipment used for managing the pupae during adult emergence and maturation are taken to the chilling rooms. Here, the collection of the adults is carried out once they have been chilled and knocked down. This is done by lowering the temperature in the chilling rooms to –3°C for a period of 45 minutes to an hour, with a relative humidity of 50–70%. Once the flies are immobile, the fly collection process begins by placing a level or shelf on the collection table. The accessories for food, water, pupae containers, and other devices are removed, and then the flies are placed onto the collection table and dumped into PARC boxes located under the table (Figure 10.9). Finally, the flies are weighed and then loaded into the release machine (FAO/IAEA/USDA 2019).

Figure 10.9 Collection table.

10.6 RELEASE METHODS

The release of sterile flies is carried out by ground or aerial means. Ground release primarily uses paper bags, plastic containers or pails, and occasionally plastic boxes, such as PARC boxes. This can be done statically or with ground vehicles. Towers, net cages, and PARC boxes are the most used with the chilling adult system for aerial release. Paper bags and cardboard boxes were first used for aerial release (Villaseñor 1985) and gradually action programs are moving to other types of holding devices.

The main modalities in aerial release involve the use of cold boxes that contain the adult flies. These boxes are equipped with a motor-driven conveyor, also called a screw auger or vibration band, for dosing the flies when they leave the aircraft. Both modalities manage a cooling system that regulates the temperature and humidity inside the transport box for chilled flies (Dowell et al. 2020). The boxes that use conveyor belts also have a vibration system that allows the flies to leave the box without mechanical damage and ensures a uniform dosage. These boxes are called "Mubarqui Smart Release Machines" (Leal-Mubarqui et al. 2014). This equipment allows the dosage to be adjusted according to the needs of the action plan, ranging from 10 flies per square km for tsetse flies to more than half a million per square km for fruit flies (Figure 10.10).

10.7 NAVIGATION SYSTEMS

For precise releases, there are two ways of guiding an aircraft: with a satellite flagger such as AgNav or SatLoc, where the area and bandwidth are defined, and through the Macx toolset, which consists of Macx web and the MacxNav flagger designed to meet the release needs with traceability for area wide-integrated pest management. The MacxNav Flagger was designed specifically for the needs of fruit fly aerial release by

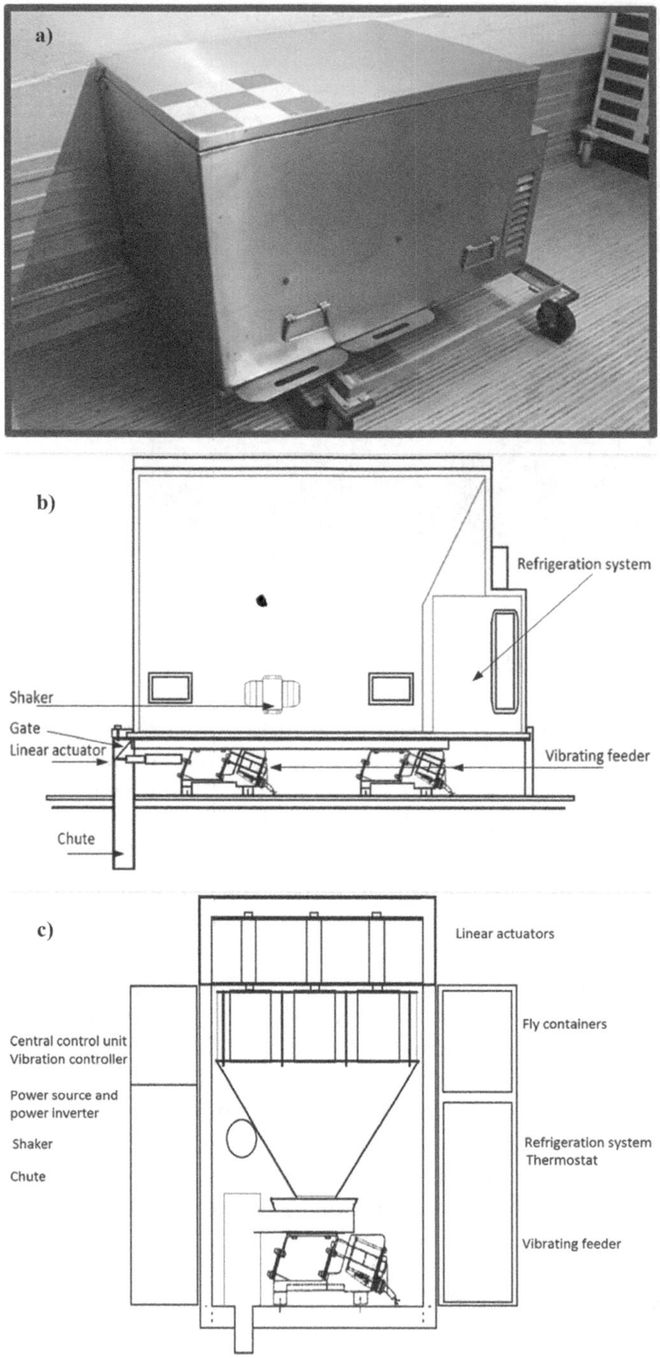

Figure 10.10 Mubarqui Smart Release Machine. (a) General view; (b) plan of the version of Mubarqui Smart Release Machine (MSRM2) used to release fruit flies; and (c) plan of the version of Mubarqui Smart Release Machine (MSRM2) used to release tsetse flies.

the Mubarqui Company for the National Campaign against Fruit Flies in Mexico. It is linked to GIS software installed on tablets on board the aircraft. These systems ensure efficient use of the available sterile insects by automatically starting and stopping the releases of sterile males when entering or leaving the release areas (Bouyer et al. 2020).

This flagger system provides navigational guidance for releasing sterile flies and offers real-time tracking for program managers. It enables them to pinpoint the exact release locations and instantly generates written reports directly from the system, eliminating the need for additional data management (Figures 10.11 and 10.12).

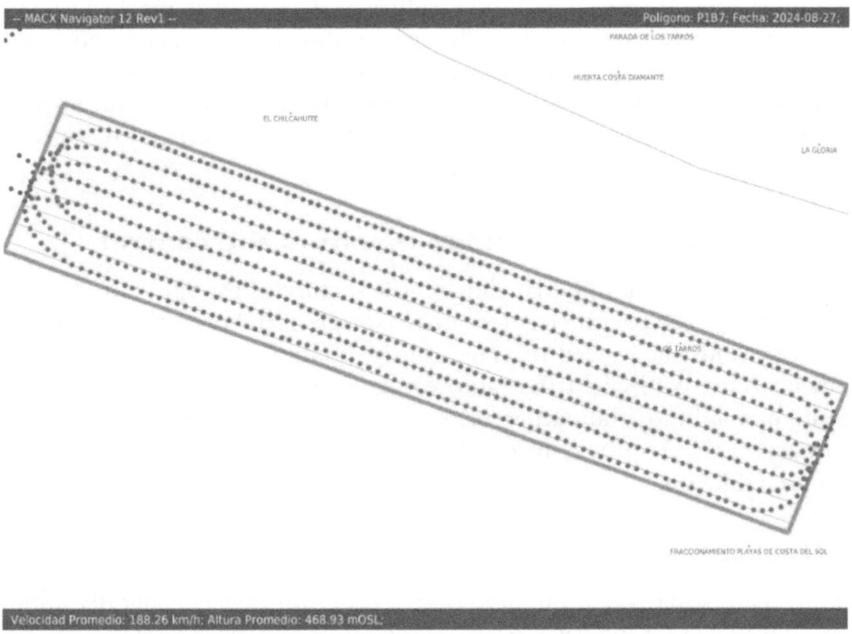

Figure 10.11 Real-time tracking report.

Figure 10.12 MacxNav Flagger.

10.8 AUXILIARY ACTIVITIES AND EQUIPMENT

10.8.1 Maintenance

All the materials of the packing systems used for the emergence of flies have a useful life according to their components. In the case of the towers, the meshes are the most fragile, which must be changed when due to use: they tear or move out of place, allowing emerged flies to escape. In addition to maintaining the meshes, it is essential to care for the metal structures, plastic components, and fabric to prevent leaks or the proliferation of microorganisms caused by food residues or dyes used to mark sterile flies. Such contamination can negatively impact fly quality during emergence and handling. To address this, specialized equipment is available for cleaning the various parts of towers and PARC boxes, including manual pressure washers, carts, and automatic washers adaptable to different tower components. Additionally, it is crucial to ensure that the flies have access to clean drinking water.

10.8.2 Temperature and humidity monitors and control

The success of the SIT not only depends on the quality of the flies but also on the care provided to them in the packing centers, the last step in the rearing and release process, where no effort must be spared to maintain the quality of the flies. For this reason, it is convenient to use equipment that automatically controls humidity and temperature, without the possibility of human error, such as automatic controls that are placed in each of the rooms that require controlled humidity and temperature conditions. If aroma therapy is used (Shelly et al. 2014), this also needs to be carefully applied and monitored.

10.8.3 Quality control

All packing centers manage a quality control program that is generally adopted from the IAEA/FAO/USDA Quality Control Manual (FAO/IAEA/USDA 2019), and the quality of the insects is monitored on a regular basis. Normally, results are within those expected from the standards established in the QC Manual no matter which packaging system is used.

10.9 CONCLUSION

In all, this review, by providing an update of the equipment for the packing, emergence, holding, and release of sterile insect requirements, with emphasis on presenting the experience from the National Fruit Fly Program in Mexico, contributes to giving insight into the alternatives available for other ongoing or starting programs.

REFERENCES

Bouyer, J., J. S. H. Cox, L. Guerrini, et al. 2020. Using geographic information systems and spatial modelling in area-wide integrated pest management programmes that integrate the sterile insect technique. In *Sterile Insect Technique, Principles and Practice in Area-Wide Integrated Pest Management*, ed. V. A. Dyck, J. Hendrichs and A. S. Robinson, 703–730. Boca Raton, FL: CRC Press.

Dowell, R. V., J. Worley, P. J. Gomes, et al. 2020. Supply, emergence and release of sterile insects. In *Sterile Insect Technique, Principles and Practice in Area-Wide Integrated Pest Management*, ed. V. A. Dyck, J. Hendrichs and A. S. Robinson, 441–448. Boca Raton, FL: CRC Press.

FAO/IAEA. 2017. *Guideline for Packing, Shipping, Holding and Release of Sterile Flies in Area-Wide Fruit Fly Control Programmes*, ed. J. L. Zavala-Lopez, and W. R. Enkerlin. Rome, Italy: FAO/IAEA.

FAO/IAEA/USDA. 2019. *Product Quality Control for Sterile Mass-Reared and Released Tephritid Fruit Flies, Version 7.0*. Vienna, Austria: International Atomic Energy Agency.

Hendrichs, J., and R. H. Prokopy. 1994. Food foraging behavior of frugivorous fruit flies. In *Fruit Flies and the Sterile Insect Technique*, ed. C. O. Calkins, W. Classen, and P. Liedo, 37–55. Boca Raton, FL: CRC Press.

Leal-Mubarqui, R. L., R. Cano-Pérez, R. R. Angulo-Kladt, et al. 2014. The smart release machine, a universal system for applying the sterile insect technique. *PLOS One* 9(7):e103077. https://doi.org/10.12371/journal.pone.0103077

Liedo, P., P. Leal, A. Cerdio, et al. 2024. Use of SIT for suppression in area-wide fruit fly management programs: The case of Mango Ataulfo in Southeast Mexico. *11th Tephritid Workers of the Western Hemisphere (TWWH) Meeting*, Jamaica.

Liedo, P., D. Orozco, L. Cruz-López, et al. 2013. Effect of post-teneral diets on the performance of sterile *Anastrepha ludens* and *Anastrepha obliqua* fruit flies. *Journal of Applied Entomology* 137(l):49–60. https://doi.org/10.1111/j.1439-0418.2010.01568.x

Papanastasiou, S. A., J. R. Carey, and N. T. Papadopoulos. 2019. Effects of early-life protein starvation on longevity and sexual performance of male medfly. *PLOS One* 14(7):e0219518. https://doi.org/10.1371/journal.pone.0219518

Pereira, R., B. Yuval, P. Liedo, et al. 2021. Improving post-factory performance of sterile male fruit flies in support of the sterile insect technique. In *Sterile Insect Technique. Principles and Practice in Area-Wide Integrated Pest Management*, ed. V. A. Dyck, J. Hendrichs, and A. S. Robinson, 631–656. Boca Raton, FL: CRC Press.

Servicio Nacional de Sanidad, Inocuidad y Calidad Agroalimentaria (SENASICA). 2020. *Reporta Agricultura la erradicación de la mosca del Mediterráneo; salva la producción y exportación hortofrutícola del país*. Comunicado. Gobierno de México.

Shelly, T. E., D. O. McInnis, C. Rodd, et al. 2014. Sterile insect technique and Mediterranean fruit fly (Diptera: Tephritidae): assessing the utility of aromatherapy in a Hawaiian coffee field. Journal of Economic Entomology 100(2):273–282.

Villaseñor, A. 1985. *Comparación de tres sistemas de liberación aérea para mosca del Mediterráneo estéril, Ceratitis capitata (Wied.). Tesis de licenciatura*. Facultad de Ciencias Agrícolas, Universidad Autónoma de Chiapas.

Zavala, J. L., J. Arredondo, E. Hernández, et al. 2020. Empaque y liberación de moscas estériles. In *Moscas de la fruta: Fundamentos y Procedimientos para su Manejo*, ed. P. Montoya, J. Toledo, and E. Hernández, 499–517. Ciudad de México: S y G editores.

Zavala, J. L., J. M. Gutiérrez, E. Cotoc, et al. 2014. New Mediterranean fruit fly emergence and release facility at Tapachula, Chiapas, México. *Proceedings of the 9th International Symposium on Fruit Flies of Economic Importance (ISSFEI)*, 12–16 May 2014, Bangkok, Thailand.

Zavala López, J. L., R. Cano Pérez, and R. L. Leal de la Garza. 2020. Centro de Empaque Móvil, un concepto de calidad aplicado a la TIE, para liberación en el sistema de adulto enfriado. *Congreso Americano de moscas de las Frutas and 10th Meeting of Tephritid Workers of the Western Hemisphere*, November 2–6, Colombia.

Chapter 11

Production of a genetic sexing strain of *Anastrepha fraterculus* (Diptera: Tephritidae) for SIT programs in Brazil

Paloma Guazzelli Della Giustina, Adalécio Kovaleski, Ana Julia Prestes, Henrique Martinelli Amaral, Luís Anselmo Lopez, Maria de Lourdes Zamboni Costa, and Thiago Mastrangelo

11.1 INTRODUCTION

The sterile insect technique (SIT) is a biological control method used to suppress or eradicate pest populations by releasing sterile males into the environment. These males, carrying dominant lethal mutations in their sperm, mate with wild females, and the eggs laid by them are unviable, which leads to a gradual reduction in the pest population over time (Dyck et al. 2021). Originally developed for the New World Screwworm (Knipling 1953), the SIT has since been gaining significant momentum globally in fruit fly management, with several operational area-wide integrated pest management (AW-IPM) programs in different countries (Enkerlin 2021; Rendon and Enkerlin 2021; Aluja et al. 2024).

Many studies have shown that the release of only sterile males optimizes the application of the SIT in three main ways (Rendon et al. 2000; Robinson 2002; Orozco et al. 2013). First, it prevents the damage caused by oviposition from sterile females, also referred to as "sterile stings," on host fruits, which could compromise their marketability (Hendrichs et al. 1995; Klassen and Curtis 2005), mainly because sterile females can retain stinging behavior, creating entry points for pathogens (Meats et al. 2002). Second, handling only males can lead to a reduction of almost half of the costs in operational programs related to radiation, packaging, marking, transport and release (Caceres et al. 2004). And third, male-only releases prevent sterile males from mating with sterile females, which could distract the males from seeking and mating with wild females, thereby reducing the induction of sterility in wild females (Robinson et al. 1986). Field and semifield cage studies have demonstrated that the male-only releases can result in significantly greater suppression of wild populations compared to conventional bisexual releases (McInnis et al. 1994; Rendon et al. 2000, 2004; Klassen and Curtis 2005; Orozco et al. 2013; Franz et al. 2021; Shelly and Manoukis 2022). Therefore, the removal of females prior to release is advantageous both economically and in terms of field efficiency (Caceres et al. 2004).

In this context, Whitten (1969) was the first to propose the development of genetic sexing strains (GSSs) using male-linked translocations, which enable sex separation or female elimination during development via genetic mechanisms like pupal color or developmental rate (Franz et al. 1994, 1996; Hendrichs et al. 1995). In tephritid fruit flies, sex determination follows an XX/XY chromosome system. GSSs can be created

DOI: 10.1201/9781003644330-14

by identifying a recessive autosomal mutation and linking the wild-type allele to the Y chromosome through a translocation, resulting in females being homozygous for the mutation and males being heterozygous and therefore showing the normal phenotype (Meats et al. 2002).

The first GSS for tephritids, developed for *Ceratitis capitata* (Wiedemann, 1824) (Diptera: Tephritidae), used pupal color as a marker controlled by the white pupae (*wp*) gene. In that strain, heterozygous males emerged from brown pupae and homozygous females from white pupae, allowing for efficient visual sex separation (Franz 2005). However, differentiation occurred only at the pupal stage, requiring the rearing of female larvae, which increased costs and logistical complexity (Klassen and Curtis 2005). A second-generation strain introduced a temperature-sensitive lethal (*tsl*) gene linked to the *wp* gene, allowing the elimination of homozygous *tsl* female embryos when exposed to 32–34°C (Caceres et al. 2000, 2004). This advancement improved the technique and significantly reduced the costs of rearing and releasing (Franz et al. 2021).

Several GSSs have been developed for other tephritid species using the same *wp* gene as a phenotypic marker (Franz et al. 2021), such as *Zeugodacus* (*Bactrocera*) *cucurbitae* (Coquillett, 1899) (Diptera: Tephritidae) (McInnis et al. 2004) and *Bactrocera dorsalis* (Hendel, 1912) (Diptera: Tephritidae) (McCombs and Saul 1995).

For the *Anastrepha* genus, the first GSSs were developed using the black pupae (*bp*) gene as a selectable marker, which acts through the absence of expression of the *ebony* gene, leading to excessive melanization (Paulo et al. 2025). A reciprocal translocation links the wild-type allele to the male-determining region, enabling sex separation, with heterozygous males emerging from brown pupae and homozygous females emerging from black pupae (Zepeda-Cisneros et al. 2014). One example is the Tapachula-7 GSS, which is successfully employed in the control of the Mexican fruit fly, *Anastrepha ludens* (Loew, 1873) (Diptera: Tephritidae), demonstrating competitiveness (Orozco et al. 2013) and dispersal patterns similar to those of wild insects (Flores et al. 2015). Recently, a novel pupal color-based GSS was developed for *A. ludens*, named GUA10, which promises to further increase the SIT performance given its genetic stability and high production efficiency (Ramírez-Santos et al. 2021).

The same trait was used for the development of a GSS strain for *Anastrepha fraterculus* (Wiedemann, 1830) (Diptera: Tephritidae), based on natural mutations from a laboratory strain of the Brazil-1 morphotype and was named GSS-89 (Meza et al. 2020). The black pupae phenotype was characterized by the black color of the pupae, as well as the very dark coloration and wing veins in the adult stage, compared to the wild-type phenotype. As it occurs in other GSS, due to the translocation, only 50% of the sperm produced by GSS males are genetically balanced, making the strain semisterile (Robinson et al. 1999; Morrison et al. 2010; Zepeda-Cisneros et al. 2014; Meza et al. 2020).

Depending on how the sexing system was genetically constructed, it can be destabilized by recombination events between the translocation breakpoint and the mutations, which can lead to the accumulation of unwanted recombinant flies in the colony and even the breakdown of the mass-reared strain (Franz et al. 2021). In mass-rearing facilities, it is usually necessary to apply specific rearing methodologies that enable stable and predictable mass production of the GSS. A key rearing technology in the production of GSSs is the filter rearing system (FRS), which is briefly based on the early detection and elimination of recombinant individuals from a smaller colony of the GSS kept under low stress conditions (Caceres et al. 2000).

Aiming at the future application of the GSS-89 in AW-IPM projects such as the MOSCASUL Biological Control and Integrated Fruit Fly Management Center (Kovaleski and Mastrangelo 2021), a pupal lot of the GSS-89 was imported to Brazil on February 2020 with the legal permit obtained from the Ministry of Agriculture of Brazil (*Permissão de Importação* MAPA No. 026/2019 to the Center of Nuclear Energy in Agriculture of the University of Sao Paulo (CENA/USP)). Initially, the lot was transferred to the "Costa Lima" Quarantine Laboratory from the *Brazilian Agricultural Research Corporation* (EMBRAPA). After adult emergence, 5-day-old flies were inspected by Dr. Marcoandre Savaris from Esalq/USP, and then the adult cages were released from quarantine and transferred to the CENA/USP in March 2020.

Here, we describe how this GSS was adapted to the rearing conditions at CENA/USP, the quality control carried out and the production levels achieved. Additionally, we describe the resolution of problems that appeared over time, such as high recombination rates at the beginning of the strain's establishment.

11.2 MATERIAL AND METHODS

The colonies of the bisexual strains (e.g., Piracicaba and Vacaria strains) (Mastrangelo et al. 2021) and the GSS-89 were maintained at CENA under controlled environmental conditions (26±2°C and 70%–80% RH) following the rearing protocols of Walder et al. (2014) and Mastrangelo et al. (2021). The standard quality control parameters were recorded for each strain according to the methodologies recommended by the FAO/IAEA/USDA (2019) (i.e., total volume of eggs produced, egg hatch (%), total number of pupae produced, pupal weight (mg), pupal recovery (%), egg-pupa recovery (%), emergence of adults (%), fliers (%) and sex ratio) over the generations. The initial colony of the GSS-89 was maintained for nine generations, and the quality control parameters were recorded as previously mentioned. Also, the percentages of recombinants of each batch of pupae, with 50–100 pupae of each color per sample, were monitored.

A one-way analysis of variance *F*-test was applied for the quality control parameter of egg hatch (%) over generations at 5% significance (ANOVA). When significant differences were detected, Tukey's honest significant difference test (α=0.05) was applied to compare the means. For the pupal weight (mg) of brown and black pupae (two days before adult emergence) and adult emergence (%) of males and females, a two-way analysis of variance *F*-test was applied at a 1% level of significance, considering the generation and sex (factorial scheme 23 by 2) as factors. The Bartlett (1937) and Shapiro and Wilk (1965) tests were performed to verify the homocedasticity assumptions and the normality of the errors, respectively. Analyses were performed in the program SAS 9.4 (SAS Institute Inc. 2013).

11.3 RESULTS AND DISCUSSION

11.3.1 Adaptation of the GSS-89 to rearing conditions in Brazil

An initial colony of the GSS-89 was successfully established at the Food Irradiation & Radioentomology Laboratory of CENA, in small cages (30×30×30 cm) with up to 1,620 flies per cage at a 1:1.7 (♂:♀) ratio (Figure 11.1). Then, the GSS was adapted

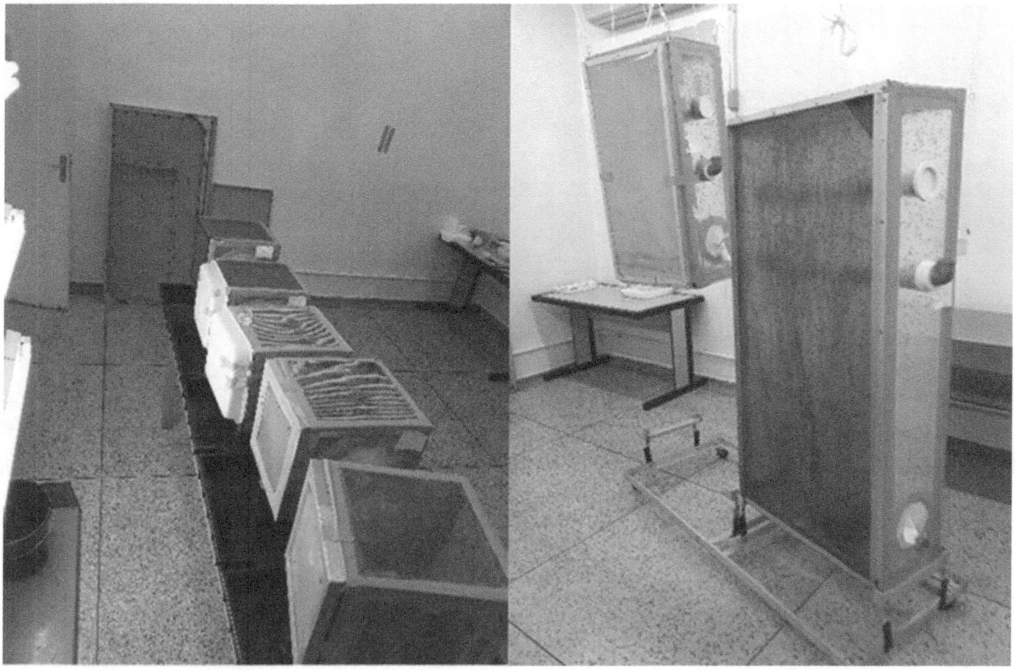

Figure 11.1 Small adult cages (30×30×30cm) used to initiate the GSS colony at CENA/USP (on the left). Medium (48×100×30cm) and large (75×150×30cm) cages for the GSS colony.

to CENA's rearing system, females adapted easily to oviposit on the vertical red or blue silicone panels, which are made of voile cloth coated externally with a thin layer of transparent silicon rubber (<0.5 mm thick). This layer was protected by a wet sponge cloth placed all along the oviposition chamber door (made of aluminum angles and a transparent polyvinyl chloride cover to which the sponge is attached) to avoid dehydration of the laid eggs (Mastrangelo et al. 2021). The larvae developed well on CENA's standard larval diet with carrageenan and corn flour, which had been used routinely for the bisexual strains (Figure 11.2). As higher volumes of pupae were obtained in later generations, medium (48×100×30 cm) and large (75×150×30 cm) cages started being used, always using an adult density of up to 0.3 flies/cm^2 (Mastrangelo et al. 2021).

Between the F_1 and F_9 generations of the GSS-89, mean egg hatch was 56.4±1.7%, a total volume of 1.1 L of pupae (≈36,300 pupae) was produced (Figure 11.3) and adult emergence was ca. 78.0±0.6%.

In the first three generations, the recombinant rates were lower than 10%. In F_2 and F_3, the percentage of recombinant females emerging from brown pupae was ca. 3.5%, but after F_4, the recombinant rates in brown pupae increased sharply from above 20% (Figure 11.4). More recombinant wild-type females were found in brown pupae than black males emerging from black pupae over generations.

The likely cause of the high presence of recombinant flies in this initial colony must have been the fact that the adult cages of the first five generations were set up exclusively with pupae (ca. 40% of brown pupae and 60% of black pupae per cage) instead of selected adults, as part of the need to reduce activities in the laboratory during the

Figure 11.2 Larval development of the GSS-89 in CENA's standard diet with corn flour and carrageenan.

Figure 11.3 Larvae of the GSS-89 pupating in vermiculite (at the left) and selected pupae from the GSS (at the right) for the assessment of quality control parameters at CENA/USP.

pandemic time. These results highlight the need to genetically clean the colony and establish a complete FRS to select 100% of adults before setting the ovipositing cages. According to Meza et al. (2020), recombinant rates should not be higher than 1% for the GSS-89.

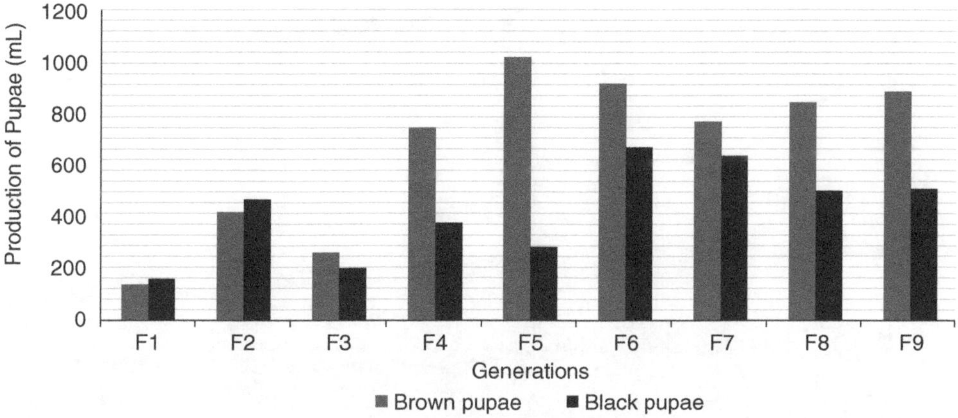

Figure 11.4 Total volume of brown and black pupae produced per generation of the initial GSS-89 colony at CENA/USP.

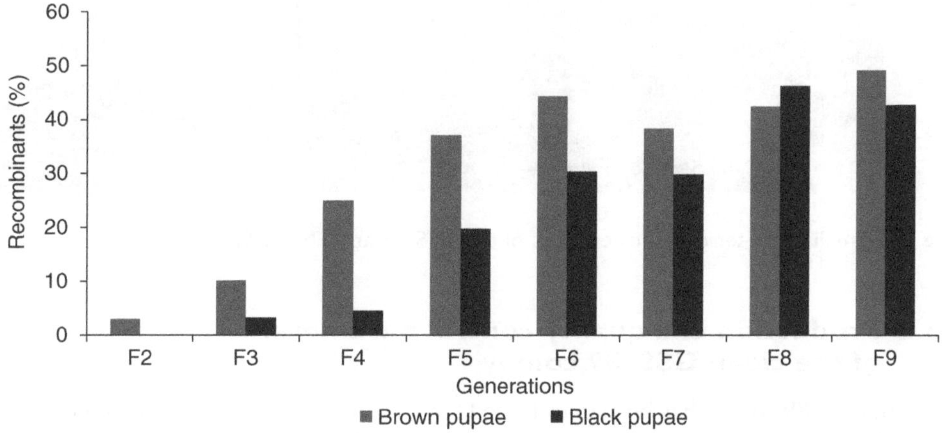

Figure 11.5 Rates of recombinant flies emerging from the initial GSS-89 colony at CENA/USP.

To clean the colony, 200 small cages (containing one male to five black females) were set in May 2021 trying to find male lines that presented the intact translocation characteristic of the GSS-89 and the same biological parameters described by Meza et al. (2020) (Figures 11.6 and 11.7).

After screening the pupal offspring and emerged adults of the 200 lines, the families in which no recombinant flies were observed were maintained isolated and then used to start a new pure colony of the GSS-89. A clean colony of the GSS (3 small cages 30(l)×30(w)×30(h) cm with 1,620 flies/cage at a 1:1.7 (♂:♀) ratio) was established with selected adults from the single male families named No. 30 and No. 83 in early 2022.

A clean colony of the GSS-89 and an FRS have been maintained since then at CENA/USP, and its production has been scaled up for several experiments and pilot projects.

Figure 11.6 Small cages set to filter new lines of the GSS-89 at CENA/USP.

11.3.2 Production and quality control parameters of the clean GSS-89 colony

The clean GSS-89 colony has been maintained for 22 generations so far (Figures 11.8–11.10). The total amount of eggs collected was ca. 1.5 L (ca. 17.3 million eggs), with a mean egg hatch of 52.8±1.5% (Figure 11.8).

The number of eggs collected varied throughout the generations mainly due to the different number of cages used in each generation and the fact that not all of the pupae produced were intended for the expansion of the mother colony. Several batches of pupae were used for other laboratory experiments involving the evaluation of different diets, sterilization doses and protocols, sexual compatibility and field dispersal tests, among others.

Until the F_3 generation, the clean colony of the GSS was maintained with 3–4 small cages (30×30×30 cm) with 1,620 flies/cage (ca. 600 males and 1,020 black females per cage). With the greater availability of pupae, medium-sized cages (50(l)×30(w)×100(h) cm) began to be introduced in the rearing, with ca. 6,650 flies/cage at a 1:1.7 (♂:♀) ratio.

From F_{16} onwards, the mother colony of GSS-89 began to be maintained in most generations with three medium-sized cages and 3–4 small cages (Figure 11.9). Estimating the individual fecundity of black females using our production data is complicated because of the different number of females used in different generations and the natural

Figure 11.7 Plastic cup adapted as cage to filter new lines of the GSS-89 at CENA/USP.

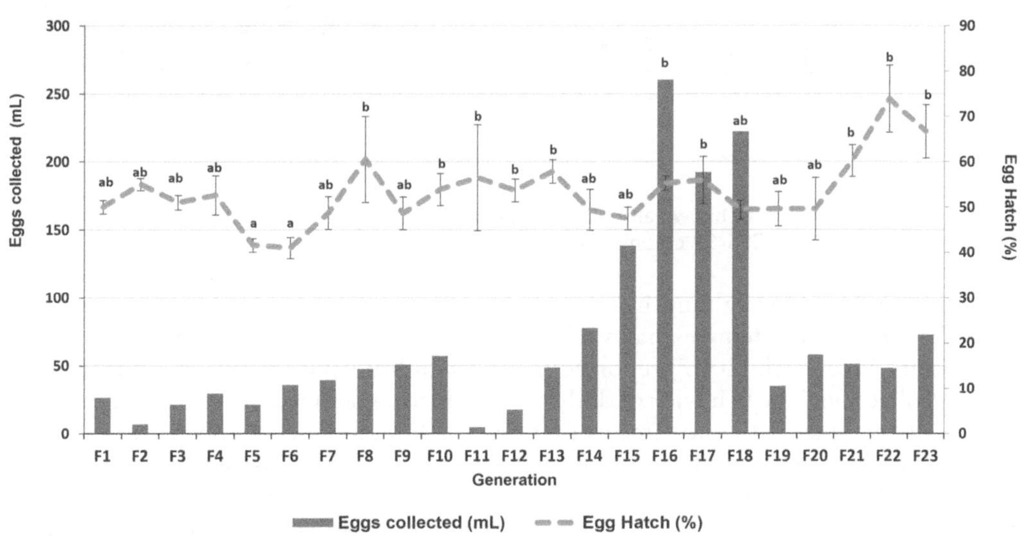

Figure 11.8 Total volume of eggs produced and mean egg hatch values produced per generation of the clean GSS-89 colony. Error bars on dashed line represent standard errors from the means of egg hatch. Different letters above the error bars indicate significant differences by Tukey's test ($P < 0.05$).

Figure 11.9 Ovipositing cages (4 medium and 4 small cages) for the 16th generation of the clean GSS-89 colony.

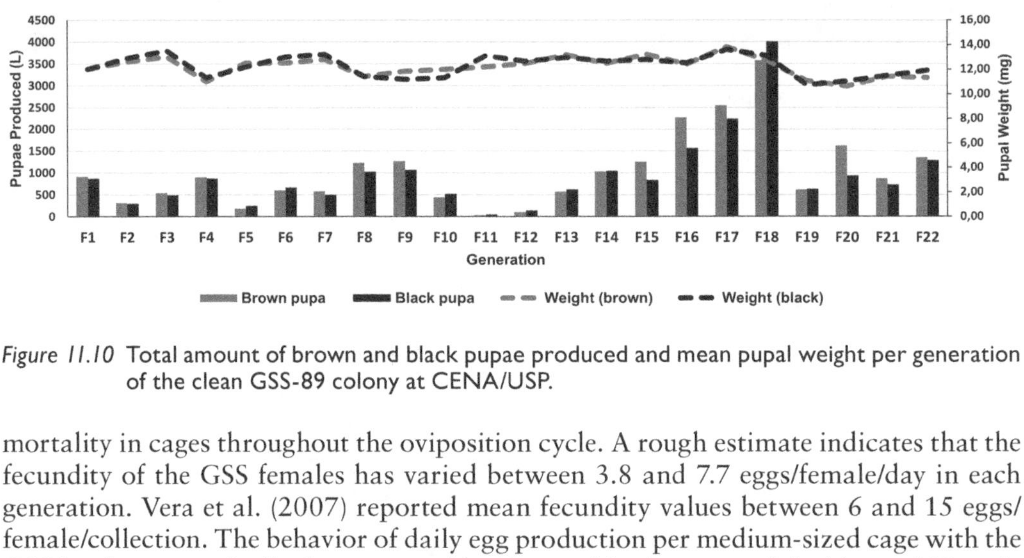

Figure 11.10 Total amount of brown and black pupae produced and mean pupal weight per generation of the clean GSS-89 colony at CENA/USP.

mortality in cages throughout the oviposition cycle. A rough estimate indicates that the fecundity of the GSS females has varied between 3.8 and 7.7 eggs/female/day in each generation. Vera et al. (2007) reported mean fecundity values between 6 and 15 eggs/female/collection. The behavior of daily egg production per medium-sized cage with the GSS has been practically the same as that observed in the cages of the bisexual Vacaria strain when kept at a density of 0.3 flies/cm^2 as reported by Mastrangelo et al. (2021), with a production peak (ca. 3.8–4.7 mL of eggs/cage) occurring between the 4th and 6th day of oviposition followed by a drop in egg production in the following days.

Egg production was lower in generations F_{11} and F_{12} due to management problems in the mother colonies and not to biological or genetic issues.

Regarding egg hatch, over 23 generations, significant differences were observed between the mean egg hatch values of the F_5 and F_6 generations, which were smaller (ca. 41%), and the mean values from generations F_8, $F_{10\text{-}13}$, $F_{16\text{-}17}$ and $F_{21\text{-}23}$ ($F = 8.2$; C.V. = 8.7%; $P < 10^{-3}$) (Figure 11.8).

Vera et al. (2007) reported rearing and quality control parameters of an *A. fraterculus* bisexual strain reared in Tucuman, Argentina, for more than 7 years, and the mean egg hatch values reported by them varied between 76% and 88.4%. Meza et al. (2020) found that the fertility of the GSS-89 was expected to be about 54%, because the (Y;A) translocation makes the GSS semisterile (i.e., only 50% of the sperm produced by males of the GSS-89 are genetically balanced). The Tapachula-7, a GSS of *A. ludens*, also presents the expected lower fecundity and fertility of a GSS compared with the standard bisexual strain of *A. ludens* mass-reared at the Moscafrut facility (Zepeda-Cisneros et al. 2014; Quintero-Fong et al. 2018). However, an increase in the percentage of egg hatch has been observed in the last generations of the GSS-89 (Figure 11.8), although no recombinant flies were found during the filtration process. A similar phenomenon has been observed already for the Tapachula-7 strain of *A. ludens* as production was scaled up at the Moscafrut facility, without any indication of loss of the genetic sexing system (Meza, personal communication).

Over generations, a total of 43.3 L of pupae were produced (≈1.4 million pupae), with 22.7 L of brown and 20.6 L of black pupae (≈750,255 brown pupae and 678,562 black pupae), with an overall mean pupal weight of 12.3 ± 0.2 mg. As expected for this strain (Meza et al. 2020), a slightly larger number of brown pupae have been produced in each generation (Figure 11.10). Considering an average egg-pupa recovery of 55.1 ± 6.4% (Giustina, unpublished data), which has been obtained in the GSS rearing using CENA's standard diet with carrageenan and corn flour, a single medium-sized cage producing a total of 20 mL of eggs (≈234,000 eggs) in 10 days of oviposition would be capable of generating at least 3.9 L of pupae (ca. 129,000 pupae).

There was no significant difference between the mean pupal weight of brown and black pupae in each generation ($F = 3.3$; $P = 0.07$) and also the interaction (generation x sex) was not significant ($F = 1.8$; $P = 0.03$). However, significant differences were noticed among generations in each sex ($F = 29.6$; $P < 10^{-3}$), but there was no clear trend of increasing or decreasing weight values as generations progressed (Table 11.1). The variations were probably caused by rearing or environmental factors, such as heating problems or high population density in larval trays (which help generate smaller pupae) in some generations. The highest pupal weight values (ca. 13.5–13.8 mg) were observed in generations F_3 and F_{17}, for both males and females, while the lowest means (range of 10–11.4 mg) were noticed in generations F_{19} and F_{20} for black and brown pupae, respectively (Table 11.1). The F_{19} and F_{20} generations occurred in January-February 2024, a period when it was difficult to maintain the temperature in the rearing rooms due to high summer temperatures and less labor.

Considering the minimum pupal weights of fruit flies accepted by the FAO/IAEA/USDA (2019), the GSS-89 pupae reared on CENA's diet weighted more than *A. suspensa* (10 mg), but less than *A. ludens* (17.1 mg) and *A. obliqua* (13.5 mg). Vera et al. (2007, 2014) reported mean pupal weights for *A. fraterculus* around 13.1 and 13.7 mg.

The mean emergence of adults from brown and black pupae over generations was 82.4 ± 1.9% and 73.9 ± 2.1%, respectively (Table 11.2). The interaction (generation × sex) was significant for the emergence of adults ($F = 4.9$; $P < 10^{-3}$) with regard to sex alone ($F = 244.8$; $P < 10^{-3}$) and among generations ($F = 50.2$; $P < 10^{-3}$) (Table 11.2). The emergence of black females was significantly lower than that of males in almost all generations. This may be a consequence of the fact that GSS-89 females are homozygous for

Table 11.1 Pupal weight (mg) of brown and black pupae (2 days before adult emergence) of the GSS-89 maintained at CENA/USP over generations

Generation	Sex	
	♂	♀
F_1	12.0±0.2 cdefgh	12.0±0.1 cdefg
F_2	12.6±0.2 bcd	12.9±0.3 abcd
F_3	13.0±0.2 abc	13.5±0.3 a
F_4	11.0±0.3 hi	11.3±0.3 fgh
F_5	12.5±0.2 bcde	12.2±0.15 bcdef
F_6	12.5±0.2 bcde	13.0±0.22 abcd
F_7	12.8±0.3 abcd	13.2±0.34 ab
F_8	11.4±0.3 efghi	11.4±0.4 efgh
F_9	11.8±0.5 defgh	11.2±0.5 fgh
F_{10}	12.0±0.3 cdefgh	11.3±0.1 fgh
F_{11}	12.2±0.4 befg	13.1±0.4 abc
F_{12}	12.5±0.2 bcde	12.7±0.2 abcd
F_{13}	13.2±0.3ab	13.0±0.4 abcd
F_{14}	12.5±0.2 bcde	12.7±0.2 abcd
F_{15}	13.2±0.8 ab	12.8±0.7 abcd
F_{16}	12.4±0.2 bcdef	12.5±0.2 abcde
F_{17}	13.8±0.2 a	13.6±0.2 a
F_{18}	12.7±0.1 abcd	13.1±0.1 abc
F_{19}	11.0±0.6 ghi	10.7±0.7h
F_{20}	10.6±0.5 i	11.0±0.4 gh
F_{21}	11.4±0.6 efghi	11.5±0.5 efgh
F_{22}	11.3±0.5 fghi	11.9±0.4 defg
F_{23}	12.7±0.4 abcd	13.1±0.2 ac
ANOVA	F= 13.9; C.V.=3.1%; P< 10^{-3}	F= 17.8; C.V.=2.9%; P< 10^{-3}

Means (± SE) followed by the same letters in the columns do not differ significantly by Tukey's test (P>0.01). ANOVA, analysis of variance; C.V., coefficient of variation.

the black pupae (*bp*) mutation and deletions at the *ebony* locus, which plays an important role in inhibiting black melanization in adult fruit flies, can have pleiotropic effects on both embryo viability and adult development in tephritids (Paulo et al. 2025).

Considering the minimum pre-irradiation adult emergence percentages of fruit flies accepted by the FAO/IAEA/USDA (2019), the overall means from the GSS-89 males at CENA were slightly smaller than the values recommended for *A. suspensa* (85%), *A. ludens* (91%) and *A. obliqua* (92%). Vera et al. (2007) showed mean adult emergence values between 77.1% and 88.6%, while Walder et al. (2014) reported a mean emergence of 85% for the Piracicaba strain.

The overall flight ability ranged between 73.5% and 98%, and the mean sex ratio (i.e., no. black pupae/ total no. of brown and black pupae produced) was 0.48 (Figure 11.10). The percentage of flyers was higher than 70%–90%, which is above or similar to the minimum pre-irradiation values accepted by the FAO/IAEA/USDA (2019) for *A. ludens* (88%), *A. obliqua* (89%) and *A. suspensa* (75%).

Table 11.2 Adult emergence (%) (means±SE) of males and females of the GSS-89 maintained at CENA/USP over generations

Generation	*Sex* ♂	♀	*ANOVA*
F_1	70.0±0.5 ij	70.0±0.4 cdef	$F=0.1$; C.V.=0.64%; $P=0.9$
F_2	82.0±0.7 defg A	68.0±0.8 def B	$F=520$; C.V.= 1.1%; $P<10^{-3}$
F_3	83.8±1.0 cde A	75.0±1.0 bcde B	$F=116$; C.V.= 1.3%; $P<10^{-3}$
F_4	69.3±1.7 ij	67.0±1.4 ef	$F=3.3$; C.V.=2.3%; $P=0.15$
F_5	97.5±1.4 a A	88.0±1.2 a B	$F=79.6$; C.V.= 1.4%; $P<10^{-3}$
F_6	88.1±0.6 abcde A	80.0±0.8 abc B	$F=196$; C.V.=0.8%; $P<10^{-3}$
F_7	90.2±1.9 abcd A	83.0±1.8 ab B	$F=22.7$; C.V.=2.1%; $P=0.009$
F_8	90.1±1.2 abcd A	83.0±1.4 ab B	$F=44.5$; C.V.= 1.5%; $P=0.002$
F_9	85.3±5.8 bcde	70.6±5.7 cdef	$F=9.8$; C.V.=7.4%; $P=0.035$
F_{10}	78.1±0.9 fghi A	68.8±1.8 def B	$F=64.1$; C.V.= 1.9%; $P=0.001$
F_{11}	80.0±0.3 efgh A	73.8±0.2 bcde B	$F=887.1$; C.V.=0.3%; $P<10^{-3}$
F_{12}	83.5±0.2 cdefg A	76.1±0.2 bcde B	$F=2053$; C.V.=0.3%; $P<10^{-3}$
F_{13}	83.5±0.3 cdefg A	79.1±0.4 abc B	$F=232$; C.V.=0.4%; $P<10^{-3}$
F_{14}	61.4±0.6 j	60.6±0.7 fgh	$F=2.3$; C.V.=0.7%; $P=0.21$
F_{15}	71.0±1.1 hij	66.0±2.4 efg	$F=10.8$; C.V.=2.7%; $P=0.03$
F_{16}	92.8±1.3 abc	87.2±3.0 a	$F=8.8$; C.V.=2.6%; $P=0.04$
F_{17}	92.3±1.5 abc	83.8±2.9 ab	$F=20.3$; C.V.=2.6%; $P=0.01$
F_{18}	93.8±0.9 ab A	83.0±1.3 ab B	$F=139$; C.V.= 1.3%; $P<10^{-3}$
F_{19}	74.0±4.02 ghi A	54.0±3.9h B	$F=38.5$; C.V.=6.2%; $P<10^{-3}$
F_{20}	80.0±1.5 efgh A	56.0±2.9 gh B	$F=162$; C.V.=3.4%; $P<10^{-3}$
F_{21}	75.6±9.03 fghi	62.0±7.6 fgh	$F=4$; C.V.= 12.1%; $P=0.12$
F_{22}	82.0±7.6 defg	78.0±8.5 abcd	$F=0.37$;C.V.= 10.1%; $P=0.58$
F_{23}	91.8±3.7 abc	88.0±4.1 a	$F=1.42$; C.V.=4.3%; $P=0.29$
ANOVA	$F=26.1$; C.V.=3.8%; $P<10^{-3}$	$F=28.9$; C.V.=4.4%; $P<10^{-3}$	

Means (± SE) followed by the same lowercase letter in the columns or uppercase letter in the lines do not differ significantly by Tukey's test or the Student's *t*-test, respectively ($P>0.01$). ANOVA, analysis of variance; C.V., coefficient of variation.

As a result of the implementation of an FRS to maintain the integrity of the sexing system, the current recombinant rates are below 0.05%. This allowed scaling up the GSS-89 colony for the conduction of several bioassays and sending batches of irradiated brown pupae to pilot projects in the southern region of Brazil.

11.4 CONCLUSIONS

The GSS-89 of *A. fraterculus* has adapted successfully to the rearing conditions used at CENA in Brazil. The problems faced due to the accumulation of recombination events, found in the establishment of the strain during the pandemic period, were successfully overcome. However, to avoid the presence or accumulation of recombinant flies in the mother colony, a strict FRS must be maintained throughout the generations.

ACKNOWLEDGMENTS

We would like to thank Dra. Simone de Souza Prado and Dr. Luiz Alexandre N. de Sá (*in memoriam*) from EMBRAPA Environment for all the assistance during the import and quarantine of flies at the "Costa Lima" Quarantine Laboratory. We thank student trainees from Esalq/USP for support in the mass rearing of *Anastrepha* flies. Also, we acknowledge the support provided for this research via the FAO/IAEA Coordinated Research Projects "Improvement of Colony Management in Insect Mass-rearing for SIT Applications" (CRP No. D42017) and "Improving rearing, handling, and field components for fruit fly SIT application" (CRP No. D41029).

REFERENCES

Aluja, M., S. M. Ovruski, F. R. Mello Garcia, et al. 2024. Fruit fly (Tephritidae) management in the neotropical region: History, state of the art, and perspectives. In *Management of Fruit Flies in the Americas*, ed. F. R. Mello Garcia, 11–66. Switzerland: Springer Nature.

Bartlett, M. S. 1937. Properties of sufficiency and statistical tests. *Proceedings of the Royal Statistical Society, Series A* 160: 268–282.

Caceres, C., J. P. Cayol, W. Enkerlin, et al. 2004. Comparison of Mediterranean fruit fly (*Ceratitis capitata*) (Tephritidae) bisexual and genetic sexing strains: Development, evaluation and economics. In *International Symposium on Fruit Flies of Economic Importance*, 367–381. Stellenbosch, South Africa: Isteg Scientific Publications.

Caceres, C., K. Fisher, and P. Rendon. 2000. Mass rearing of the medfly temperature sensitive lethal genetic sexing strain in Guatemala. In *Area-Wide Control of Fruit Flies and Other Insect Pests*, ed. K. H. Tan, 551–558. Penang, Malaysia: Penerbit Universiti Sains.

Dyck, V.A., J. Hendrichs, and A. S. Robinson. 2005. *Sterile Insect Technique: Principles and Practice in Area-Wide Integrated Pest Management*. Dordrecht: Springer.

Enkerlin, W. R. 2021. Impact of fruit fly control programmes using the sterile insect technique. In *Sterile Insect Technique: Principles and Practice in Area-Wide Integrated Pest Management*, ed. V. A. Dyck, J. Hendrichs, and A. S. Robinson, 979–1006. Boca Raton: CRC Press.

Flores, S., S. Campos, E. Gómez, et al. 2015. Evaluation of field dispersal and survival capacity of the genetic sexing strain Tapachula-7 of *Anastrepha ludens* (Diptera: Tephritidae). *Florida Entomologist* 98(1): 209–214.

Food and Agriculture Organization of the United Nations/International Atomic Energy Agency/ United States Department of Agriculture [FAO/IAEA/USDA]. 2019. *Product Quality Control for Sterile Mass-Reared and Released Tephritid Fruit Flies, Version 7.0*. Vienna, Austria: IAEA.

Franz, G. 2005. Genetic sexing strains in Mediterranean fruit fly, an example for other species amenable to large-scale-rearing for the sterile insect technique. In *Sterile Insect Technique: Principles and Practice in Area-Wide Integrated Pest Management*, ed. V. A. Dick, J. Hendrichs, and A. S. Robinson, 427–452. Dordrecht: Springer.

Franz, G., K. Bourtzis, and C. Caceres. 2021. Practical and operational genetic sexing systems based on classical genetic approaches in fruit flies, an example for other species amenable to large-scale rearing for the sterile insect technique. In *Sterile Insect Technique: Principles and Practice in Area-Wide Integrated Pest Management*, ed. V. A. Dyck, J. Hendrichs, and A. S. Robinson, 576–599. Boca Raton: CRC Press.

Franz, G., E. Gencheva, and P. Kerremans. 1994. Improved stability of genetic sex-separation strains for the Mediterranean fruit fly, *Ceratitis capitata. Genome* 37: 72–82.

Franz, G., P. Kerremans, P. Rendon, et al. 1996. Development and application of genetic sexing systems for the Mediterranean fruit fly based on a temperature sensitive lethal. In *Fruit Fly Pests: A World Assessment of Their Biology and Management*, ed. B. A. McPheron, and J. G. Steck, 185–191. Delray Beach, FL: St. Lucia Press.

Hendrichs, J., G. Franz, and P. Rendon. 1995. Increased effectiveness and applicability of the sterile insect technique through male-only releases for control of Mediterranean fruit flies during fruiting seasons. *Journal of Applied Entomology* 119: 371–377.

Klassen, W., and C. F. Curtis. 2005. History of the sterile insect technique. In *Sterile Insect Technique: Principles and Practice in Area-Wide Integrated Pest Management*, ed. V. A. Dyck, J. Hendrichs, and A. S. Robinson, 3–36. Dordrecht: Springer.

Knipling, E. F. 1953. Possibilities of insect control or eradication through the use of sexually sterile males. *Journal of Economic Entomology* 48: 456–462.

Kovaleski, A., and T. Mastrangelo. 2021. MOSCASUL Programme: First steps of a pilot project to suppress the South American fruit fly in southern Brazil. In *Area-Wide Integrated Pest Management. Development and Field Application,* ed. J. Hendrichs, R. Pereira, and M. J. B. Vreysen, 215–230. Boca Raton, FL: CRC Press.

Mastrangelo, T., A. Kovaleski, B. Maset, et al. 2021. Improvement of the mass-rearing protocols for the South American fruit fly for application of the sterile insect technique. *Insects* 12: 622. https://doi.org/10.3390/insects12070622

McCombs, S.D., and S. H. Saul. 1995. Translocation-based genetic sexing system for the oriental fruit fly (Diptera: Tephritidae) based on pupal color dimorphism. *Annals of the Entomological Society of America* 88: 695–698.

McInnis, D. O., S. Tam, C. Grace, et al. 1994. Population suppression and sterility rates induced by variable sex ratio, sterile insect releases of *Ceratitis capitata* (Diptera: Tephritidae) in Hawaii. *Annals of the Entomological Society of America* 87: 231–240.

McInnis, D. O., S. Tam, R. Lim, et al. 2004. Development of a pupal color-based genetic sexing strain of the melon fly, *Bactrocera cucurbitae* (Coquillett) (Diptera: Tephritidae). *Annals of the Entomological Society of America* 97: 1026–1033.

Meats, A., P. Maheswaran, M. Frommer, et al. 2002. Towards a male-only release system for SIT with the Queensland fruit fly, *Bactrocera tryoni*, using a genetic sexing strain with a temperature-sensitive lethal mutation. *Genetica* 116: 97–106.

Meza, J. S., K. Bourtzis, A. Zacharopoulou, et al. 2020. Development and characterization of a pupal-colour based genetic sexing strain of *Anastrepha fraterculus* sp. 1 (Diptera: Tephritidae). *BMC Genetics* 21: 1–9.

Morrison, N. I., G. Franz, M. Koukidou, et al. 2010. Genetic improvements to the sterile insect technique for agricultural pests. *Asia-Pacific Journal of Molecular Biology and Biotechnology* 18: 275–295.

Orozco, D., M. R. Hernandez, J. S. Meza, et al. 2013. Do sterile females affect the sexual performance of sterile males of *Anastrepha ludens* (Diptera: Tephritidae)? *Journal of Applied Entomology* 137: 321–326.

Paulo, D.F., T. N. M. Nguyen, C. M. Ward, et al. 2025. The genetic basis of the black pupae phenotype in tephritid fruit flies. *Communication Biology* 8: 60. https://doi.org/10.1038/s42003-025-07489-y

Quintero-Fong, L., J. Toledo, L. Ruiz-Montoya, et al. 2018. Demography of a genetic sexing strain of *Anastrepha ludens* (Diptera: Tephritidae): Effects of selection based on mating performance. *Agricultural and Forest Entomology* 20:1–8. https://doi.org/10.1111/afe.12223

Ramírez-Santos, E., P. Rendon, G. Gouvi, et al. 2021. A novel genetic sexing strain of *Anastrepha ludens* for cost-effective sterile insect technique applications: Improved genetic stability and rearing efficiency. *Insects* 12(6):499. https://doi.org/10.3390/insects12060499

Rendon, P., and W. Enkerlin. 2021. Area-wide fruit fly programmes in Latin America. In *Area-Wide Integrated Pest Management, Development and Field Application*, ed. J. Hendrichs, R. Pereira, and M. J. B. Vreysen, 161–198. Boca Raton, FL: CRC Press.

Rendon, P., D. O. McInnis, D. R. Lance, et al. 2000. Comparison of medfly male only and bisexual releases in large-scale field trials. In *Area-Wide Control of Fruit Flies and Other Insect Pests*, ed. T. K. Hong, 517–525. Malaysia, Penang: Penerbit Universiti Sains.

Rendon, P., D. O. McInnis, D. R. Lance, et al. 2004. Medfly (Diptera: Tephritidae) genetic sexing: Large-scale field comparison of males-only and bisexual sterile fly releases in Guatemala. *Journal of Economic Entomology* 97: 1547–1553.

Robinson, A. S. 2002. Mutations and their use in insect control. *Mutation Research/Reviews in Mutation Research* 511: 113–132.

Robinson, A. S., U. Cirio, G. H. S. Hooper, et al. 1986. Field cage studies with a genetic sexing strain in the Mediterranean fruit fly, *Ceratitis capitata. Entomologia Experimentalis et Applicata* 41(3): 231–235.

Robinson, A. S., G. Franz, and K. Fisher. 1999. Genetic sexing strains in the medfly, *Ceratitis capitata*: Development, mass-rearing and field application. *Trends in Entomology* 2: 81–104.

SAS Institute Inc. 2013. *SAS Software Version 9.4*. Cary, NC: SAS Institute Inc.

Shapiro, S. S, and M. B. Wilk. 1965. An analysis of variance test for normality (complete samples). *Biometrika* 52: 591–611.

Shelly, T. E., and N. C. Manoukis. 2022. Mating competitiveness of *Bactrocera dorsalis* (Diptera: tephritidae) males from a genetic sexing strain: Effects of overflooding ratio and released females. *Journal of Economic Entomology* 115(3): 799–807.

Vera, T., S. Abraham, A. Oviedo, et al. 2007. Demographic and quality parameters of *Anastrepha fraterculus* (Diptera: Tephritidae) maintained under artificial rearing. *Florida Entomologist* 90: 53–55.

Vera, M. T., A. Oviedo, S. Abraham, et al. 2014. Development of a larval diet for the South American fruit fly *Anastrepha fraterculus* (Diptera: Tephritidae). *International Journal of Tropical Insect Science* 34: S73–S81.

Walder, J. M. M., R. Morelli, K. Z. Costa, et al. 2014. Large scale artificial rearing of *Anastrepha sp.1 aff. fraterculus* (Diptera: Tephritidae) in Brazil. *Scientia Agricola* 71: 281–286. https://doi.org/10.1590/0103-9016-2013-233

Whitten, M.J. 1969. Automated sexing of pupae and its usefulness in control by sterile insects. *Journal of Economic Entomology* 62: 272–273.

Zepeda-Cisneros, C. S., J. S. Meza Hernandez, V. Garcia-Martinez, et al. 2014. Development, genetic and cytogenetic analyses of genetic sexing strains of the Mexican fruit fly, *Anastrepha ludens* Loew (Diptera: Tephritidae). *BMC Genetics* 15: 1–11.

Section IV

Natural enemies and biological control

Chapter 12

Abundance and diversity of tephritid parasitoids in different agroecological and climatic regions of Madagascar

Henriette Rasolofoarivao, Lala Harivelo Ravaomanarivo, and Hélène Delatte

12.1 INTRODUCTION

Tephritid fruit fly species are known to be one of the major pests of many horticultural crops in the Indian Ocean (White and Elson-Harris 1992; Ryckewaert et al. 2010; De Meyer et al. 2012; Rasolofoarivao et al. 2022a). The tropical continental island of Madagascar is a hot spot of biodiversity where 74 species in the family Tephritidae have been described (Hancock 1985; Rasolofoarivao et al. 2022a). Within those 74 species, 32 are frugivorous species whose larvae attack fruits and vegetables of economic importance. Five species (*Bactrocera dorsalis* Hendel, *Ceratitis malgassa* Munro, *Neoceratitis cyanescens* Bezzi, *Dacus demmerezi* Bezzi, and *Dacus vertebratus* Bezzi) have been reported as major pests of various crops on the island (Franck and Delatte 2020; Rasolofoarivao et al. 2022a), three of which (*C. malgassa*, *N. cyanescens*, and *D. demmerezi*) are endemic to Madagascar. Detected in 2010, *B. dorsalis* is the most recent invader, severely affecting fruit production (De Meyer et al. 2012; Rasolofoarivao et al. 2022a; Deschepper et al. 2022). Other exotic species of Tephritidae are present on the island, such as *Ceratitis cosyra* Walker and *Ceratitis capitata* (Wiedemann) (Dubois 1965; De Meyer 2000). Nevertheless, high populations and severe economic damage had never been reported for either of these two species, even in the first records made by Dubois (1965) around the 1950s.

Fruit flies are naturally exposed to various predators, the most important being ants (Formicidae) and rove beetles (Staphylinidae), which are known to cause high larval mortality in sub-saharan Africa (Vayssières et al. 2016; Mohamed et al. 2016). Several species of Hymenoptera are well-known as parasitoids and play an important role in the regulation of tephritid pests of the genus *Bactrocera* and *Zeugodacus* in sub-Saharan Africa (Mohamed et al. 2016). Madl and van Achterberg (2014) listed 577 species of Braconidae in the island of Madagascar and the islands of the south-western part of the Indian Ocean, including 29 species of Opiinae, that are endemic to Madagascar. However, the biology of most of these species is unknown, and they may attack other non-tephritid hosts (Fischer and Madl 2008). In the 1960s, four wasp species (*Opius concolor* Szépligeti, *Opius oophilus* Fullaway (syn. *Fopius arisanus* Sonan), *Opius longicaudatus* Ashmead (syn. *Diachasmimorpha longicaudata* Ashmead), and *Dirhinus giffardii* Silvestri (Table 12.1) were introduced to Madagascar (Antananarivo, Brickaville and Ampasimbe; East coast) from Hawaii and France to control *C. malgassa*. In 1969, a few years after their introduction, some specimens of *F. arisanus* and very few of *D. longicaudata* were sampled in the same areas. *Fopius arisanus* is a

DOI: 10.1201/9781003644330-16

Table 12.1 Species of parasitoids of fruit fly species (Tephritidae) present in Madagascar (Dubois 1965; Ravaomanarivo 1996; Fischer and Madl 2008; Madl and Van Achterberg 2014)

Parasitoid species	*Family/subfamily*	*Endemic/ introduced*	*Distribution outside Madagascar*
Diachasmimorpha longicaudata	Braconidae	Introduced	Asia, America, Africa, Mauritius, Reunion
Fopius arisanus	Braconidae	Introduced	America, Australia, Africa, Indian Ocean islands
Opius concolor	Braconidae	Introduced	Africa, Europe (Italy, Israel, France), Middle East, USA (Hawaii, California)
Psytallia distinguenda Granger	Braconidae	Endemic	Mascarenes
Psytallia insignipennis	Braconidae	Endemic	Mascarenes
Psytallia phaeostigma Wilkinson	Braconidae	Introduced	Africa (e.g., Ethiopia, South Africa, and Kenya), Mayotte, Mauritius, Reunion, Hawaii
Dirhinus giffardii	Chalcididae	Introduced	Nigeria, Hawaii, Israel, Pakistan
Pachycrepoideus vindemniae	Pteromalidae	Introduced	America, Africa
Muscidifurax raptor rs	Pteromalidae	Introduced	America, the Caribbean, Africa, Middle East, Europe, Australia, and many Pacific islands

polyphagous, solitary, egg-pupal braconid endoparasitoid (Bautista et al. 1998) that is native to tropical areas in South and Southeast Asia (Fullaway 1951). In 1976 and 1977, further introductions of generalist parasitoids (Pteromalidae) of dipteran species were made. These were the pupal parasitoids *Pachycrepoideus vindemniae* Rondani and *Muscidifurax raptor* Girault and Sanders (Table 12.1). In the 1980s, these two species were collected from citrus orchards in the highlands (Ambohijafy) and seemed to have become established (Ravaomanarivo 1996). To date, several parasitoid species from East Africa were introduced for biological control (Fischer and Madl 2008); however, some of these species failed to establish, and information regarding the target insect remains unknown. Furthermore, to the best of our knowledge, no survey has been conducted in more than 10 years to identify tephritid parasitoids and their potential impact on the native tephritid community from Madagascar.

To understand the current situation regarding the possible regulation of fruit fly populations by hymenopteran parasitoids, we studied the diversity and distribution of tephritid parasitoids and host plants of these tephritid species found on the island.

12.2 MATERIALS AND METHODS

12.2.1 Fruit sampling and rearing in the laboratory

We followed the same sampling and fruit incubation protocols as in Rasolofoarivao et al. (2022a). Fruits were collected from 240 sites across 6 provinces, 17 regions, and 48 districts of Madagascar (Figure 12.1). Sampling sites were geolocated, each sample (a piece of fruit or a batch of fruit) was weighed and incubated on river sand in the

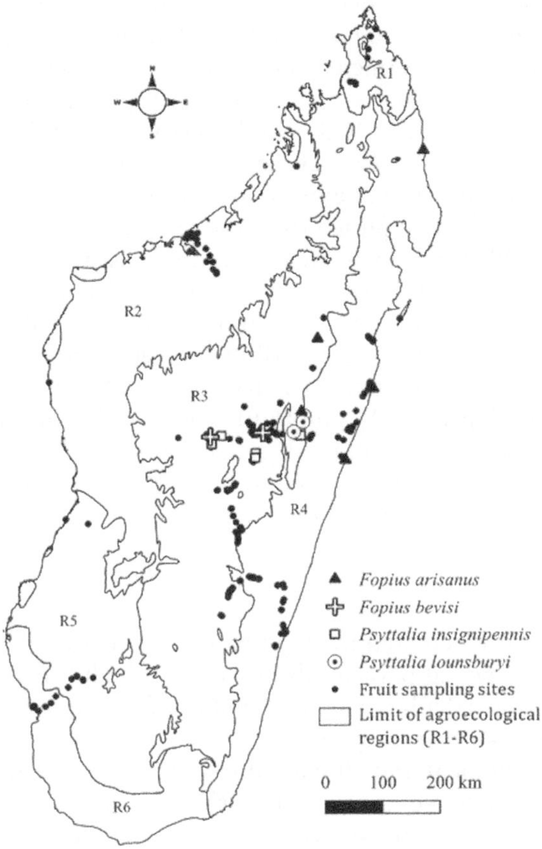

Figure 12.1 Distribution of parasitoid species across agroecological and climatic regions (R1–R6) of Madagascar.

Antananarivo laboratory until pupation. After emergence, tephritids were identified following White and Elson-Harris (1992), De Meyer (2000), and Virgilio et al. (2014) keys. Adults were counted, and some specimens were killed in 96% ethanol and stored at –20°C. Full details of each tephritid host plant species (weight and number of samples per host plant) are provided in Table 12.1. Samples from 2016 to 2018 are the same as described in Rasolofoarivao et al. (2022a), while the samples from 2019 to 2022 are from the present study. Sampling covered all six different agroecological zones (Cornet 1974) of the island (Figure 12.1).

12.2.2 Parasitoid identification

The species of parasitoids were identified using Wharton's key (Wharton and Gilstrap 1983; Wharton et al. 1999) and the online identification key of Wharton and Yoder (2018). Observations of parasitoids and emerged fruit flies were made using a stereomicroscope (SMZ 745T Nikon) equipped with an Amscope MU1000 digital camera mounted on an ocular tube. Additionally, we sequenced two standard DNA barcode fragments for each identified species in order to confirm morphological identifications. The DNeasy Blood and Tissue Kit (Qiagen, Hilden, Germany) was used to extract DNA from adult specimens. Universal primers CHO and LCO (Vrijenhoek 1994) were used to amplify part of the mitochondrial Cytochrome Oxidase I (COI) gene. Polymerase chain reaction (PCR) conditions were as follows: an initial step of 15 minutes at 95°C, followed by 40 cycles of 30 seconds at 95°C, 50 seconds at 49°C, and 30 seconds at 72°C, and a final extension of 10 minutes at 72°C. A portion of the 28S ribosomal gene was then amplified from the same DNA sample using primers D2F and D3R (Park et al. 2000). The PCR cycle program consisted of an initial denaturation step for 15 minutes at 95°C, followed by 40 cycles of denaturation for 30 seconds at 95°C, annealing for 30 seconds at 56.5°C, and extension for 30 seconds at 72°C, and a final extension for 10 minutes at 72°C. All PCRs were performed in a final volume of 20 µL containing 4 µL DNA extract, 5 µL H_2O, 10 µL 2X type-it master mix (Qiagen, Germany), 0.5 µL of each primer (10 pmol/µL). The resulting products were then sent to Macrogen Europe for sequencing. Barcode sequences were aligned using Geneious R10 (https://www.geneious.com) and subjected to BLAST to relate the species to the closest taxa available in the GenBank database.

12.2.3 Data analysis

The parasitism rate was calculated by dividing the number of recovered parasitoids by the total number of recovered parasitoids and emerged adult fruit flies in each sample. The infestation rate was calculated as the number of pupae per kg of fruit. A generalized linear model (GLM) was used to analyze the effect of the interaction between host plant and agroecological and climatic division on parasitism rate. Using a similar GLM, parasitism rates were compared across three altitudinal gradients (0–499 masl, from 500 to 1,000 masl, and above 1,000 masl) and six agroecological regions (R1–R6). A Pearson correlation test was conducted to examine the relationship between the parasitism rate and the variation in the tephritid infestation level among tephritid host plants at the sites where parasitism was detected. A network analysis was performed using the igraph package in R to visualize the relationship between plant species, fruit flies, and their parasitoids. All analyses (significance threshold less than 0.05) were performed in R (R Development Core Team 2021).

12.3 RESULTS

12.3.1 Fruit fly species and parasitism level

A total of 41,304 fruits (743.15 kg) from 61 families and 200 species were sampled between 2016 and 2022. Out of the 200 plant species that were collected, 66 were found to be host of Tephrididae (Table 12.2). No adult fruit fly was obtained from pupae of three plant species (*Carica papaya* Linnaeus, *Artocarpus heterophyllus* Lamarck, and *Rubus rosifolius* Smith). Out of the 12 fruit fly species found, 5 were found to be associated with parasitoids (Figure 12.2).

Parasitism levels were significantly different between plants (GLM, deviance=5.398, P=0.001) and ranged from 0.96 (*Cucurbita pepo* Linnaeus) to 19.63% (*Prunus persica* Linnaeus) for all parasitoid species, except for *S. jambosa* (71.88%) (Table 12.3).

Table 12.2 List of plant species sampled, recognized as hosts of Tephritidae in Madagascar in 2016–2022

Plant family	*Plant species (scientific name)*	*Plant species (common name)*	*Total no. of fruits*	*Total weight (kg)*
Anacardiaceae	*Mangifera indica* Linnaeus	Mango	468	99.064
Anacardiaceae	*Sclerocarya birrea* (A. Richard) Hochstetter	Marula	224	6.987
Anacardiaceae	*Spondias dulcis* Solander ex Parkinson	Jew plum	205	12.702
Annonaceae	*Annona senegalensis*	Wild custard - apple	15	7.658
Annonaceae	*Annona squamosa* Linnaeus	Sugar - apple	25	5.965
Annonaceae	*Xylopia sp*		200	0.111
Asparagaceae	*Dracaena reflexa* Lamarck	Dragon tree	455	0.415
Caricaceae	*Carica papaya*	Papaya	35	16.554
Combretaceae	*Terminalia catappa* Linnaeus	Tropical almond	289	5.039
Cucurbitaceae	*Citrullus colocynthis* (L.) Scharder	Bitter apple	57	5.785
Cucurbitaceae	*Citrullus lanatus* (Thunb.) Matsum and Nakai	Water melon	29	18.428
Cucurbitaceae	*Citrullus sp*		24	2.355
Cucurbitaceae	*Cucumis sativus* Linnaeus	Cucumber	198	24.706
Cucurbitaceae	*Cucumis sp*		17	5.244
Cucurbitaceae	*Cucumus melo* Linnaeus	Muskmelon	30	4.682
Cucurbitaceae	*Cucurbita maxima* Duschesne	Pumpkin	178	40.688
Cucurbitaceae	*Cucurbita pepo*	Squash	237	22.973
Cucurbitaceae	*Luffa cylindrica* M. Roemer	Sponge gourd	55	9.0738
Cucurbitaceae	*Momordica charantia* Linnaeus	Bitter gourd	60	0.592
Cucurbitaceae	*Sechium edule* (Jacq.) Swartz	Chayote	281	40.359
Cucurbitaceae	*Trichosanthes cucumerina* Linnaeus	Snakegourd	40	2.96
Ebenaceae	*Diospyros kaki*	Japanese persimmon	204	12.05
Lauraceae	*Persea americana* Miller	Avocado	69	17.685
Loganiaceae	*Strychnos spinosa* Lamarck	Monkey orange	18	5.845
Meliaceae	*Non identified*		15	0.046

(*Continued*)

Table 12.2 (Continued) List of plant species sampled, recognized as hosts of Tephritidae in Madagascar in 2016–2022

Plant family	*Plant species (scientific name)*	*Plant species (common name)*	*Total no. of fruits*	*Total weight (kg)*
Moraceae	*Artocarpus altilis* (Parkinson) Fosberg	Breadfruit	31	10.831
Moraceae	*Artocarpus heterophyllus*	Jackfruit	27	42.554
Moraceae	*Ficus polita* Vahl	Heart-leaved fig	61	1.411
Myrtaceae	*Psidium cattleyanum* Sabine	Strawberry guava	798	5.322
Myrtaceae	*Psidium friedrichsthalianum* Niedenzu	Coronilla	15	1.215
Myrtaceae	*Psidium guajava*	Common guava	737	31.535
Myrtaceae	*Syzygium jambos*	Rose-apple	508	8.332
Myrtaceae	*Syzygium malaccense* (L.) Merril and L.M.Perry	Malay apple	57	2.394
Oxalidaceae	*Averrhoa carambola* Linnaeus	Star fruit	134	7.908
Podocarpacae	*Podocarpus gaussenii* Woltz	Bastard yellowwood	352	1.413
Rosaceae	*Eriobotrya japonica* (Thunb.) Lindley	Loquat	1320	13.049
Rosaceae	*Malus domestica* (Suckow) Borkhausen	Apple	59	2.84
Rosaceae	*Prunus domestica* Linnaeus	Plum	107	2.202
Rosaceae	*Prunus persica*	Peach	476	18.062
Rosaceae	*Pyrus communis* Linnaeus	Pear	27	3.343
Rosaceae	*Rubus myrianthus Baker*	Blackberry	682	0.42
Rosaceae	*Rubus rosifolius*	Thimbleberry	128	0.095
Rubiaceae	*Coffea sp*	Coffea	1005	1.268
Rutaceae	*Citrus paradisi* Macfadyen	Grapefruit	18	11.03
Rutaceae	*Citrus reticulata*	Mandarin	178	10.395
Rutaceae	*Citrus sinensis*	Orange	428	41.714
Sapindaceae	*Litchi chinensis* Sonnen	Lychee	309	3.978
Sapotaceae	*Mimusops coriacea* (A. de Candolle) Miquel		28	1.042
Solanaceae	*Capsicum annum* Linnaeus	Bell pepper	172	1.646
Solanaceae	*Capsicum baccatum* Linnaeus		24	0.209
Solanaceae	*Capsicum chinense* Jacquin	Habanero-type pepper	99	0.382

(Continued)

Table 12.2 (Continued) List of plant species sampled, recognized as hosts of Tephritidae in Madagascar in 2016–2022

Plant family	*Plant species (scientific name)*	*Plant species (common name)*	*Total no. of fruits*	*Total weight (kg)*
Solanaceae	*Capsicum frutescens* Linnaeus	Pepper pili	762	0.373
Solanaceae	*Capsicum sp*	Pepper	461	1.071
Solanaceae	*Lycopersicum esculentum* Miller	Tomato	708	14.431
Solanaceae	*Lycopersicum esculentum cerasiforme* Miller		257	1.392
Solanaceae	*Solanum betaceum* Cavanilles	Tree tomato	109	3.987
Solanaceae	*Solanum indicum* Linnaeus	African eggplant	567	4.027
Solanaceae	*Solanum macranthum* Linnaeus	Potato tree	29	1.772
Solanaceae	*Solanum mauritianum* Scopoli	Bugweed	1519	2.398
Solanaceae	*Solanum melongena*	Egg plant	250	10.358
Solanaceae	*Solanum muricatum* Aiton	Melon pear	38	3.42
Solanaceae	*Solanum nigrum*	Black nightshade	3856	1.948
Solanaceae	*Solanum pseudocapsicum* Linnaeus	Jerusalem cherry	63	0.117
Solanaceae	*Solanum seaforthianum* Andrews	Brazilian nightshade	216	0.137
Solanaceae	*Solanum torvum* Swartz		363	0.678
Solanaceae	*Solanum tuberosum* Linnaeus	Potato	13	0.013

Plants are ordered alphabetically by family and scientific name.

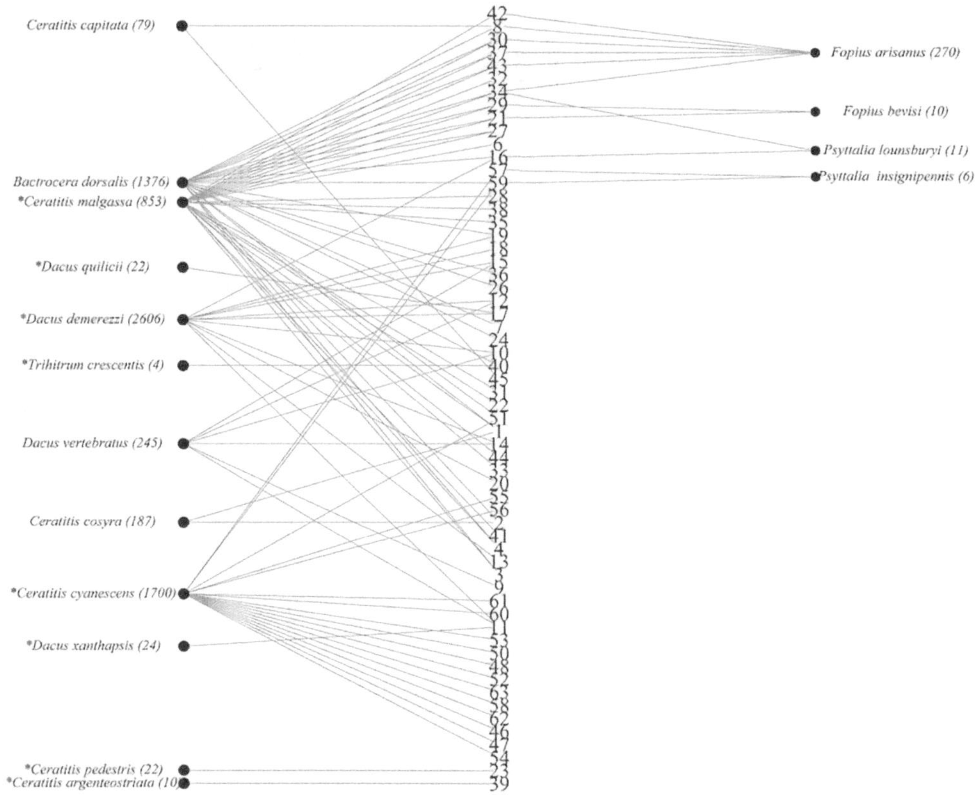

Figure 12.2 Network of interaction between tephritid parasitoids, tephritids (fruit flies), and host plants of these tephritid species (n: number of adult/species). * Endemic fruit fly species of Madagascar.

No correlation was detected between the parasitism rates and the infestation indexes or the weight of the samples (Pearson test, $P>0.05$).

12.3.2 Diversity of parasitoid species

Four parasitoid species were found: *F. arisanus*, *Fopius bevisi* Brues, *Psyttalia insignipennis* Granger, and *Psyttalia lounsburyi* Silvestri. The identification of *F. bevisi* was based only on morphological characters. Fragments of the COI and (or) 28S ribosomal barcodes of 58 specimens were successfully sequenced. Sequences were submitted to the Barcode of Life Data Systems. *Fopius arisanus* was the most abundant species, accounting for 90.30% (270 individuals) of the collected adult parasitoids. The frequency of the other three species was as follows: *F. bevisi* 3.34% (10 individuals), *P. lounsburyi* 3.68% (11 individuals), and *P. insignipennis* 2.01% (6 individuals) (Table 12.4).

For the fruit fly species/parasitoid samples associated, *F. arisanus* emerged from fruits that produced three fruit fly species (*C. capitata*, *C. malgassa*, and *B. dorsalis*), *F. bevisi* from two species (*C. malgassa* and *B. dorsalis*), *P. insignipennis* from *N. cyanescens*, and *P. lounsburyi* from *B. dorsalis* and *D. demmerezi* (Table 12.4). We were not able

Table 12.3 Parasitism rate from fruit fly-infested fruit

Plant family	*Plant species*	*No fruit*	*Total weight (kg)*	*Infestation rate (no. of pupae/kg)*	*Parasitism rate (%) (all species of parasitoids)*	*Parasitism rate (%) (only F. arisanus)*
Combretaceae	*Terminalia catappa*	289	5.039	203.21	10.88	10.88
Cucurbitaceae	*Cucurbita pepo*	237	22.973	70.30	0.96	0.00
Ebenaceae	*Diospyros kaki*	204	12.05	3.90	3.45	0.00
Moraceae	*Artocarpus altilis*	31	10.831	0.09	(-)	(-)
Myrtaceae	*Psidium guajava*	737	31.535	22.13	1.97	0.00
Myrtaceae	*Syzygium jambosa*	508	8.332	119.06	71.88	71.88
Rosaceae	*Eriobotrya japonica*	1320	13.049	24.45	1.06	(+)
Rosaceae	*Prunus persica*	248	9.65	123.01	19.63	23.32
Rutaceae	*Citrus reticulata*	178	10.395	11.54	15.15	15.15
Rutaceae	*Citrus sinensis*	428	41.714	8.87	3.66	3.66
Solanaceae	*Solanum melongena*	250	10.358	30.22	2.31	0.00
Solanaceae	*Solanum nigrum*	455	0.484	204.55	4.35	0.00

(–) only one *Fopius arisanus* adult emerged. (+) we have no complete data on the number of emerged individuals. Two hymenopteran adults emerged from *Prunus persica*, but were damaged and could not be identified.

Table 12.4 Diversity of parasitoid species and tephritid species emerged from infested fruits. No: number of adult/species

	Fruit fly emergence (No.)					Parasitoid emergence (No.)			
Plant species	*Ceratitis capitata*	*Ceratitis malgassa*	*Neoceratitis cyanescens*	*Bactrocera dorsalis*	*Dacus demmerezi*	*Fopius arisanus*	*Fopius bevisi*	*Psyttalia insignipennis*	*Psyttalia lounsburyi*
Terminalia catappa	49	10	0	416	0	57	0	0	0
Cucurbita pepo	0	0	1	1	931	0	0	0	9
Diospyros kaki	0	26	0	2	0	0	1	0	0
Artocarpus altilis	0	0	0	0	0	1	0	0	0
Psidium guajava	0	265	0	182	0	0	9	0	0
Syzygium jambosa	0	49	0	5	0	138	0	0	0
Eriobotrya japonica	0	88	0	98	0	0	0	0	2
Prunus persica	0	141	0	76	0	66	0	0	0
Citrus reticulata	0	28	0	0	0	5	0	0	0
Citrus sinensis	0	76	0	3	0	3	0	0	0
Solanum melongena	0	0	169	0	0	0	0	4	0
Solanum nigrum	0	0	44	0	0	0	0	2	0

to assign the parasitoid species to the tephritid host exactly, as more than one fruit fly species developed in some of the incubated fruit samples.

Fopius arisanus was found emerging from fruits of six plant species, four of which are known to be associated with *B. dorsalis* (*Terminalia catappa* Linnaeus, *Syzygium jambosa* Linnaeus Alston, *P. persica*, and *Citrus sinensis* Linnaeus Osbeck). The other three parasitoid species were found on two plant species each (Figure 12.2).

12.3.3 Geographical distribution of parasitoids

Parasitoids were found in only 20 of the 240 sampling sites (Figure 12.1). Highly significant differences in parasitism rates were observed between altitudinal gradients (deviance=7.68, df=2, P=0.004) and agroecological regions (deviance=27.193, df=5, P<0.0001). *Fopius arisanus* was found in both the highlands (R3) at medium altitudes and at high altitudes (ranging from 762 to 1,288 masl), as well as in the eastern region (R4) at low altitudes ranging from 1 to 13 masl. *Fopius bevisi, P. insignipennis*, and *P. lounsburyi* were distributed throughout the central highlands (R3) at more than 864 masl.

12.4 DISCUSSION

Fopius arisanus was the most abundant parasitoid in our samples, and it was found emerging from six plant species, parasitizing three fruit fly species: *C. capitata*, *C. malgassa*, and *B. dorsalis*. It was mainly found in fruits with a high infestation of *B. dorsalis* such as those within the Myrtaceae family, except for *Citrus reticulata* Blanco, where only *C. malgassa* was found. These results are consistent with the literature, where *F. arisanus* has been described emerging from fruit flies associated with the Myrtaceae in Hawaïi (Vargas et al. 1993), or in Australia (Clarke et al. 2022) from at least 84 other plant species, belonging to 35 families (Rousse et al. 2005; Rousse 2007). It has also been reported as the main parasitoid of *C. capitata* in Hawaii after its introduction on the island (Haramoto and Bess 1970; Vargas et al. 2001a). This study showed its ability to parasitize fruit flies from two genera in Madagascar, including an endemic species, increasing its known range of hosts and confirming its large range of fruit fly species. It is important to note that due to the wide host range of *F. arisanus*, it could play a role in regulating tephritid species in Madagascar, where 32 frugivorous species have been recorded (De Meyer et al. 2012; Rasolofoarivao et al. 2022a). Yet, the distribution of *F. arisanus* in Madagascar is rather restricted, occurring mainly in the eastern lowlands and central parts of the island. Approximately 60 years after its release, *F. arisanus* is still found in the few areas where it was originally introduced. Following the last release in 1965 (about 13,000 females were released at two sites), *F. arisanus* slowly dispersed from the original release site and has only been recovered in two regions. Conversely, after its release in Kenya, *F. arisanus* has covered a total area of 50.34 m^2 in five years (Agboka et al. 2022). In both countries, the climatic and agroecological areas of release were different. Therefore, knowing this potential dispersal capacity of *F. arisanus*, if new releases were to be considered in Madagascar, it might be important to release *F. arisanus* in multiple sites to improve its dispersal and establishment.

This limited range may also be attributed to various factors, such as environmental or abiotic conditions that are unfavorable for the establishment of this introduced species. According to Nanga et al. (2021), *F. arisanus* completes its developmental cycle between 15°C and 30°C, but stops and dies at temperatures above 32°C. The optimal female flight temperature is around 20°C (Rousse et al. 2009), and the optimal embryonic development is between 22°C and 25°C, with a minimum of 10.3°C (Groth et al. 2017). The extreme climatic conditions in Madagascar, which range from 5.5°C (minimum temperature) in the austral winter in the highlands to 41.5°C (maximum temperature) in the summer in the south-western lowlands, may have hindered the establishment of *F. arisanus* on these parts of the island.

In Africa, parasitism rates of up to 70% have been observed in *B. dorsalis* using laboratory populations and up to 40% in the field (Mohamed et al. 2010). Populations of *F. arisanus* in Hawaii were able to parasitize 65%–70% of *B. dorsalis* (Haramoto and Bess 1970; Vargas et al. 1993). In Madagascar, *B. dorsalis* spread throughout the island in 10 years after its arrival. However, *F. arisanus* did not follow the same pattern despite the arrival of its primary Asian host (Vargas et al. 2001b), and was only found in the areas of initial releases in the 1960s. Currently, parasitism rates do not exceed 23.32%. Regarding the results of such a low distribution and parasitism rates, one cannot exclude the hypothesis of the long (almost 50 years) coevolution of *F. arisanus* and the resident fruit fly community in Madagascar (mostly with the *Ceratitis* genus), which may also have been responsible for the differences between Africa or Hawaii and Malagasy parasitism by *F. arisanus*. This parasitoid was introduced into these countries from populations reared on *B. dorsalis* in countries where *B. dorsalis* was also a new invader compared to Madagascar where no *Bactrocera* species were present. In Reunion, a similar observation was made where *F. arisanus* was introduced to control *Bactrocera zonata* Saunders in the 2000s in the absence of *B. dorsalis* (Rousse et al. 2005). When *B. dorsalis* arrived 17 years after its establishment (Moquet et al. 2021), parasitism rates on *B. dorsalis* were also low, as observed in Madagascar (Moquet et al. 2023).

This is the first time that *F. bevisi* has been found in Madagascar. *Fopius bevisi* is an African larval, koinobiont, solitary, endophagous parasitoid of the genus *Ceratitis*, originally described in South Africa (Greathead 1971; Clausen 1978). A large diversity of fruit fly parasitoids has been described in Madagascar (Dubois 1965; Fischer and Madl 2008; Madl and Van Achterberg 2014), but *F. bevisi* was never mentioned, although it may have been undetected due to very low populations. It was found in only two fruit species (*Diospyros kaki* Linnaeus and *Psydium guajava* Linnaeus) of all tephritid host fruit species collected in Madagascar. We suspect that it parasitizes both *C. malgassa* and *B. dorsalis*, but preferentially *C. malgassa*. It was introduced from Africa to Hawaii for biological control programs against *C. capitata* and *B. dorsalis* but its acclimatization was unsuccessful in those countries (Wharton and Gilstrap 1983; Mohamed et al. 2016). In Kenya, it has been reported in wild fruits of the genus *Strychnos* infested with *Ceratitis querita* Munro (Clausen 1978) and in a single sample of fruit of *Annona senegalensis* Persoon, attacked by *C. cosyra* (Copeland et al. 2006). Other fruit fly species such as *Dacus ciliatus* Loew have been reported as hosts of *F. bevisi* (Mohamed et al. 2016). Several species of the tribe Ceratitini are present in Madagascar, including *C. capitata*, which probably invaded the island in the 1950s (Dubois 1965). Very small populations of this species were found in this study, mostly restricted to the eastern lowlands of the island (Rasolofoarivao et al. 2022a). *Fopius bevisi* may have been involved in the regulation of medfly populations in Madagascar. Nevertheless, further experiments on its tephritid host species preferences should be conducted to better understand its role in regulating tephritid species in Madagascar and elsewhere.

Psyttalia insignipennis was found emerging from two plant species infested by *N. cyanescens, Solanum melongena* Linnaeus, and *Solanum nigrum* Linnaeus (Table 12.4). It is a koinobiont endoparasitoid, which parasites larvae and emerges from the host puparium (Wharton et al., 1999). *Psyttalia insignipennis* is endemic and restricted to Madagascar and the Mascarene islands (Wharton et al. 1999). Although it was initially described as a parasitoid of the endemic tomato fly *N. cyanescens* (Brenière 1965; Dubois 1965), it has been reported in other tephritid species, such as *C. capitata* and *Ceratitis catoirii* Guérin-Méneville on Reunion Island (Wharton et al. 1999). Low parasitism of this species (2.31%–4.35%) was observed in Madagascar. In Reunion Island, there are few data on *P. insignipennis* attacking *N. cyanescens*, suggesting that parasitism by *P. insignipennis* seems to be occasional (Ryckewaert et al. 2010).

Neoceratitis cyanescens is the most serious threat to Solanaceae crops, with 16 plant species known to be their hosts in Madagascar, with infestations reaching up to 58% (Rasolofoarivao et al. 2022b). Unfortunately, even in the area of endemicity of *P. insignipennis* and *N. cyanescens,* population control of *N. cyanescens* is not effective enough to limit the damage to Solanaceae.

As with *F. bevisi*, this is the first study to record *P. lounsburyi* in Madagascar. It was found emerging from pupae coming from two plant species, *C. pepo* and *Eriobotrya japonica* (Thunberg) Lindley. It emerged together with fruits infested by *D. demmerezi* (*C. pepo*) and *B. dorsalis* (*E. japonica*). It is a solitary egg-pupal, endoparasitoid, koinobiont species (Daane et al. 2008) and is known as an African parasitoid of *Bactrocera oleae* Rossi (Copeland et al. 2004). It was also found developing in *C. capitata* (Thaon et al. 2009). The native range of *P. lounsburyi* is sub-Saharan Africa, but it has been exported to other countries (e.g., USA (California) and France) as part of biological control programs against the olive fruit fly *B. oleae* (Daane et al. 2008; Malausa et al. 2010) and in Hawaii against *C. capitata* (Hallouti et al. 2020). The reduced impact of *B. oleae* and *C. capitata* in fruit has been largely attributed to the action of this introduced parasitoid in France with observed parasitism rates of up to 57% (Malausa et al. 2010). However, it was found in Madagascar emerging most likely from two different species of Tephritidae and from a very different genus (*Dacus*), so further studies on its tephritid host range could be interesting, despite the low parasitism rate observed (between 0.96% and 1.06%).

12.5 CONCLUSION

The level of parasitism and parasitoid diversity observed (only four braconid species described, parasitism levels ranging from 0.96% to 19.63%, from 12/66 plant species and 2/6 agroecological regions covered) was surprisingly low in a continental island with high agricultural value and with great biodiversity and endemism. The low parasitism levels observed could be due to several non-exclusive hypotheses, such as (i) a past event of competition when *F. arisanus* was first introduced, knowing its capacity to outcompete other parasitoids (Mahat and Clarke 2021); (ii) the low fruit fly infestations observed in sampled fruits; (iii) the increased forest fires recorded in Madagascar in recent years that might have reduced large numbers of beneficial insects; or (iv) cultural practices, in which the use of pesticides targeting pest species could have a negative impact on parasitoid populations.

Finally, the low level of parasitism found in this study may not support the role of the indigenous parasitoid community in effectively controlling fruit flies. Considering the promising results of *F. arisanus* against *B. dorsalis* in Hawaii, Kenya, Reunion, and other countries where *F. arisanus* is not endemic, it seems relevant to propose to try to release new strains/populations of *F. arisanus* reared on *B. dorsalis* and in large areas. However, any new introduction to the island should be followed by studies on its acclimatization, its spread, and impact on fruit fly populations and other endemic species of parasitoids.

ACKNOWLEDGEMENTS

The authors are grateful to the master II students (L. Ravalison and N. Razafiarivelo) from the University of Antananarivo, Département d'Entomologie for their help. We would like to acknowledge the Direction de la Protection des Végétaux (DPV) of Madagascar for lending a laboratory to carry out fruit fly rearing. We thank Haran J. from UMR CBGP for confirming the determination of *F. bevisi* on taxonomic criteria; Joelle Sadeyen from UMR PVBMT for technical assistance for the molecular part; and Samuel Nibouche for the barcoding analysis. This work was funded by the European Union (FEDER INTERREG VI), Conseil Régional de la Réunion and CIRAD, under EPIBIO-OI «EPIdémiosurveillance et BIOcontrole dans le Sud-Ouest de l'Océan Indien» project and undertaken in the framework of the research platform "Biocontrôle et épidémiosurveillance vegetal en ocean Indien" (https://www.dp-biocontrole-oi.org/).

REFERENCES

Agboka, K. M., H. E. Z. Tonnang, E. M. Abdel-Rahman et al. 2022. A systematic methodological approach to estimate the impacts of a classical biological control agent's dispersal at landscape: Application to fruit fly *Bactrocera dorsalis* and its endoparasitoid *Fopius arisanus*. *Biological Control* 175:105053.

Bautista, R. C., E. J. Harris, and P. O. Lawrence. 1998. Biology and rearing of the fruit fly parasitoid *Biosteres arisanus*: clues to insectary propagation. *Entomologia Experimentalis et Applicata* 89:79–86.

Brenière, J. 1965. Liste des parasites et prédateurs des principaux insectes nuisibles aux cultures à Madagascar. *Agronomie Tropicale* 20:344–349.

Clarke, A. R., C. Harris, B. J. Kay, et al. 2022. Opiine parasitoids (Hymenoptera: Braconidae) and biological control of fruit flies (Diptera: Tephritidae) in Australia: Past, present and future. *Annals of Applied Biology* 180(1): 44–72.

Clausen, C. P. 1978. Introduced parasites and predators of arthropod pests and weeds: A world review. Curculionidae. *US Department of Agriculture. Agriculture Handbook* 480:259–276.

Copeland, R. S., R. A. Wharton, Q. Luke, et al. 2006. Geographic distribution, host fruit, and parasitoids of African fruit fly pests *Ceratitis anonae*, *Ceratitis cosyra*, *Ceratitis fasciventris*, and *Ceratitis rosa* (Diptera: Tephritidae) in Kenya. *Annals of the Entomological Society of America* 99:261–278.

Copeland, R. S., I. White, M. Okumu, et al. 2004. Insects associated with fruits of the Oleaceae (Asteridae, Lamiales) in Kenya, with special reference to the Tephritidae (Diptera). *Bishop Museum Bulletin in Entomology* 12:135–164.

Cornet, A. 1974. Essai de cartographie bioclimatique à Madagascar. *Notice Explicative de l'ORSTOM* 55:1–28.

Daane, K. M., K. R. Sime, X. G. Wang, et al. 2008. *Psyttalia lounsburyi* (Hymenoptera: Braconidae), potential biological control agent for the olive fruit fly in California. *Biological Control* 44:79–89.

De Meyer, M. 2000. Systematic revision of the subgenus *Ceratitis* MacLeay s.s. (Diptera, Tephritidae). *Zoological Journal of the Linnean Society* 128:439–467.

De Meyer, M., S. Quilici, A. Franck, et al. 2012. Records of frugivorous fruit flies (Diptera: Tephritidae: Dacini) from the Comoro archipelago. *African Invertebrates* 53:69–77.

Deschepper, P., S. Vanbergen, Y. Zhang, et al. 2022. *Bactrocera dorsalis* in the Indian Ocean: A tale of two invasions. *Evolutionary Applications* 16:48–61.

Dubois, J. 1965. La mouche des fruits malgache (*Ceratitis malagassa* Munro) et autres insectes des agrumes, pêchers et pruniers à Madagascar. *Fruits* 20:435–460.

Fischer, M., and M. Madl. 2008. Review of the opiinae of the Malagasy subregion (Hymenoptera: Braconidae). *Linzer biologische Beiträge* 40:1467–1489.

Franck, A., and H. Delatte. 2020. *Plantes-hôtes des mouches des fruits: Comores,* Madagascar, Maurice, Mayotte, Réunion, Seychelles. Saint-Pierre: CIRAD.

Fullaway, D. T. 1951. Review of the Indo-Australasian parasites of the fruit flies (Tephritidae). *Proceedings of the Hawaiian Entomological Society* 14:243–250.

Greathead, D. J. 1971. A review of biological control in the Ethiopian Region. *Technical Communications. Commonwealth Institute of Biological Control* 5:162.

Groth, M. Z., A. E. Loeck, S. D. Nornberg, et al. 2017. Biology and thermal requirements of *Fopius arisanus* (Sonan,1932) (Hymenoptera: Brachonidae) reared on *Ceratitis capitata* eggs (Wiedemann) (Diptera: Tephritidae). *Neotropical Entomology* 46:554–560.

Hallouti, A., A. Zahidi, E. H. Lamharchi, et al. 2020. First report of *Psyttalia lounsburyi*, a new larval parasitoid of the Mediterranean fruit fly *Ceratitis capitata* in Moroccan Argane Forest. *Journal of Entomology* 17:58–67.

Hancock, D. L. 1985. Trypetinae (Diptera: Tephritidae) from Madagascar. *Journal of the Entomological Society of Southern Africa* 48:283–301.

Haramoto, F. H., and H. A. Bess. 1970. Recent studies on the abundance of the Oriental and Mediterranean fruit flies and the status of their parasites. *Proceedings of the Hawaiian Entomological Society* 20:551–566.

Madl, M., and C. van Achterberg. 2014. A catalogue of the Braconidae (Hymenoptera: Ichneumonoidea) of the Malagasy subregion. *Linzer biologische Beiträge* 46:5–220.

Mahat, K., and A. R. Clarke. 2021. Competition between *Diachasmimorpha kraussii* and *Fopius arisanus* in *Bactrocera tryoni*: does native parasitoid-host association matter? *BioControl* 66:297–306.

Malausa, J. C., A. Auguste-Maros, S. Cheyppe-Buchmann, et al. 2010. Introductions of the African parasitoid *Psyttalia lounsburyi* in South of France for classical biological control of *Bactrocera oleae*. *IOBC/wprs Bulletin* 59:163–170.

Mohamed, S. A., S. Ekesi, and R. Hanna. 2010. Old and new host-parasitoid associations: Parasitism of the invasive fruit fly *Bactrocera invadens* (Diptera: Tephritidae) and five African fruit fly species by *Fopius arisanus*, an Asian opiine parasitoid. *Biocontrol Science and Technology* 20:183–196.

Mohamed, S. A., M. M. Ramadan, and S. Ekesi. 2016. In and out of Africa: Parasitoids used for biological control of fruit flies. In Fruit Fly Research and Development in Africa-Towards a Sustainable Management Strategy to Improve Horticulture, ed. S. Ekesi, S. Mohamed, and M. De Meyer, 325–368. Switzerland: Springer International Publishing.

Moquet, L., B. Jobart, R. Fontaine, et al. 2023. Tri-trophic interactions among *Fopius arisanus*, Tephritid species and host plants suggest apparent competition. *Ecology and Evolution* 13:e9742.

Moquet, L., J. Payet, S. Glenac, et al. 2021. Niche shift of tephritid species after the Oriental fruit fly (*Bactrocera dorsalis*) invasion in La Réunion. *Diversity and Distributions* 27:109–129.

Nanga, S. N., S. Kekeunou, A. F. Kuate, et al. 2021. Temperature-dependent phenology of the parasitoid *Fopius arisanus* on the host *Bactrocera dorsalis*. *Journal of Thermal Biology* 100:103031.

Park, Y. H., S. Sensoy, C. Wye, et al. 2000. A genetic map of cucumber composed of RAPDs, RFLPs, AFLPs, and loci conditioning resistance to papaya ringspot and zucchini yellow mosaic viruses. *Genome* 43:1003–1010.

R Development Core Team. 2021. R: *A Language and Environment for Statistical Computing*. R Foundation for Statistical Computing.

Rasolofoarivao, H., M. De Meyer, and H. Delatte. 2022a. Diptera: Tephritidae, fruit flies, laliboankazo. In The New Natural History of Madagascar, ed. S. M. Goodman, 1080–1085. Princeton: Princeton University Press.

Rasolofoarivao, H., L. R. Ravaomanarivo, and H. Delatte. 2022b. Host plant ranges of fruit flies (Diptera: Tephritidae) in Madagascar. *Bulletin of Entomological Research* 112:1–12.

Ravaomanarivo, L. R. 1996. *Contribution à l'étude de la biologie de Ceratitis (Ceratitis) malgassa Munro et à l'étude systématique des espèces du genre Ceratitis (Diptères, Tephritidae) à Madagascar*. Thèse de doctorat, Université d'Antananarivo.

Rousse, P. 2007. *Fopius arisanus, le droit à l'erreur: spécificité parasitaire et sélection de l'hôte chez un parasitoïde ovo-pupal de mouches des fruits Tephritidae*. Thèse de doctorat, Université de la Réunion, La Réunion, France.

Rousse, P., F. Gourdon, M. Roubaud, et al. 2009. Biotic and abiotic factors affecting the flight activity of *Fopius arisanus*, an egg-pupal parasitoid of fruit fly pests. *Environnemental Entomology* 38:896–903.

Rousse, P., E. Harris, and S. Quilici. 2005. *Fopius arisanus*, an egg-pupal parasitoid of Tephritidae. Overview. *Biocontrol News and Information* 26:59–69.

Ryckewaert, P., J. P. Deguine, T. Brévault, et al. 2010. Fruit flies (Diptera: Tephritidae) on vegetable crops in Reunion Island (Indian Ocean): State of knowledge, control, methods and prospects for management. *Fruits* 65:113–130.

Thaon, M. M., A. Blanchet, and N. Ris. 2009. Contribution à l'optimisation de l'élevage du parasitoïde *Psyttalia lounsburyi*. *Cahier des Techniques de l'INRA* 66:21–31.

Vargas, R. I., S. L. Peck, G. T. McQuate, et al. 2001a. Potential for areawide integrated management of Mediterranean fruit fly (Diptera: Tephritidae) with a Braconid parasitoid and a novel bait spray. *Journal of Economic Entomology* 94:817–825.

Vargas R. I., M. Ramadan, T. Hussain, et al. 2001b. Comparative demography of six fruit fly (Diptera: Tephritidae) parasitoids (Hymenoptera: Braconidae). *Biological Control* 25:30–40.

Vargas, R. I., J. D. Stark, G. Uchida, et al. 1993. Opiine parasitoids (Hymenoptera, Braconidae) of Oriental fruit fly (Diptera, Tephritidae) on Kauai Island, Hawaii Islandwide relative abundance and parasitism rates in wild and orchard guava habitats. *Environmental Entomology* 22:246–253.

Vayssières, J. F., J. Offenberg, A. Sinzogan, et al. 2016. The use of weaver ants in the management of fruit flies in Africa. In Fruit Fly Research and Development in Africa-Towards a Sustainable Management Strategy to Improve Horticulture, ed. S. Ekesi, S. Mohamed, and M. De Meyer, 389–434. Switzerland: Springer International Publishing.

Virgilio, M., I. White, and M. De Meyer. 2014. A set of multi-entry identification keys to African frugivorous flies (Diptera, Tephritidae). *ZooKeys* 428:97–108. https://keys.lucidcentral.org/keys/v3/fruitflies/ (accessed January 25, 2025).

Vrijenhoek, R. 1994. DNA primers for amplification of mitochondrial cytochrome c oxidase subunit I from diverse metazoan invertebrates. *Molecular Marine Biology and Biotechnology* 3:294–299.

Wharton, R., and F. Gilstrap. 1983. Key to and status of opiine braconid (Hymenoptera) parasitoids used in biological control of *Ceratitis* and *Dacus sl* (Diptera: Tephritidae). *Annals of the Entomological Society of America* 76:721–742.

Wharton, R., S. Quilici, B. Hurtrel, et al. 1999. The status of two species of *Psyttalia* Walker (Hymenoptera: Braconidae: Opiinae) reared from fruit-infesting Tephritidae (Diptera) on the Indian Ocean Islands of Reunion and Mauritius. *African Entomology* 7:85–90.
Wharton, R., and M. Yoder. 2018. Parasitoids of fruit-infesting Tephritidae. https://paroffit.org.
White, I. M., and M. M. Elson-Harris. 1992. *Fruit Flies of Economic Significance: Their Identification and Bionomics*. Wallingford, UK: CABI International.

Section V

Area-wide IPM and action programs

Chapter 13

Fifty-five years of a successful National Fruit Fly Program in Chile

Ricardo Rodríguez Palomino

13.1 EARLY HISTORY

The National Fruit Fly Detection System (NFFDS) in Chile has been a successful program for over 55 years, maintaining Chile as a fruit fly-free area. The first detection of the Mediterranean fruit fly (*Ceratitis capitata* Wied.) (Medfly) in Chile was in 1963, and four eradications have been carried out in Pica, the last in 2017. The NFFDS uses a network of 16,000 specific traps for tephritid species and protects about 350,000 hectares for fruit production in the fruit fly-free area. Chilean fruit growing is the primary agricultural export with an annual average value of US$ 5,000 million returns in recent years. Strict quarantine controls at each checkpoint, a permanent surveillance system in the country, and a corrective action protocol for Medfly are some of the methods used to maintain Chile as a fruit fly-free area. The NFFDS is successful in maintaining Chile as a fruit fly-free area based on five powerful pillars: (i) professional, technical and administrative human team of excellence and committed with the Fruit Fly Program; (ii) strict quarantine control at each entry checkpoint; (iii) permanent surveillance system in the country; (iv) corrective action protocol for the Mediterranean fruit fly; and (v) incursion verification of Medfly in the country and of a very occasional entry of *Bactrocera dorsalis* (Hendel) in Easter Island. The NFFDS and its five essential supports are responsible for the maintenance of Chile's status as a country free of fruit flies of economic importance for over 55 years. This freedom allowed Chile to export fruits without restrictions for fruit flies of economic importance (Figure 13.1).

When agricultural and forestry exports started in Chile in the 1960s, one of the marketing slogans was that Chile "is a phytosanitary island": isolated by the Pacific Ocean, the Andes Mountains, the Atacama Desert and low temperatures in the south. But this changed dramatically with the globalization of trade, tourism, climate change and today with immigrants.

The following historical milestones that helped keep Chile as a country free of fruit flies of economic importance are as follows. The first step was the foundation of the vegetable health police service (today plant protection division—SAG in 1923). In 1934, the first eradication of *Anastrepha fraterculus* (Wied) occurred at Arica city and Azapa Valley (both close to the border with Peru). In 1935, a thesis on fruit flies written by a student from the University of Chile warned of the danger that these insects represent for the fruit development if they enter Chile.

The first detection of the Medfly in Chile was in 1963 at the Oasis of Pica (Figure 13.2), located in the desert of Atacama, 2,000 km north from the main fruit

DOI: 10.1201/9781003644330-18

Figure 13.1 First official detections of *Ceratitis Capitata* in South America: Brazil: 1901; Uruguay: 1932; Argentina: 1934; Bolivia: 1942; Peru: 1957; Venezuela: 1960; Chile: 1963; Ecuador: 1978; Colombia: 1986.

production area. Subsequently, four eradications were carried out in Pica; the last eradication was in 2017.

The NFFDS was created in 1980 as part of the *eradication and exclusion project of fruit flies in Chile*, with the purpose to obtain pest-free areas (PFAs) for fruit flies, mainly for Medfly. From time to time, different actions were implemented to fight against fruit flies that are intercepted in Chile. The project was approved as a state policy (Law No. 3.557), and its implementation was developed by the Agriculture and Livestock Service (SAG), which is a unit of the Ministry of Agriculture (Olalquiaga and Lobos 1993; Rodriguez and Lobos 2010; Shelly et al. 2014).

After the recognition of Chile as a fruit fly-free country, there was a considerable increase in fruit exports (Figure 13.3).

Fruit is the primary agricultural export for Chile, with an annual average value in recent years valued at US$5,000 million. The industrial fruit sector generates around 2.5 million jobs, permanent and seasonal. The annual budget for the National Fruit Fly Program is approximately US$7 million. It is estimated that the PFA for fruit flies in Chile returns US$714 to the country for every dollar invested.

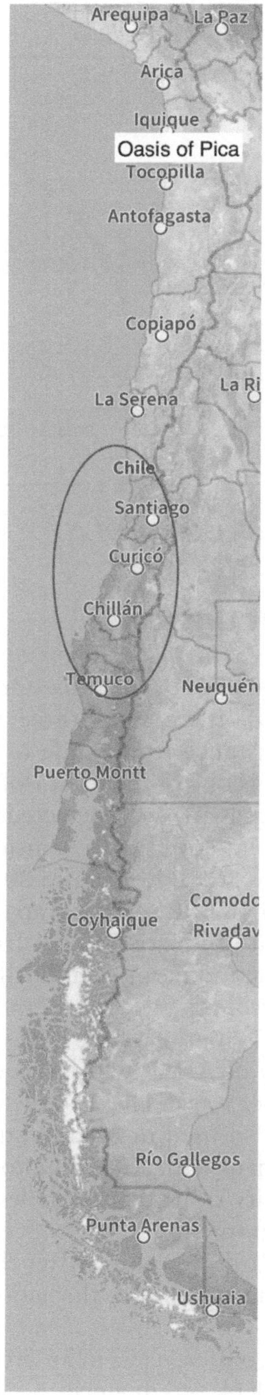

Figure 13.2 Detection of the first Medfly at the Oasis of Pica. The circle denotes the main fruit production area in Chile.

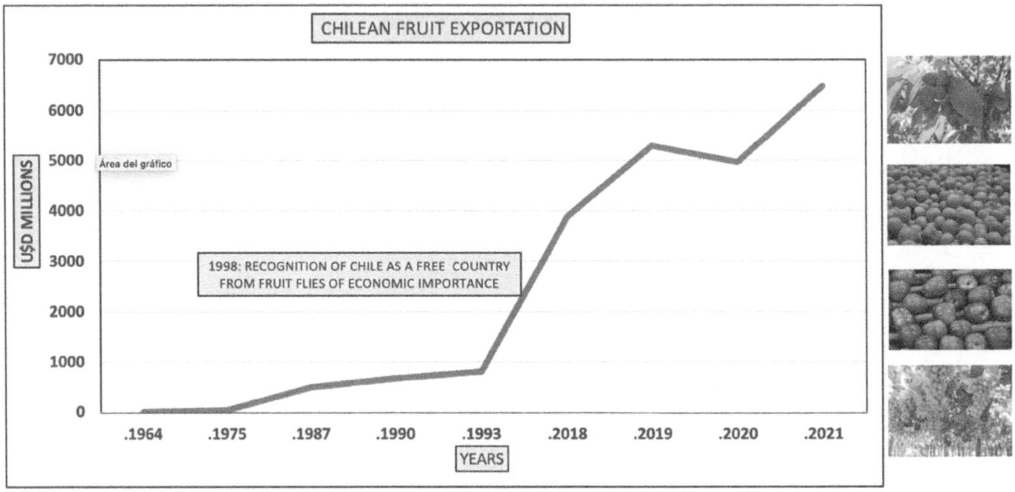

Figure 13.3 Fruit exports for Chile from 1964 to 2021.

13.2 OTHER FRUIT FLY INCURSIONS

Occasionally, there are incursions of *B. dorsalis* (Oriental fruit fly—OFF) on Easter Island. During December 2010, there were three OFF detections by the NFFDS and all were eradicated. The distance from Chile to Easter Island is about 3,800 km (2,375 miles), and Easter Island is one of the most isolated islands in the Pacific Ocean. In 1972, *Bactrocera tryoni* (Froggatt) was detected in Easter Island and eradicated (Figure 13.4). This campaign employed as a consultant Dr. M. A. Bateman from Australia (Olalquiaga and Lobos 1993, Rodríguez 2010).

The NFFDS has a network of 16,000 specific traps for tephritid species and protects about 350,000 hectares for fruit production in the fruit fly-free area (the whole country). The surveillance is constant and permanent during the year, for early detection of the possible entry of fruit flies from the genera: *Ceratitis, Anastrepha, Bactrocera* and *Rhagoletis*. In 1992 with the contribution of the IAEA and FAO, a sterile insect production center was built in the city of Arica, which has a maximum production capacity of 50 million irradiated males per week. These sterile insects were released preventively in the city of Arica and were also used in Medfly outbreaks in other places of Chile. Additionally, there is an agreement with Argentina to import irradiated Medflies from Mendoza to increase the quantity of irradiated males to be released in an eradication and as a preventive release program.

The program has some standard responses. After the detection of one fruit fly specimen (either male or not inseminated female), the actions are as follows:

- Intense surveillance for two life cycles, in an area of 64 km^2.
- Additional 930 traps are deployed in a grid.
- 200 meters around the detection site, systematic host fruit sampling is carried out (in every property).
- No quarantine is implemented.

Figure 13.4 There have been incursions of *Bactrocera* spp. into Easter Island.

The conditions for transitioning to eradication are as follows:

- Two or more fruit fly specimens are detected, or
- One inseminated female, or
- Immature stages (eggs, larvae, pupae).

When eradication is triggered, the actions are as follows:

- Control actions and regulated (quarantine) areas are implemented in a radius of 7.2 km.
- Intensive surveillance extended to 196 km^2, for three life cycles.
- Additional 1,802 traps are deployed in a grid.
- Host fruit sampling is carried out 800 m around each detection site.
- SIT release rate between 1,500 and 2,000 sterile males per hectare.
- Pesticide applied—GF-120.
- Soil treatment with a pesticide (Diazinon).
- Harvest of host fruit.

Note: Chile as a country does not lose its fruit fly-free condition, only the regulated area (quarantine area).

The NFFDS successfully maintains Chile as a fruit fly-free area and is based on five powerful supporters:

1. Professional, technical and administrative human team of excellence and committed to the Fruit Fly Program (Figure 13.5).

   There are many phytosanitary problems with "clandestine" passengers, such as insects that sometimes are not intercepted at the border line control. During 2019, approximately 6,000,000 people entered the international airport from Santiago; some passengers illegally carry fruit, some of which may be infested with fruit flies (Figure 13.6).
2. When entering Chile, passengers must sign a statutory declaration that no fruit or products of animal origin are being carried into Chile. There is an immediate fine (between US$30 and US$11,000) if the statement is falsified or not made and prohibited products are detected.
3. Strict quarantine control at each checkpoint; there are approximately 101 control points with the presence of SAG inspectors, dogs and X-ray machines (Figure 13.7).
4. There is a permanent surveillance system in the country with specific traps for the target fruit flies, and host fruit is harvested and then analyzed in search for fruit fly larvae or eggs (Figure 13.8).

Figure 13.5 Fruit fly eradication actions.

Figure 13.6 Schematic of "clandestine" passengers.

5. There is a contingency protocol to respond quickly and effectively to any detection of exotic fruit flies of economic importance in the country; the principal objective of the response plan is to eradicate them in the shortest time possible (Figure 13.9). Additionally, there is a handbook with a protocol for the detection of fruit flies of economic importance (Del Pino-Gracia and Garrido 1996; Tassan et al. 1982).
6. Chile has signed agreements on subjects related to fruit flies of economic importance with neighboring countries: Peru and Argentina.

13.3 RESULTS AND CONCLUSIONS

Chile has a successful system for fruit fly detection and response. This NFFD system and its five pillars help maintain the fruit fly status of Chile and protect Chilean exports. The NFFDS uses a network of 16,000 specific traps to protect 350,000 hectares of fruit production in the fruit fly-free area. The system provides for early detection of Medfly in Chile and of a very occasional entry of *B. dorsalis* in Easter Island (last detection 2011). Additionally, the system detected *Anastrepha fraterculus* during 2017 at Iquique, a city located in the Atacama Desert. These early detections of incursions provided the first warning of possible incursions and allowed Chile to respond in a timely manner to eradicate the incursions. The NFFDS uses a combination of surveillance, quarantine,

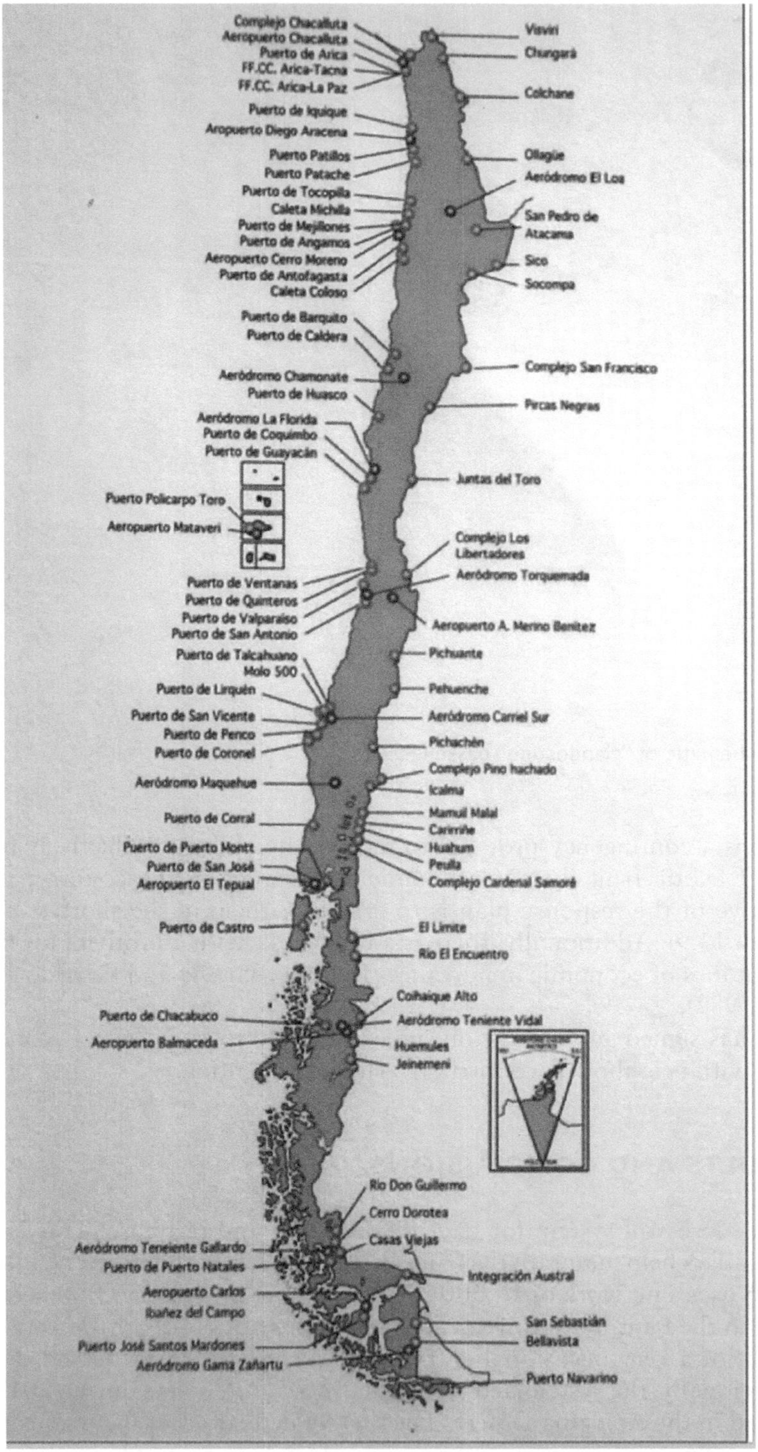

Figure 13.7 Checkpoints in Chile.

Figure 13.8 Surveillance system in Chile.

Figure 13.9 Management protocols.

and control actions to eradicate fruit flies when detected. Additionally, strict quarantine control at checkpoints, including SAG inspectors, dogs and X-ray machines, helps prevent the entry of fruit flies into Chile. The NFFDS protects Chile's primary agricultural export, generating around 2.5 million jobs and an annual average value of US$5 billion. The NFFDS has been operational for over 55 years, which allows Chile to export fruits without restrictions for fruit flies of economic importance (Olalquiaga 1968, 1979).

ACKNOWLEDGMENTS

I thank Ana María Salazar Suárez, Diana Pérez-Staples, Betzabé Verónica Murrieta, Abraham Moreno Mejía, Dinesh Rao and Bernard Dominiak for help with the manuscript and figures. I thank our present colleagues, the International Atomic Energy Agency and the Organization Committee for the 11th International Symposium on Fruit Flies of Economic Importance.

REFERENCES

Del Pino-Gracia, A., and A. Garrido. 1996. Evaluación de puesta de *Ceratitis capitata* Wied., con temperaturas variables en campo y constantes en laboratorio. *Boletín de Sanidad Vegetal. Plagas* 22 (2) (February): 401–410. https://www.mapa.gob.es/ministerio/pags/biblioteca/revistas/pdf_plagas/BSVP-22-02-401-410.pdf (articulo en español)

https://cdn.portalfruticola.com/2018/01/7bfaab8d-guia_deteccion_moscas.pdf

Olalquiaga, G. 1968. *Lucha contra la mosca del Mediterráneo en Chile.* Ministerio de Agricultura, Servicio Agrícola y Ganadero.

Olalquiaga, G. 1979. *Erradicación de la mosca del Mediterraneo en la provincia de Los Andes – V Región de Valparaíso.* Ministerio de Agricultura, Servicio Agrícola y Ganadero.

Olalquiaga, F. G., R. T. Bobadilla., H. T. Del l'Orto., L. H. Ramírez., S. F. Santa Cruz, and T. O. Miranda. 1966. La mosca del Mediterráneo en Chile *(Ceratitis capitata)*. Ministerio de Agricultura, Santiago de Chile. https://dtisartec.senasica.gob.mx:8080/biblioteca/libros/libros/Olalquiaga%20F.G.%20y%20otros%201966.pdf

Olalquiaga, F. G., and C. A. Lobos. 1993. La mosca del Mediterráneo en Chile, introducción y erradicación. Ministerio de Agricultura, Universidad Austral de Chile, Santiago de Chile. https://agris.fao.org/search/en/providers/123819/records/64735e5c53aa8c896309c02a

Rodríguez, R., and C. A. Lobos. 2010. *Plan de acciones correctivas para la Mosca del Mediterráneo Ceratitis capitata (Wied.).* Ministerio de Agricultura, Servicio Agrícola y Ganadero.

Shelly, T., N. Epsky, E. B. Jang, J. R. Flores, and R. Vargas. 2014. *Trapping and the Detection, Control, and Regulation of Tephritid Fruit Flies.* Springer, Dordrecht Heidelberg New York London.

Tassan, R. L., T. K. Palmer, K. S. Hagen, A. Cheng, G. Feliciano, and T. L. Blough. 1982. Mediterranean fruit fly cycle estimations for California eradication program. In Fruit Flies of Economic Importance, ed. R. Cavalloro. 564–569. Rotterdam: A. A. Balkema.

Chapter 14

Current status of the National Fruit Fly Detection Programme in Chile

Jocelyn Yevenes Flores, Jeremías Nitor Carrillo, Nayadeth Álvarez Ugalde, Valeska Contreras Pizarro, Ruth Castro Feria, Paulina Concha Egaña, and Marcelo Contreras Ortiz

14.1 INTRODUCTION

The increase in fresh fruit exports over the last 20 years has positioned Chile as the leading fruit exporter in the southern hemisphere and the world's leading exporter of table grapes and blueberries. The national fruit production area covers approximately 373,000 ha between Atacama and Los Lagos. The sector produces approximately five million tonnes of fruit. In January 2024, fresh fruit export volumes recorded an increase of 14.4% over the previous season; this increase was accompanied by an 8.2% increase in the value transported, reaching a total of US$1,891,729,213. This value included a wide range of fresh fruits, such as grapes, cherries, blueberries and apples.

Chile is currently the only country in Latin America that remains free of economically significant fruit flies belonging to the family Tephritidae. The Mediterranean fruit fly, *Ceratitis capitata* (Wied.) (Medfly) was detected and described for the first time in 1963 in the province of Tamarugal, Tarapacá Region (Olalquiaga and Bobadilla, 1966). Since then, Chile established a permanent management programme against this pest, which progressively eradicated from certain regions (Olalquiaga, 1979; Lobos, 1991; Olalquiaga and Lobos, 1993) until successfully eradicating from the entire country in 1995. However, Chile continually faces the potential threat of incursions (Atenas, 1990), mainly from the Mediterranean fruit fly *Ceratitis capitata* (Wied.) (Medfly) and other invasive flies of quarantine importance (Lobos, 2010). Despite this, the country has some essential advantageous natural features that hinder the natural spread of these pests. These include the Andes Mountain range in the east of the country, thousands of square kilometres of desert in the north, the Pacific Ocean in the west and, finally, an extremely cold subpolar climate in the south.

Despite Chile's isolation, the entry of tephritids takes place through fruit hidden in passengers' luggage or in the cabins of vehicles and illegal entry through unauthorised border crossings. Chile has a rigorous quarantine system consisting of qualified personnel, a canine brigade and X-ray machines at border controls at each entry point. The only exception to this isolation was the province of Arica, on the border with Peru, which is only a few kilometres away from Peru. In 2004, after 10 years of intense work with the Peruvian phytosanitary authorities, the province of Arica was declared free of Mediterranean fruit flies. Therefore, this area was incorporated into the free zone, and Chile regained its status as a fruit fly-free country. Currently, the Agriculture and Livestock Service (SAG) applies a preventive release programme through the rearing

DOI: 10.1201/9781003644330-19

and release of sterile flies (SIT) in specific areas of the Arica and Parinacota, Tarapacá (Pica), Valparaíso and Metropolitan regions; these areas are characterised by a higher probability of incursion and establishment of the pest.

In 1980, the Chilean Ministry of Agriculture (through Servicio Agrícola y Ganadero) developed the Project "Suppression, Eradication, and Exclusion of the Fruit Fly in Chile" to maintain and improve favourable phytosanitary conditions. The main objectives of this project were to keep economically significant fruit flies out of the country and eradicate any outbreaks detected in Chile (Servicio Agrícola y Ganadero [SAG], 1982).

As a result of the activities carried out by SAG's fruit fly programme, Chile can be proud of its fruit fly-free status. There are no species from the genera *Ceratitis*, *Bactrocera*, *Anastrepha* and *Dacus* in the country. Only five species of *Rhagoletis* are present in Chile, and all are related to solanaceous hosts. These species are *Rhagoletis tomatis* (Foote), *Rhagoletis nova* (Schiner), *Rhagoletis conversa* (Bretes), *Rhagoletis panela* (Foote) and *Rhagoletis brancici* (Frias) (Gonzalez, 2005). The fruit fly programme operates through the National Fruit Fly Detection System (NFDSS), which includes the detection of the pests by trapping and fruit sampling in different areas located from the Arica and Parinacota region to the Carlos Ibáñez del Campo region (Aysén) (Servicio Agrícola y Ganadero, 2013).

Being a fruit fly-free zone is an essential advantage for Chilean fresh fruit exports, as no post-harvest quarantine treatments or other restriction measures are necessary. In

Table 14.1 Number of Mediterranean fruit fly detections from 2020 to 2023

Year	*Region*	*Single captures*[a]	*Spouts*[b]	*Total*
2020	Valparaíso	1	1	2
2021	Tarapacá	1	3	4
	Valparaíso	0	2	2
	Metropolitan	0	1	1
2022	Arica and Parinacota	5	3	8
	Tarapacá	2	1	3
	Antofagasta	1	1	2
	Atacama	1	1	2
	Coquimbo	1	0	1
	Valparaíso	2	0	2
	Metropolitan	5	7	12
	O'Higgins	1	0	1
	Ñuble	1	0	1
2023	Arica and Parinacota	1	2	3
	Antofagasta	1	4	5
	Atacama	0	1	1
	Coquimbo	1	2	3
	Valparaíso	3	1	4
	Metropolitan	3	4	7

[a] Single capture: Detection of a single fruit fly specimen, which triggers intensive surveillance activities.

[b] Spout: Detection of an inseminated female or more than one adult specimen, or immature stages of the same fruit fly species within a specific area.

addition, for over 20 years, Argentina and Peru have collaborated through binational agreements to create buffer zones and eradicate the Mediterranean fruit fly from areas close to the border with Chile. This programme has been beneficial for all three countries. Discussions are currently underway to sign a similar agreement with Bolivia.

Despite the above, a few fruit flies are detected in the country annually (Table 14.1). To deal with this, Chile relies on a science-based response plan (IPRFFSP, 2006) approved by its trading partners, activated upon detection of any stage of fruit fly. Therefore, depending on the nature of the detection, there are two courses of action:

Single fly capture: This is defined as one specimen, male or female (not inseminated). When this occurs, an intensive delimiting surveillance is conducted, including high-density trapping deployed over an area of 64 km^2 and systematic fruit sampling on the backyards of each property within a 200-m radius around the detection. Surveillance lasts for two life cycles (Tassan et al. 1982).

Outbreak (multiple captures): This is defined as two or more adult specimens (either in one or more traps or events), one inseminated female, or immature stages. In this case, SAG carries out an eradication campaign, which includes intensive delimitation surveillance in an area of 196 km^2, with trapping and fruit sampling. Additional control activities are applied, such as fruit destruction, soil drenching, bait spraying and sterile male releases in a 200-m radius from the discovery site. In addition, regulated areas are established at 7.2 km from each detection point, and this activity is maintained until the completion of three biological cycles following the detection of the last adult in the regulated area (Servicio Agrícola y Ganadero, 2004).

14.2 THE NATIONAL FRUIT FLY DETECTION SYSTEM (NFDSS)

The main objective of the NFDSS is to detect early entry of exotic fruit fly and minimise the risk of passive spread. The programme extends from the border with Peru (18° south latitude, I Region) to southern Chile (47° south latitude, XI Region). This detection system includes all towns and cities in that area and rural areas where fruit fly hosts are present.

Based on the NFDSS Technical Guidelines (Servicio Agrícola y Ganadero, 2013), the system has 16,767 traps installed in the territory. The traps and attractants used are listed in Table 14.2. These traps are superimposed on the plan of the area analysed, a

Table 14.2 Attractant and trap type according to the fruit fly target species

Fruit fly target species	*Attractant*	*Trap*
Ceratitis capitata (males)	Trimedlure	Jackson
Ceratitis capitata (females)	Hydrolysed protein + borax Biolure[a]	McPhail Multilure
Anastrepha spp. and others	Hydrolysed protein	McPhail
Bactrocera spp. (mainly *Dorsalis* Complex)	Methyl eugenol	Steiner
Bactrocera spp. (mainly *B. tryoni*)	Cuelure	Steiner
Zeugodacus cucurbitae	Cuelure	Steiner

[a] Its use is in outbreak.

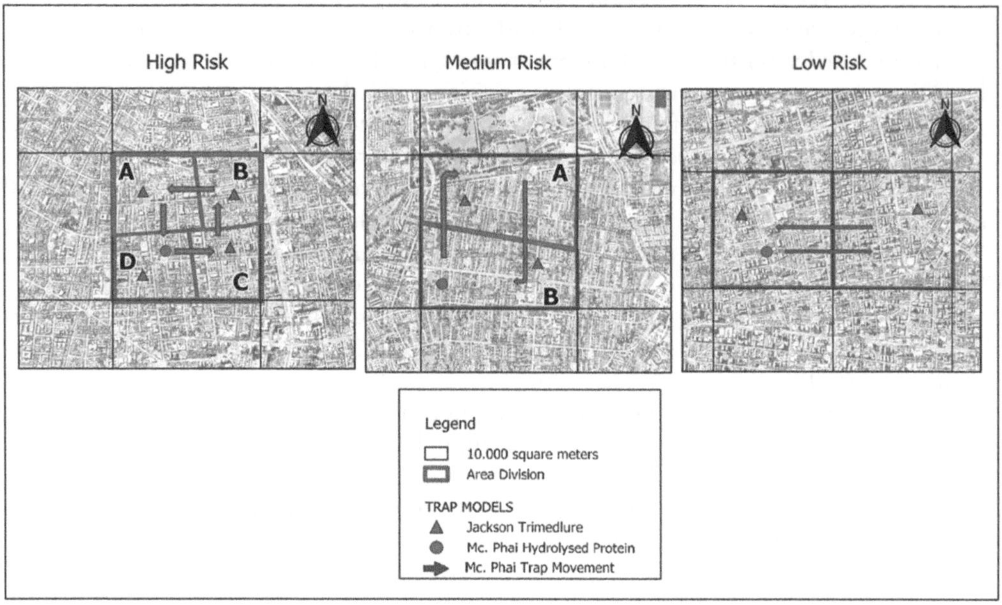

Figure 14.1 Trap distribution according to the risk of detecting a fruit fly adult.

grid of 100-ha (1 km²) quadrants (Figure 14.1). Each 100-ha quadrant is divided into two sub-quadrants of 50 or four sub-quadrants of 25 ha. The subdivision of quadrants allows for a systematic and homogeneous coverage of the areas. These quadrants facilitate the location and monitoring of the traps, organising the surveillance of each trap installed in the area, thus ensuring that all the trappable areas are monitored (López, 1991). This system facilitates the programmed movement of the traps according to the phenology of the available hosts, maintaining surveillance throughout the calendar year (Stubbs, 1986). The 100-ha quadrants were defined for each region of the country using the fruit fly programme's Geographic Information System (GIS), which assigns each fruit fly trap site a unique, unrepeatable and unmodifiable number. The datum is in WGS84.

14.2.1 Deployment of traps in urban areas

In urban areas of the country, where fruit fly host species are abundant in gardens and streets, traps are placed according to a grid system with uniform distribution. This system considers trap lines to ensure adequate coverage in these highly populated areas (Lobos, 1991) (Table 14.3).

For the distribution of the traps in the field, a grid of 100 ha (1 km²) quadrants must be superimposed on the plan of the analysed area. Therefore, each 100 ha quadrant can be divided into two sub-quadrants of 50 ha or four sub-quadrants of 25 ha, thus facilitating the work of distributing the traps. The 100-ha quadrants for each region of the country are previously defined by the GIS of the fruit fly programme. Each quadrant is assigned a unique, unrepeatable and not adjustable specific number. The datum of this layer is in WGS84 (Servicio Agrícola y Ganadero, 2013).

Table 14.3 Trap density for urban areas

Attractant	*High-risk area*	*Medium-risk area*	*Low-risk area*
Trimedlure	1 t/25 ha	1 t/50 ha	1 t/100 ha
Methyl eugenol	1 t/250 ha	1 t/1,000 ha	1 t/2,500 ha
Cuelure	1 t/250 ha	1 t/1,000 ha	1 t/2,500 ha
Hydrolysed protein + borax	1 t/100 ha	1 t/100–150 ha	1 t/150–500 ha

Each sub-quadrant shall have one TML trap in high-risk areas, relocated four times a year but always within the allocated area (25 ha). Each square shall have one protein trap relocated within the assigned area (100 ha). In addition, fixed traps are placed in high-risk areas (hot spots) that have yet to be relocated (Figure 14.1).

14.2.2 Deployment of traps in rural areas

Surveillance traps in rural areas (less populated localities, commercial orchards with a regular insecticide control programme, or areas with scattered commercial orchards) are placed along paths, giving easy access to host trees, according to Table 14.4.

According to the instructions for Planning Activities and the Technical Guidelines in the NFDSS, the country has a sensitive network of the 16,767 traps for timely detection as listed in Table 14.5 in the regions of Arica and Parinacota in the north of Chile and Carlos Ibáñez del Campo, Aysén.

14.3 DNA ANALYSIS

Upon detection of a *C. capitata* specimen, DNA analysis is carried out to determine the possible origin of this incursion. This analysis is based on the PCR/RFLP technique using four markers, which allows for the identification of different haplotypes (Lobos, 1995). Additionally, implemented since 2000, this technique has allowed for the determination of the routes with the highest risk of Medfly entry. The most common haplotypes detected are AABB, BBBB and AACB. According to the results, the most likely sources are the Andean countries, which have the highest frequency of the AABB haplotype, while the Atlantic coast countries tend to record the BBBB and AACB haplotypes. However, although the AABB haplotype is more commonly found in Central America than in South America, it is rarely detected in the samples analysed, suggesting a broader risk not limited to neighbouring countries.

14.4 THE USE OF THE STERILE INSECT TECHNIQUE (SIT)

In 1987, an experimental release was carried out using Medfly material from Hawaii to eradicate the pest in Arica, northern Chile. Between 1988 and 1990, the SIT was applied using pupae from the United States (Hawaii), Guatemala and Mexico. In 1992, with the support of the International Atomic Energy Agency (IAEA) and the Food and

Table 14.4 Trap density for rural areas (trap/ha)

Attractant	*High-risk area*	*Medium-risk area*	*Low-risk area*
Trimedlure	1 t/50–100 ha	1 t/100–150 ha	1 t/150–300 ha
Methyl eugenol	1 t/1,000 ha	1 t/1,500 ha	1 t/2,000 ha
Cuelure	1 t/1,000 ha	1 t/1,500 ha	1 t/2,000 ha
Hydrolysed protein + borax	1 t/150 ha	1 t/300 ha	1 t/500 ha

Table 14.5 Total number of fruit fly traps installed in Chile

Region	*Attractant*					*Total*
	TML	*Protein*	*3 Components*	*Cuelure*	*Methyl*	
Arica and Parinacota	161	124	486	8	7	786
Tarapacá	339	128	86	10	10	573
Antofagasta	718	152	0	6	6	882
Atacama	386	119	0	16	16	537
Coquimbo	1919	544	39	27	28	2,557
Valparaíso	2867	867	0	128	129	3,991
Metropolitan	3214	924	5	136	136	4,415
B. O'higgins	920	271	0	31	31	1,253
From Maule	425	177	1	22	0	625
Ñuble	221	105	0	15	0	341
Bio-Bio	92	21	0	5	0	118
Araucanía	171	72	0	0	0	243
De los Rios	122	57	0	0	0	179
From Los Lagos	168	75	0	0	0	243
C.Ibáñez Del Campo-Aysén	17	7	0	0	0	24
Total	**11,740**	**3,643**	**617**	**404**	**363**	**16,767**

Agriculture Organisation of the United Nations (FAO), the Sterile Insect Production Centre (CPIE) was built in the Lluta Valley, Arica Province, to mass rear a bisexual strain. By 1995, the definitive eradication of *C. capitata* from Arica was achieved, allowing Chile to declare its territory free of fruit flies by integrating SIT as a control tool.

In 1996, the first genetic sexing strain was introduced for the release of males only. Currently, the SIT is used as a preventive measure in Arica, Pica, Los Andes and Santiago. The insects are produced at the CPIE, with 500–1,000 males per ha being released twice a week as a preventive measure. Control activities, supervised through the Product Quality Control for Sterile Mass-Reared and Released Tephritid Fruit Flies (FAO/IAEA/USDA, 2019), ensure that the material meets or exceeds international standards. The strain released is the Vienna-8 (*wp*, *tsl*). The insects are released in paper bags on predefined routes, which include 80 bag opening points.

14.5 CONCLUSIONS

The fruit industry in Chile is a fundamental pillar of the national economy, therefore maintaining fruit fly-free status is a government priority. Chile has effective natural isolation that prevents the pest from arriving by direct flight/dispersal from neighbouring countries. However, Chile faces the challenge of constant biological pressure due to the high prevalence of the pest in other countries in the region, which generates transient Mediterranean fruit fly events.

Against this background, efforts are organised on three levels:

- **Beyond borders:** Increasing work and exchange of knowledge and experience on economically significant fruit flies through bilateral agreements, internships and training with neighbouring countries such as Peru, Argentina, Bolivia and Brazil.
- **At the border:** A strict system of inspections and quarantine must be strengthened in the different regions of the country due to the high pressure of illegal entry of agricultural products and unauthorised border crossings, where fruit fly-infested host species are entering. Because of this situation, efforts are focused on increasing inspection points and road controls.
- **Within the country:** Despite having a susceptible detection system that allows early detection of fruit fly specimens entering the national territory, scientists are working to increase this sensitisation capacity. To this end, regulators have considered classifying the territory into two risk categories: high and medium, depending on the region. Additionally, scientists are looking to increase the density of food traps for more effective detection. This strategy is based on the increasing pressure of the pest's entry and the changes in the country's climatic conditions, which are becoming increasingly favourable for establishing these species (Hill, 1986).

In addition, the SIT is used preventively in the region of Arica and Parinacota, which borders Peru. Also, SIT is used in the Metropolitan Region, Pica (Tarapacá Region), and the commune of Los Andes (Valparaíso Region), particularly in the land port sector. Currently, work is underway to increase the hectares treated with SIT by 100,000 ha.

Finally, DNA analysis is carried out to identify the risk routes and probable origin of the specimens detected, contributing to this pest's effective prevention and control.

REFERENCES

Atenas, R. L. 1990. *Informe campaña de erradicación de la mosca del Mediterráneo (Ceratitis capitata (Wied.)) en la Ciudad de Los Andes, V Región* [Unpublished internal document]. Santiago, Chile: Ministerio de Agricultura, Servicio Agrícola y Ganadero (SAG).

FAO/IAEA/USDA. 2019. *Product Quality Control for Sterile Mass-Reared and Released Tephritid Fruit Flies, Version 7.0*. Vienna, Austria: International Atomic Energy Agency.

Gonzalez, J. 2005. *Guía para la identificación de Tephritidae de Chile y reconocimiento de géneros de importancia cuarentenaria*. Santiago, Chile: Servicio Agrícola y Ganadero, Departamento de Protección Agrícola, Proyecto Moscas de la Fruta.

Hill, A. R. 1986. Choice of insecticides in Steiner traps affects the capture rate of fruit flies (Diptera: Tephritidae). *Journal of Economic Entomology* 79(2): 533–536.

IPRFFSP – International Panel for Review of Fruit Fly Surveillance Programs. 2006. *Review of the Fruit Fly Surveillance Programs in the United States*. Riverdale, MD: USDA/APHIS/PPQ/ Fruit Fly Program.

Lobos, A. C. 1991. *Informe campaña de erradicación mosca del Mediterráneo (Ceratitis capitata* (Wied.) *[Diptera: Tephritidae]) de la Ciudad de Coquimbo, IV Región (Informe interno no publicado)*. Santiago, Chile: Ministerio de Agricultura, Servicio Agrícola y Ganadero.

Lobos, C. 1995. *Guía para la detección de moscas de frutas (Diptera: Tephritidae)* (1ª ed). Santiago, Chile: Ministerio de Agricultura de Chile, Servicio Agrícola y Ganadero.

Lobos, C. 2010. *Guía para la Detección de Moscas de la Fruta de Importancia Económica (Diptera: Tephritidae)* (4ª ed.). Santiago, Chile: Ministerio de Agricultura de Chile, Servicio Agrícola y Ganadero.

López, L. M. 1991. Evaluación de trampas para moscas del Mediterráneo. In *V Curso Internacional sobre moscas de la fruta, Módulo III, métodos de detección*, 11–17. Metapa de Domínguez, Chiapas, México: Centro Internacional de Capacitación en Moscas de la Fruta.

Olalquiaga, G. 1979. *Erradicación de la mosca del Mediterráneo en la Provincia de Los Andes*. Santiago, Chile: Ministerio de Agricultura, Servicio Agrícola y Ganadero.

Olalquiaga, G., and R. Bobadilla. 1966. *La mosca del Mediterráneo en Chile (Ceratitis capitata)*. Santiago, Chile: Boletín técnico Nº 20- Ministerio de Agricultura de Chile, Dirección de Agricultura y Pesca.

Olalquiaga, G., and C. Lobos. 1993. *La Mosca del Mediterráneo en Chile, Introducción y Erradicación (Monografía)*. Santiago, Chile: Servicio Agrícola y Ganadero.

Servicio Agrícola y Ganadero (SAG). 1982–1987. *Instructivos, circulares, charlas y notas técnicas referentes al Sistema Nacional de Detección de moscas de la fruta en Chile* [Documentos no publicados de uso interno]. Ministerio de Agricultura de Chile.

Servicio Agrícola y Ganadero. 2004. *Plan de emergencia para moscas de la fruta*. Publicación Interna del SAG.

Servicio Agrícola y Ganadero. 2013. *Directrices técnicas del Sistema Nacional de Detección de Mosca de la Fruta (SNDMF)* [Documento general]. Programa Nacional Moscas de la Fruta, Ministerio de Agricultura.

Stubbs, E. J. 1986. *Procedimientos para detección de plagas exóticas y moscas de las frutas*. Organismo Internacional Regional de Sanidad Agropecuaria (OIRSA), Depto. de Sanidad Vegetal.

Tassan, R. L., K. S. Hagen, A. Cheng, T. K. Palmer, G. Feliciano, and T. L. Bough. 1982. Mediterranean fruit fly life cycle estimations for the California eradication program. In *Fruit Flies of Economic Importance. Proceedings of the CEC/IOBC International Symposium CEC/IOBC Symposium, Athens*, ed. R. Cavalloro, 564–570. Rotterdam, the Netherlands: A. A. Balkema.

Chapter 15

National Fruit Fly Control and Eradication Program (Procem) in Argentina

Esteban Garavelli, María Florencia Vazquez, and Paola Fedyszak

15.1 NATIONAL FRUIT FLY CONTROL AND ERADICATION PROGRAM (PROCEM)

Fruit flies (Tephritidae) represent one of the main pests for the Argentine fruit and vegetable industries. Fruit flies cause a significant direct impact on production (direct losses) and also impose restrictions on national and international trade and the need for phytosanitary measures for the mobilization of host products. There are two species in Argentina of economic importance: the Mediterranean fruit fly (*Ceratitis capitata* Wied.) and the South American fruit fly (*Anastrepha fraterculus* Wied.). The creation of the Argentine Institute of Plant Health and Quality (IASCAV, for its Spanish acronym) in January 1992 led to the integration and strengthening of the provincial actions developed until then against the pest and culminated in the creation of the National Fruit Fly Control and Eradication Program (Procem, for its Spanish acronym), through ex-IASCAV Resolution No. 134/94. Currently, Procem is under the coordination of the National Service for Agrifood Health and Quality (SENASA, for its Spanish acronym) with decentralized execution through provincial and municipal organizations, according to the governance criteria of each region. Additionally, technical approaches are adapted to the incidence of the pest in each region.

Following an initial analysis of the agroecological conditions and the productive potential of the fruit-growing areas in Argentina, it was determined that the geographic isolation conditions of the Central and Southern Oases of the province of Mendoza and the Productive Valleys of the Patagonian Region presented conditions favorable to undertake an eradication project (ISPM 10 1999). The natural isolation consists of semidesert areas with xerophilous vegetation (without hosts to fruit flies) that surround the irrigated productive valleys, the Andes Mountains as the western limit and the Atlantic Ocean as the eastern limit for the Patagonian Region. The climatic conditions of these regions make winter survival of the pest difficult, in addition to the scarce number of winter hosts (citrus fruits), which is practically limited to urban areas (ISPM 37 2016).

Fruit growing is the main economic activity in the Patagonian Region and the Central and Southern Oases of Mendoza, contributing significantly to the generation of added value and employment. In the international market, Argentina's position assumes relevance due to its counter-seasonal character in relation to the northern hemisphere. This favorable trade position drove the implementation of eradication actions, with the aim of improving the conditions for international market access of fruits and vegetables.

DOI: 10.1201/9781003644330-20

In other regions, such as northeastern Argentina, where eradication is not considered feasible due to agroecological conditions, a suppression strategy is applied with the aim of reducing the direct damage caused by the pest.

15.2 PROGRAM OVERVIEW

15.2.1 Goals

The National Fruit Fly Control and Eradication Program (Procem) has the following general objectives:

- Protect the socioeconomic heritage of fruit and vegetable resources and agroecosystems threatened by the establishment of populations of fruit flies and related economically important dipterans.
- Reduce direct economic damage to fruit and vegetable production (yield and quality) and indirect damage (quarantine restrictions linked to access to internal and external markets) caused by *C. capitata* and *A. fraterculus*, through integrated pest management based on environmentally friendly control methods.
- Maintaining international recognition and the health of fruit produced in fruit fly-free areas.

The specific objectives of the program are as follows:

- Standardize surveillance and phytosanitary procedures at the national level, in all areas under the program.
- Knowledge transfer to producers and other actors involved in the fruit and vegetable chains.

15.2.2 Current program description and components

The Procem bases its activities on the official surveillance system, which consists of trapping and fruit sampling. The control actions implemented by the Program depend on the results of these activities, which may include chemical control, cultural–mechanical practices, mass trapping or bait stations and the sterile insect technique (SIT).

15.2.2.1 Surveillance

The purpose of surveillance activities is to

- Determine the presence or absence of the pest in the adult or immature stages (larvae, pupae and eggs);
- Determine spatial distribution, seasonal variation, relative abundance and other indicators.

In this regard, two types of actions are carried out:

Trapping: The objective of trapping activities is to determine the presence or confirm the absence of adult flies and to analyze the population dynamics through the FTD index (Flies per Trap per Day) and distribution. Generally, traps are checked weekly. Trap distribution is 1 trap/km^2 in rural areas and 2–5 traps/km^2 in urban areas and risk points (quarantine checkpoints, airports and fruit concentration markets).

Jackson-type traps with a sexual attractant and McPhail or Multilure-type traps with food attractants are used. Also, specific attractants are used to confirm the absence of economically important species of the *Bactrocera* genus, which are absent in Argentina. Currently, more than 11,000 traps are in operation in the different regions under the program.

Fruit sampling: The objective of fruit sampling is to detect the presence of immature stages of the pest. Samples are taken from fruits that show symptoms of infestation (directed sampling).

Both surveillance procedures are aligned with the International Standards for Phytosanitary Measures (ISPM) (ISPM 26 2006).

15.2.2.2 Control

Control activities aim to

- Mitigate the impact of the pest by reducing its population levels.
- Carry out preventive control actions and/or eradication of outbreaks in free and/or low pest prevalence areas.

Chemical control: Biological bait insecticides are used, which are environmentally friendly and harmless to humans. They are used in urban areas, in backyards, and in large rural areas where fruit and vegetable production is taking place.

Cultural–mechanical control: This is based on the collection of fruits that are on the ground or that remain unharvested on the trees. Then, the collected material is denatured and/or destroyed.

Mass trapping or bait stations: They are based on placing a high density of traps or bait stations with an attractant and an insecticide to capture or kill as many adults as possible. It is recommended that the attractant used be directed especially at females, which will reduce the general infestation of the fruit. It is used in both rural and urban areas as a preventive or suppression control method.

SIT: This tool consists of the large-scale release of sterile males of *C. capitata* that do not produce offspring when mated with wild females. In this way, the biological cycle of the species is interrupted, allowing wild populations to be reduced in areas where the pest is present at low levels. Therefore, SIT minimizes the risk of pest multiplication and survival of those fertile wild flies that could enter the protected areas recognized as "pest-free areas".

Since 2019, a technological advance in the SIT has been implemented, in line with the main programs worldwide. This consisted of the adoption of the Chilled Adult Release Technique (FAO/IAEA/USDA 2019) for the preventive release of sterile flies in pest-free areas, which offers numerous comparative advantages over traditional release in paper bags. The chilled adult technique has been gradually adopted by the province of Mendoza and the Patagonian Region. This technique provides greater competitiveness and efficiency of the released insects that are subjected to less stress and have better conditions before release (hydration and food). Also, this technique represents an advantage for the environment, since the use of paper bags is eliminated.

Mass-rearing facilities: For the production of these sterile flies, the program has two mass-rearing facilities. One facility is located in the province of Mendoza, with an estimated potential production capacity of 600 million pupae of *C. capitata* per week. Additionally, work is being done currently on modular production lines for *Anastrepha fraterculus,* the spotted-wing fly, *Drosophila suzukii* (Matsumura), and other pests of interest for the development of SIT. These pests include the cattle screwworm, *Cochliomyia hominivorax* (Coquerel) and Mosquito, *Aedes aegypti* Linnaeus. The second facility is located in the province of San Juan, with a potential production capacity of 40 million pupae of *C. capitata* per week, in addition to having a production of the parasitoid *Diachasmimorpha longicaudata* (Ashmead) in the experimental stage.

15.2.2.3 Quarantine system

The main objective of the Quarantine Protection System is to prevent the entry of infested fruits into protected regions. It is made up of a series of Quarantine Control Posts (land checkpoints and airports) located at the entry points to those areas where the program is implemented with eradication objectives. At the land posts (phytosanitary barriers), all vehicles entering are controlled, inspected and disinfected. This process is carried out 24 hours a day, 365 days a year.

The entry of host plant products is regulated by SENASA Resolution No. 472/14, which establishes the application of quarantine treatments (methyl bromide or cold) carried out in centers officially authorized. Currently, there are 21 treatment centers authorized for this purpose.

15.2.3 Current intervention areas and their phytosanitary condition

Currently, the area covered by the program is approximately one million hectares. The phytosanitary conditions (pest situation) achieved to date in the different areas are presented in Figure 15.1.

The current regional programs are described below.

15.2.3.1 Procem Patagonia

Area of intervention: Procem Patagonia is implemented in the productive valleys of the provinces of Río Negro, Neuquén, Chubut and Santa Cruz.

Main production: pome fruit (pears and apples), cherries, stone fruits and grapes for winemaking.

Phytosanitary condition: Fruit fly-free area. This condition was achieved in 2006 and was recognized internationally by the USA, China and Chile, among others (Figure 15.2).

Objectives: Maintaining the phytosanitary condition as a fruit fly-free area and its international recognition. This status provides comparative advantages when exporting host fruits produced in the region. Based on this phytosanitary condition, fruit producers in the region avoid costly quarantine treatments when exporting to markets with restrictions, and the products maintain their high quality.

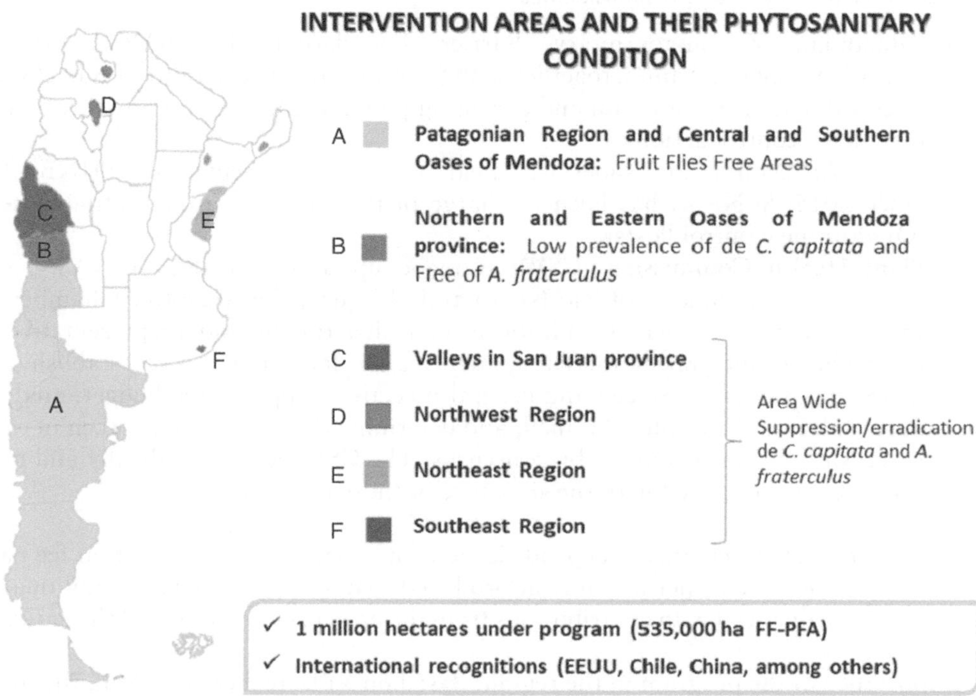

Figure 15.1 Fruit fly phytosanitary status in the different areas of fruit production in Argentina.

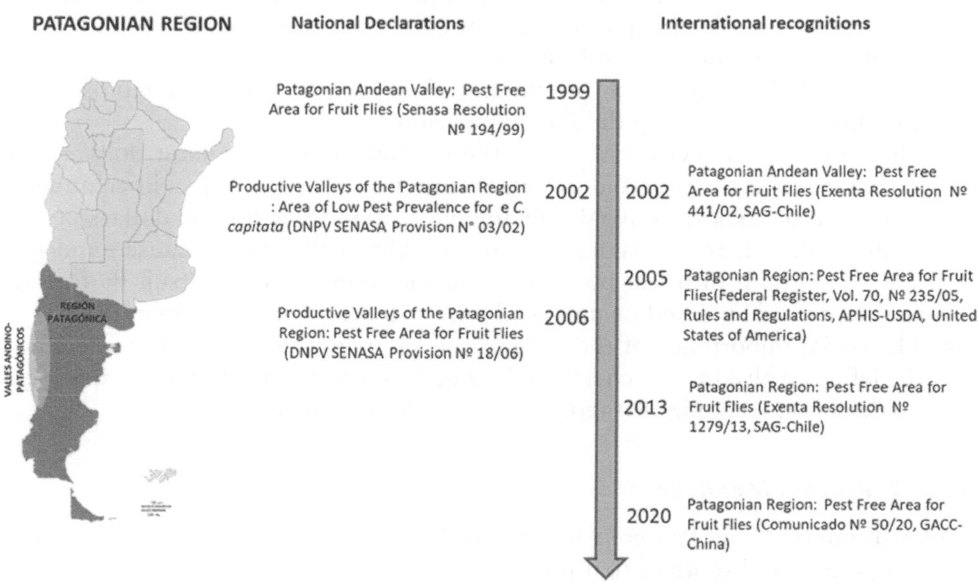

Figure 15.2 Pest-free areas in Patagonia – National declarations and international recognition.

15.2.3.1.1 Actors, roles and responsibilities

- Patagonian Zoophytosanitary Barrier Foundation (FUNBAPA, for its Spanish acronym): defines, together with SENASA, the field actions. It is in charge of coordinating the program and executing pest surveillance and preventive phytosanitary control actions.
- **SENASA:** defines and supervises field actions implemented by FUNBAPA. Since 2015, SENASA has been in charge of the Patagonian quarantine system (Quarantine Control Posts).
- **Plant Health Commission (CSV)**: is made up of FUNBAPA, SENASA, the Federation of Producers of Río Negro and Neuquén, the Argentine Chamber of Integrated Fruit Growers (CAFI, for its Spanish acronym) and a representative of the provinces integrating the Patagonian region. Its function is to establish and define the program strategy and general objectives, approve the budget, analyze proposals and the results obtained, and determine the actions to be taken in each case. All decisions taken are by consensus. The CSV meets periodically and proceedings are generated with the signature of those attending.

Sources of financing: Financing depends largely on a mandatory contribution fee paid by the production sector per ton of fruit produced that leaves the region. Additionally, the program is financed with contributions from the provinces involved and the national state.

Importance of the program in the region: Based on Villarreal et al. (2023), the main economic indicators are as follows:

- The benefit/cost ratio, since the recognition of the fruit fly-free area, indicates an average value of 17.6. Therefore, for every dollar spent on Procem Patagonia (FUNBAPA) and the Quarantine System (barriers), 17.6 dollars are generated in the region in terms of exports and savings on quarantine treatment and logistics when exporting through Chilean ports.
- Total FOB (Free on board) revenue for cherry exports from Patagonia reached 24 million dollars during the 2020/21 season.
- The fruit export sector has saved the equivalent of 27 million dollars in the period 2006–2016 by avoiding quarantine treatments (T107 cold treatment in transit). Fruit exporters avoid spending more than three million dollars per year. Additionally, there are savings in land freight in the USA because exports are able to arrive at different ports (without the recognition as a fruit fly-free area, exported produce could be unloaded only at the port of Philadelphia).
- The social importance of the program in the region is evident because 13% of the labor involved in the cherry and pome fruit chain is related to the destination markets that recognize Patagonia as a fruit fly-free area.

15.2.3.2 Procem Mendoza

Area of intervention: Procem Mendoza includes the productive oases of the north, east, center and south of that province.

Main production: Grapes for winemaking and fresh consumption, stone fruit (peaches, nectarines, plums), olives and cherries.

15.2.3.2.1 Phytosanitary condition

- **Northern and Eastern Oasis:** Low prevalence area of *Ceratitis capitata* since 2002.
- **Central and Southern Oasis:** Fruit fly-free area since 2006 (Figure 15.3).

Objectives: to maintain the phytosanitary condition of a fruit fly-free area and its international recognition, which grants comparative advantages when exporting host fruits produced in the region. Additionally, to advance in the eradication of the pest in areas of low pest prevalence, with the medium-long-term objective of achieving the status of fruit fly-free areas (ISPM 10 1999).

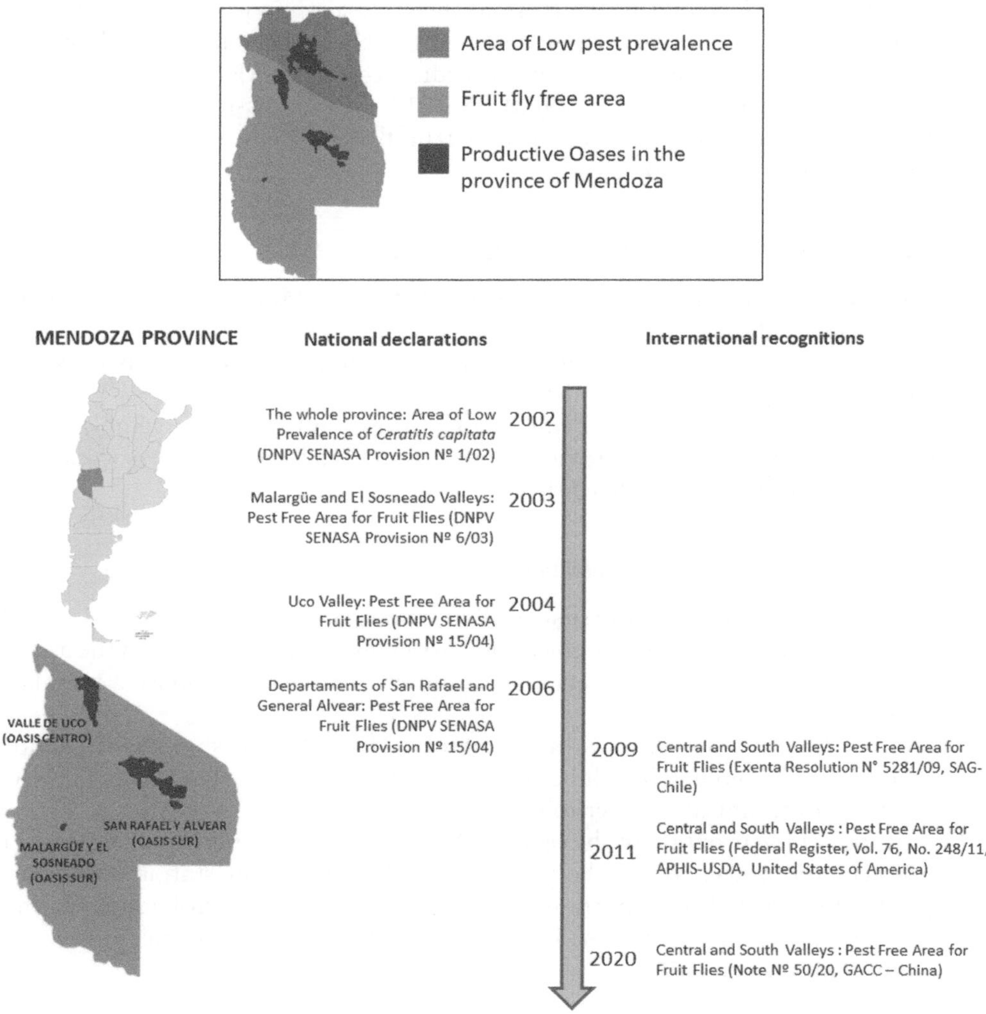

Figure 15.3 Phytosanitary conditions in Mendoza – National declarations and international recognition.

15.2.3.2.2 Actors, roles and responsibilities

- **Province of Mendoza:** the field actions (surveillance and phytosanitary control measures), the "Santa Rosa" Mass-Rearing Facility (main supplier of sterile insects in Argentina) and the Quarantine System (sanitary barriers) are carried out by the Mendoza Institute of Agricultural Health and Quality (ISCAMEN, for its Spanish acronym), dependent on the provincial government.
- **SENASA:** supervision of the operational actions of Procem Mendoza based on an annual schedule.

Sources of Financing: the financing of the field operations, the Mass-Rearing Facility and the Quarantine System depends entirely on the Provincial State government.

Importance of the Program in the Region: the strategic importance of the program for the Province is that, from its free areas, it can freely access international and local markets without quarantine restrictions regarding the pest. Also, at the local level, it implements a Systems Approach (ISPM 35 2012) in the low-prevalence areas that allows the movement of host fruits toward the fruit fly-free areas. This makes the province a strategic commercial node.

15.2.3.3 Procem NEA (Northestearn Argentina)

Area of intervention: fruit-growing area in the districts of Monte Caseros (Corrientes), Federación and Concordia (Entre Ríos).

Main production: sweet citrus fruits (oranges and tangerines) and blueberries.

Phytosanitary condition: area under suppression.

Objectives: to mitigate the direct impact of the pest on fruit and vegetable production in the region. Also, to provide an alert system to local producers to carry out phytosanitary measures efficiently.

15.2.3.3.1 Actors, roles and responsibilities

SENASA: directly carries out pest surveillance actions.

Organizations from the Provinces of Corrientes and Entre Ríos and the association of producers: implement phytosanitary measures to suppress population levels of the pest.

Sources of funding: SENASA is in charge of and funds pest surveillance actions, which include monitoring, laboratory personnel and supplies. Also, SENASA coordinates technical transfer to producers.

Importance of the program in the region: Over the years of application, the program has achieved a decrease in the fruit fly population and infestation levels in the production areas. It is important to carry out area-wide control actions because this fruit-growing area (NEA) is the main supplier of sweet citrus to the entire country all year round.

15.2.3.4 Procem San Juan

Area of intervention: productive valleys of Tulum, Zonda, Ullum, Jáchal, Iglesia and Calingasta.

Main production: grapes for fresh consumption and for winemaking, olives, pome and stone fruit.

Phytosanitary condition: area under suppression.

Objective: to mitigate the direct impact of the pest on fruit production in the province and indirectly protect areas of low pest prevalence and fruit fly-free area receptors of the fruit.

15.2.3.4.1 Actors, roles and responsibilities

- **Province of San Juan:** field surveillance actions, the "San Juan" Mass-Rearing Facility and the Quarantine System (Sanitary Barriers) are carried out by the Secretariat of Agriculture, Livestock and Agroindustry dependent on the provincial government.
- **SENASA:** supervision of the operational actions of Procem San Juan based on an annual schedule.

Sources of financing: the financing of the program, Mass-Rearing Facility and Quarantine System depends entirely on the Provincial State government.

Importance of the program in the region: by reducing the impact of the pest, opportunities are generated to promote the development of the regional economy.

15.2.3.5 Procem Jujuy

Areas of intervention: Cities of Yuto and Palma Sola, and the Region of Temperate Valleys of the province of Jujuy.

Main production: citrus, peaches and subtropical fruits.

Phytosanitary condition: area under suppression.

Objectives: to implement an alert system for citrus producers to apply control measures (Yuto and Palma Sola) and to generate temporarily fruit fly-free sites of production for early ripening peaches (Temperate Valleys) for trade in the protected areas.

15.2.3.5.1 Actors, roles and responsibilities

- Jujuy Province: directly carries out pest surveillance actions.
- SENASA: collaboration in surveillance and supervision actions.

Sources of financing: all of the program's actions are the responsibility of the provincial government.

15.2.3.5.2 Importance of the program in the region

- The alert system on fruit fly population levels gives citrus producers the possibility of carrying out control actions at the correct time.
- In the Temperate Valleys region, there is a model of replacing tobacco production with extra-early ripening peach varieties. These varieties, due to their production time, would escape the impact of the pest and, under controlled and certified

conditions, would allow their trade into pest-free areas. This situation generates a considerable commercial opportunity, and an analysis is underway to evaluate the possibility of applying a scheme of fruit fly-free sites/places of production, following the guidelines of ISPM No. 10.

15.2.3.6 Procem Southeast of the province of Buenos Aires

Area of intervention: kiwifruit-producing areas of General Alvarado and General Pueyrredón Districts.
Main production: kiwifruit.
Phytosanitary condition: area with pest surveillance actions.
Objectives: to achieve the recognition of kiwifruit-producing units as fruit fly-free sites of production, initially in the productive units in General Alvarado district, which will allow the shipment of kiwifruit to protected areas without the need for quarantine treatments.

15.2.3.6.1 Actors, roles and responsibilities

- Province of Buenos Aires, Municipalities (General Alvarado and General Pueyrredón districts) and Association of producers: carry out pest surveillance actions.
- **SENASA:** supervises pest surveillance actions and certifies the shipment of kiwifruit to protected areas on a pilot scale.

Source of financing: The surveillance system depends entirely on the union of municipal governments and producers.

Importance of the program in the region: It is expected to adapt the work area according to the regulatory framework of ISPM 10 "Establishment of pest-free places of production and pest-free production sites".

15.2.3.7 Procem Calchaquí Valleys of Salta

Area of intervention: currently, it is being implemented at the property level (at the request of some producers) in the towns of Tolombón and San Antonio, Province of Salta.
Main production: wine grapes (Tolombón) and walnut (San Antonio).
Phytosanitary condition: suppression at the production unit level.
Goals: contribute to the reduction of qualitative and quantitative losses in wine production and small walnut producers.

15.2.3.7.1 Actors

- SENASA conducts surveillance and taxonomy identification actions in the laboratory.
- National Institute of Agricultural Technology (INTA, for its Spanish acronym) conducts surveillance actions and recommendations for phytosanitary control measures.

Funding sources: survey supplies are provided by SENASA while personnel are shared between SENASA and INTA.

Importance of the program in the region: improving the phytosanitary conditions of the region to boost regional economies. Reducing the adverse impact on family economies and supporting walnut production.

15.2.4 Other Procem's actions

15.2.4.1 Actions against Drosophila suzukii

Although *Drosophila suzukii* is a dipteran species that does not belong to the family Tephritidae, it causes similar damage and therefore was incorporated into Procem's activities. The pest was detected in Argentina in 2014 (Cichón et al. 2015), and given its biology and capacity as a pest species to cause damage to healthy fruit, a national response was initiated. Procem developed surveillance actions through the implementation of a network of specific traps throughout the national territory (from Salta to Santa Cruz) in susceptible crops. The greatest impact was reported in the Patagonian Region, mainly in the Andean Region in fine fruit production and also in cherry production in eastern Patagonia and Mendoza. Therefore, the coordinated actions for its management were concentrated there.

Area of intervention: initially, traps were installed at different points throughout the country to determine its presence. Currently, surveillance is concentrated in the Patagonian Region and Mendoza.

15.2.4.1.1 Actors, roles and responsibilities

- SENASA: surveillance, coordination and identification in laboratories.
- Association of producers, municipalities, research organizations: to design and coordinate an intervention strategy to reduce population levels and economic impact of the pest.

Source of funding: SENASA provides supplies for surveillance, personnel for trap inspection and identification and information systems for recording information.

Importance of the actions: the surveillance activities determined the presence and distribution of the pest in the country and allowed to focus on the actions on the sites with higher presence and potential for damage to the fruit. The Action Plan in the Andean Region was led by SENASA and executed by different local organizations, with the aim of training the producers to adopt management measures that mitigate the impact on fine fruits (raspberries, blackberries, blueberries, cherries, strawberries and other berries).

15.2.4.2 Specific surveillance and traceability systems of Procem

SIG-PROCEM: Procem has a custom-made IT surveillance system developed to meet the needs of this program. It uses a web platform linked to a mobile App. The traps in the Procem trapping network incorporate a QR code, which is read through the App. This action records the geolocation and operation on the trap. It has the ability for offline work and is synchronized with the IT system at the end of the day. Captured flies are labeled with the trap information and sent to the laboratory for processing. Then, in the laboratory, the species and quantities captured are identified and data is recorded on the system. This allows data to be processed and information to be generated on the dynamics of the pest through graphs and weekly indicators. It should be noted that recently an artificial intelligence module is being implemented to recognize adult specimens of *C. capitata* directly through a photo of the Jackson trap inserts with the mobile application. This significantly shortens the time to obtain results.

Traceability system: This system consists of a web platform and an integrated mobile App that can trace the movements of regulated products. To do this, labels with QR codes are incorporated into each package of the products to be moved between different geographical areas. These QR labels contain information that reflects the origin of the product and its movement history. Additionally, information can be updated throughout the commercial chain with each intervention of an authorized agent, allowing the linking of the sanitary documents that cover each stage, for example, the Electronic Plant Transit Document (DTV-e) or the Quarantine Treatment Certificates. Also, labels allow for stock control at the batch level and the recording of the geolocation of the agents' activities. The web platform generates stock reports and graphs, providing valuable information when inspecting the different marketing chains.

15.2.4.3 International partnerships

- **IPPC (International Plant Protection Convention):** Procem adopts the guidelines established by the ISPM, specific to fruit flies, which enables Procem to facilitate international negotiations on host fruits. In 2023, a member of Procem´s National Coordination participated in the revision of ISPM 26, being part of the expert working group.
- **IAEA (International Atomic Energy Agency):** This organization is the international leader in the SIT. Also, IAEA promotes and coordinates technical cooperation projects for the development, training and improvements in fruit fly programs. Procem participated actively in numerous training sessions, received consultancies, support in supplies and implemented the latest technologies in accordance with the agency's manuals (e.g., the chilled adult release technique).
- **COSAVE (Southern Cone Plant Health Committee):** Procem participates in the Network of References for Fruit Fly (former Technical Group), which made it possible to generate a regional plan for strengthening capacities in the area of this pest and *D. suzukii*. In addition, face-to-face training, exchange meetings and discussions on relevant technical topics were held. Additionally, this committee works with the Inter-American Coordination Group on Plant Health (GICSV). The GICSV is a collaborative arrangement of the Regional Plant Protection Organizations of the Western Hemisphere. This group was created with the purpose of promoting joint actions aimed at preventing the introduction and spread of pests in plants, plant products and other regulated articles and encouraging the adoption of appropriate measures for their control.

15.2.4.4 Current regulations and standards

The programs in Argentina are aligned with many international and national standards and guidelines. The national regulatory framework includes the following regulations:

- **Former IASCAV Resolution No. 134/94:** approves the National Program for the Control and Eradication of Fruit Flies.
- **SENASA Resolution No. 515/01:** establishes categories of areas according to the phytosanitary condition of the pest and the level of quarantine protection implemented in each area. It establishes the different quarantine control posts for entry into the protected areas.
- **DNPV provisions No. 14/07 and 06/08:** approve Systems Approaches for host products produced in areas of low prevalence of *C. capitata* and free of *A. fraterculus*, for their industrialization in fruit fly-free areas.
- **DNPV Provision No. 12/2016:** approves Systems Approaches for fruits originating from areas of low prevalence of *C. capitata* and free of *A. fraterculus* for fresh consumption in fruit fly-free areas.
- **SENASA Resolution No. 472/14:** procedure for the phytosanitary authorization and operation of quarantine treatment centers with methyl bromide and cold.

15.3 CONCLUSIONS

Since the implementation in 1994 of the National Fruit Fly Control and Eradication Program (Procem), under the leadership of SENASA, there was a vision of integrating and strengthening provincial actions for the control of the two quarantine species of Tephritidae present in Argentina, *C. capitata* and *A. fraterculus*. Great progress has been achieved including the declaration of pest-free and low pest prevalence areas and the initiation of monitoring and suppression activities in several regions.

Procem's strategic plan, made up of a permanent surveillance system based on trapping and fruit sampling, a control component based on area-wide integrated pest management, coupled with quarantine actions and strategic communication campaigns, allowed Argentina to keep and gain strategic markets for its high-quality and valuable fruits and vegetables.

The experience gained in particular regions has served as milestones for other regions to start specific actions to hamper the damage of the pest. Most importantly, the joint work based on a public-private management model and the strategic alliances that strengthen governance and different phytosanitary conditions have been crucial to sustain the activities and integrate all the components required for the success of the actions undertaken. The recognition of pest-free areas has contributed significantly to the development of regional economies and encouraged other regions to start their own activities. Lastly, the incorporation of *D. suzukii* into the activities of Procem allowed a quick response to a new pest that threatens different regions, crops and economies.

REFERENCES

Cichón, L., S. Garrido, and J. Lago. 2015. Primera detección de *Drosophila suzukii* (Matsumura,1931) (Diptera: Drosophilidae) en frambuesas del Valle de Río Negro, Argentina. *Resúmenes del IX Congreso Argentino de Entomología*, Posadas (Argentina).

FAO/IAEA/USDA. 2019. *Product Quality Control for Sterile Mass-Reared and Released Tephritid Fruit Flies, Version 7.0*. Vienna, Austria: International Atomic Energy Agency.

ISPM 10. 1999. *Requirements for the Establishment of Pest Free Places of Production and Pest Free Production Sites*. Rome: IPPC, FAO.

ISPM 26. 2006. *Establishment of Pest Free Areas for Fruit Flies (Tephritidae)*. Rome: IPPC, FAO.

ISPM 35. 2012. *Systems Approach for Pest Risk Management of Fruit Flies (Tephritidae)*. Rome: IPPC, FAO.

ISPM 37. 2016. *Determination of Host Status of Fruit to Fruit Flies (Tephritidae)*. Rome: IPPC, FAO.

Villarreal, P., A. Mongabure, A. Borges, and C. Gomez Segade. 2023. *Evaluación del impacto económico del Programa Nacional de Control y Erradicación de Mosca de los Frutos Procem Patagonia. Reedición y actualización 2022*. Viedma: FunBaPa Ediciones.

Chapter 16

Eradicating Queensland fruit fly (*Bactrocera tryoni*) in Perth, Western Australia

History, challenges and future directions

Sonya Broughton, Touhidur Rahman, Darryl Hardie, David A. Cousins, Richard B. Johnston, Rosalie McCauley, Vincent Lanoiselet, and Rob Baxter

16.1 INTRODUCTION

Australia's $16,253.8 billion horticultural industry (Horticulture Innovation 2024) is threatened by two fruit fly species of economic concern, Queensland fruit fly (Qfly) (*Bactrocera tryoni* Froggatt) and Mediterranean fruit fly (Medfly) (*Ceratitis capitata* Wiedemann). Qfly is a native species present in Queensland, New South Wales, Victoria and the Northern Territory, while *C. capitata* is established in Western Australia (WA) and is found mainly within the south-west of WA (Dominiak and Mapson 2017). South Australia and Tasmania are free from *C. capitata* and Qfly, while WA maintains area freedom from Qfly. The east-west distribution of *C. capitata* and Qfly is a trade position for domestic and international trade negotiations (Dominiak and Mapson 2017). As WA is geographically isolated from populations of Qfly, phytosanitary measures are applied to prevent the spread of Qfly in fruit fly host produce from eastern Australia.

The Department of Primary Industries and Regional Development (DPIRD) carries out surveillance for early detection of an incursion, maintains national and international market access and demonstrates area freedom from Qfly. Maintenance of area freedom is guided by International Standards for Phytosanitary Measures (ISPM) 4, Requirements for the establishment of pest-free areas (FAO 2017), ISPM 26 Establishment of pest-free areas for fruit flies (Tephritidae) (FAO 2015) and Australia's National Fruit Fly Management Protocol. DPIRD deploys a Lynfield/cue-lure trap monitoring array in Carnarvon, Geraldton, the Perth metropolitan area, Gingin and the south-west horticultural precinct.

In 2020, a benefit–cost analysis of WA's Qfly eradication strategy estimated that over a simulated 10-year period, damage costs to commercial growers under a nil management policy were $38.0 million per year, comprising $31.9 million in lost producer surplus due to competition from domestic imports and $6.1 million from direct effects of Qfly on WA producers (Cook 2000). The break-even cost of eradication was estimated at $13.4 million (Aus). Since 1989, DPIRD has detected and successfully eradicated *B. tryoni* eight times; all detections have been within the Perth metropolitan area (Table 16.1).

DOI: 10.1201/9781003644330-21

Table 16.1 Eradication responses to *B. tryoni* in Western Australia

Year	*Location/trigger*	*Eradication techniques*	*Duration of response*	*Cost*
1989	Nedlands/Dalkeith breeding populations detected in February	Foliar baiting (malathion/ yeast autolysate), fruit stripping, MAT, SIT	24 weeks	$18.85M
1995	Victoria Park/5 males detected in 14 days in July/August	Foliar baiting (GF-120)/ fruit stripping	12 weeks	$0.37M
2011	Highgate/single female fly detected in February	Foliar baiting (GF-120)/ fruit stripping	13 weeks	
2015	Alfred Cove/5 males detected in 14 days in November	Foliar baiting (GF-120)/ fruit stripping	12 weeks	$0.76M
2018	Como/16 males detected in 14 days in April	Foliar baiting (GF-120)/ fruit stripping	27 weeks	$0.84M
2018	Fremantle/single female fly detected in January	Foliar baiting (GF-120)/ fruit stripping	13 weeks	
2020	Nedlands/Dalkeith/5 males detected in May in 14 days, breeding populations	Foliar baiting (malathion/ yeast autolysate), fruit stripping, MAT, LK, SIT	44 weeks	$12.66M

MAT, male annihilation technique; SIT, sterile insect technique; LK, Lure and Kill.

16.2 SURVEILLANCE

Since the early 1990s, Australia adopted the Lynfield trap as the standard fruit fly surveillance trap. Lynfield traps are a dry trap constructed from a 1 L plastic container (Cowley et al. 1990). Cue-lure attracts male Qfly and is combined with the organophosphate malathion on a cotton-dental wick. Wicks are suspended from the inside of the trap on a metal paper clip.

Traps are inspected weekly during summer and fortnightly during winter. Figure 16.1 provides a summary of the number of Qfly detected from 1992 to 2024. The high-risk period for Qfly incursions is from January to June, with the highest number of detections occurring in April. After the 1989 Qfly outbreak, 2,200 permanent Qfly traps were established in the Perth metropolitan area and towns in the south-west (Sproul et al. 1992). In urban areas, traps are deployed on a 400 m array and 1 km array in orchards and hung 1.5 m above the ground (Bateman 1991; Sproul et al. 1992). Also, early warning trapping is conducted at first ports of call with methyl eugenol, cue-lure and Capilure® in Lynfield traps to monitor for exotic fruit fly species such as *B. dorsalis* (Hendel), *Zeugodacus cucurbitae* (Coquillett) and *Ceratitis rosa* (Karsch).

16.3 DIAGNOSTICS

Adult Qfly is easily distinguished from *C. capitata* using wing patterns and colouration of the thorax and abdomen. Larval identification is more difficult as species of Drosophilidae, metallic-green tomato fly *Lamprolonchaea brouniana* (Bezzi) (Lonchaeidae) and *C. capitata* also may be present in fruits. In the 1989 response, identification of larvae required

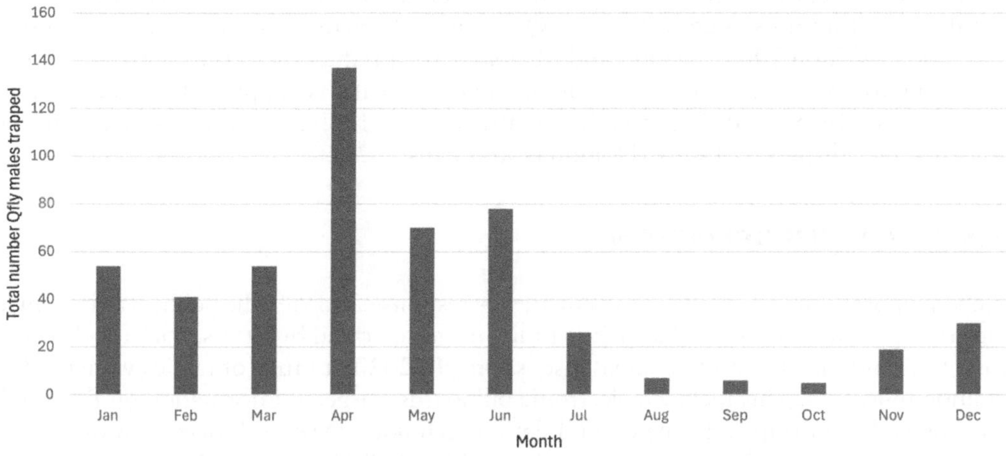

Figure 16.1 Total number of male Qfly trapped by month from 1991 to 2024.

rearing larvae in fruits, which is a labour-intensive process (Sproul et al. 1992), or electrophoresis (Dadour et al. 1992). Since 2020, loop-mediated isothermal amplification (LAMP) assays have been used for rapid identification of Qfly (Blacket et al. 2020). The LAMP assay is reported by Blacket et al. (2020) to be sensitive to low levels of target DNA (>0.016 ng/µL) and performs equally well as qPCR.

16.4 TRIGGERS

Qfly detections are reported in Table 16.1. If two or more male Qfly are trapped, an intensive larval survey is conducted and supplementary cue-lure traps are deployed on properties within 200 m of the detection site(s). The national standard that triggers an outbreak response is the detection of five male flies within two weeks, or the detection of larvae on locally grown fruit or the detection of one female Qfly (Dominiak et al. 2011). Detection of adult males triggered a response in four of the eight outbreaks; capture of a single female Qfly in two of eight outbreaks. Larvae were detected in two responses although larval detections on their own have not triggered an outbreak in WA to date.

16.5 QUARANTINE AREA AND SUSPENSION ZONE

A quarantine area (QA) is declared once an outbreak trigger has been met. A Quarantine Area Notice is used to prevent the spread of *B. tryoni* outside the QA by restricting the movement of fruit fly host plants and fruits from the QA. The QA consists of a delimitation zone, which is 200 m around the discovery point where a Qfly has been detected in a trap or larvae in fruit, a 1.5 km corrective action zone, where corrective actions such as foliage baiting and mass trapping are applied, and a 15 km controlled movement zone or suspension zone. In 1991, Australian regulators chose a 15 km radius (706.9 km^2) as the suspension area, requiring that all fruit fly produce within this area to be treated with an approved disinfestation treatment before movement out of the QA (Dominiak and Fanson 2020). Several authors suggest that, as Qfly has a short dispersal distance (Dominiak et al. 2003; Weldon 2005; Weldon and Meats 2010), a reduction in the controlled movement zone to 4 km^2 (area 50.3 km^2) for isolated outbreaks is supported (Dominiak and Fanson 2020). Meats and Edgerton (2008) attribute the 1989 programme's success to the short-distance dispersal of Qfly (Dominiak and Fanson 2020).

16.6 PUBLIC INFORMATION

Public support is important during an emergency response. In 1989, the public was informed and educated through the media such as radio, newspaper and by letters, with the distribution of 300,000 leaflets and personal discussions. In 2020, a range of tactics were used in addition to leaflets. This included electronic billboards, targeted information on Facebook and QR codes on traps and male annihilation technique (MAT) devices to provide further information with the use of a smartphone. Also, DPIRD developed free tree care tags printed on recycled plastic for plant nurseries to attach to Qfly host plants for sale. The tags include information on how to manage fruit from Qfly host trees and fruit disposal.

16.7 DATA RECORDING

In the 1989 response, trap information including location was entered onto paper in the field. Now, traps are marked with a unique barcode and global positioning system location and additional details including the host type are entered into a mobile phone in the field using "MAX" (MAXimum disease/pest management), a software package developed by Agriculture Victoria, Australia, for use in managing emergency responses. Data management has progressed from pencil and paper to Excel Spreadsheets, and in 2015, the software application MyPestGuide™ developed by DPIRD was used by field operations staff to record data on mobile phones. Currently, MyPestGuide™ is used by the public to report any suspect fruit flies as well as other pests and diseases of biosecurity concern. Staff currently use iPads to record field information into MAX. This provides response managers with up-to-date data for more informed decision-making. Additionally, Global Information Systems are used to identify areas where Qfly may be breeding and record and visualise trap locations.

16.8 FRUIT STRIPPING

At locations where Qfly had been detected in traps or fruit (infested premises), all fruit fly host fruits were stripped from the premises and from all premises within 200 m of the infested premise. Also, intensive larvae searches of all premises immediately surrounding the detection were carried out.

16.9 FOLIAR BAIT SPRAYS

Protein is required by adult fruit flies to become sexually mature (Fletcher 1987). Foliar bait sprays incorporate protein and a toxicant, killing adult flies when they ingest the bait. Malathion has been replaced by GF-120® NF Naturalure® in Qfly eradication campaigns. GF-120® contains spinosad (a mixture of spinosyn A and spinosyn D), derived from soil bacteria and other ingredients (water, sugars and plant proteins and extracts) and is the only product currently commercially available and regarded to be environmentally friendly and generally accepted by the public. GF-120® is diluted with water and applied 1.5 m from the ground using back-pack sprayers at the rate of 4–6 spots in front yards and 5–6 spots in backyards, to achieve approximately 120 bait spots per hectare. Baits are reapplied weekly or more often in the event of rain. Once an outbreak is declared, foliar baits are applied first to trees on roadsides and park areas until access to private properties and businesses can be gained. Obtaining sufficient supplies of GF-120 can be challenging, particularly during the COVID-19 pandemic as limited stock was maintained in Australia. As part of preparedness for future outbreaks, DPIRD has been testing other commercially available insecticides and entomopathogens as possible replacements.

16.10 MALE ANNIHILATION TECHNIQUE (MAT)

MAT devices combine a cue-lure and a toxicant on a suitable carrier. Male Qfly are attracted to cue-lure on the MAT and are killed, thus upsetting the male:female ratio in the Qfly population. In the 1990s, fibreboard particle board blocks (50 × 50 mm) were used as the carrier and manufactured by the Agriculture Protection Board (Bradshaw and Scott 1992). In the 2000s, Amulet™ C-L bait stations containing cue-lure replaced fibreboard. Amulet™ C-L is a commercially available cardboard wafer (65 mm x 40 mm) using fipronil (0.34% fipronil) instead of malathion as the toxicant.

MAT devices are preferentially hung by the tie inside the tree canopy of fruiting trees at a rate of 16 per hectare. Each Amulet™ C-L device was marked with an individual barcode, and the site and date of application were recorded in MAX to facilitate replacement and removal. In field trials in Hawaii, Amulet™ C-L dispensers were effective for at least 77 days (Vargas et al. 2005) and are replaced every 3 months, depending on the duration of the response. In 2020, 16,912 MAT stations were deployed across the QA.

Prior to the release of sterile fruit flies or surveillance for the reinstatement of area freedom, all MAT devices are retrieved and destroyed. From 1989 to 2011, the MAT activity was delivered by DPIRD. However, many operational activities are now outsourced to private labour companies under DPIRD supervision, with over 300 operational staff recruited for the 2020 response. Also, hire vehicles were used to support response activities, with 153 vehicles used during 2020. The use of commercial providers is an increasing trend for biosecurity activities in Australia.

16.11 LURE AND KILL (LK)

LK is another technique used to disrupt fruit fly mating. Since 2020, DPIRD used Biotrap (BioTrap Australia Pty Ltd) in Qfly responses. Biotrap is a commercially available dry trap with a clear polyvinyl chloride lid and central hole and a yellow trap base containing five vertically inclined entry holes, four around the circumference and one in the centre (Dominiak et al. 2023). The food attractant is produced from a stabilised protein concentrate, ammonium compounds and a thickening agent (xanthan gum) and is female biased, trapping mated and virgin flies and male fruit flies. Solid DDVP (dichlorvos) cubes are used as the toxicant.

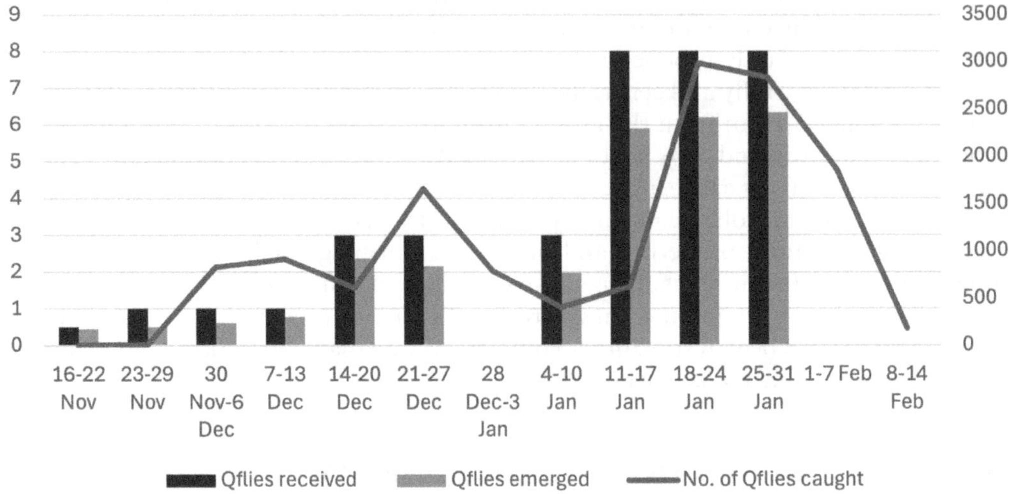

Figure 16.2 Release and recapture of sterile Qfly during the 2020 response.

During the 2020 response, 1505 traps were installed in the QA. Biotrap contents were removed from the field, and the trap contents were inspected for Qfly and Medfly adults. Medfly was used as an indicator species to evaluate the effectiveness of foliar bait sprays and fruit stripping (Figure 16.2). Additionally, Capilure® traps are deployed across the Perth metropolitan area for surveillance for exotic species such as *Ceratitis rosa*. Prior to monitoring for reinstatement of area freedom, all Biotraps were removed from the field.

16.12 STERILE INSECT TECHNIQUE (SIT)

SIT was originally pioneered in WA for the eradication of the 1989 Qfly outbreak (Yeates et al. 1992; Fisher 1996). SIT is a biologically based technique requiring Qfly to be mass reared, sterilised and released in large numbers to mate with wild females. There are no viable offspring from these matings. In the 1989 response, 40 million sterile Qfly were released each week, with 474.2 million sterile Qfly released in nine months over an area of 800 km^2 (Sproul et al. 1992). The SIT facility was repurposed in 2001 to rear the temperature sensitive lethal strain of Mediterranean fruit fly, and a new facility was established in New South Wales (Fanson et al. 2014). Currently, sterile Qfly pupae are provided by the SITPlus facility in Port Augusta, South Australia (SA), which is capable of rearing 20 million Qfly per week. The facility was upgraded enabling production of 40 million/week. Pupae were marked with fluorescent dye and sterilised using x-ray irradiation to 70 Gy, and then shipped via air from SA to WA. In WA, pupae were emerged in Murbaki towers in repurposed refrigerated shipping containers.

Quality control was undertaken to determine the emergence rate and flight ability. Shipments of SIT pupae that arrived before 17 December 2020 were affected by an issue at the SITPlus facility, in which there had been a relatively high percentage of flies emerging during the dying, packing and irradiating process. Emergence and flight ability increased after 31 December 2020 to 48.3%–68.83% flight ability and 65.6%–86.7% emergence.

The process used to distinguish between sterile and wild flies is detailed in Guillén-Aguilar et al. (2016). Briefly, when flies emerge from their puparium, an organ on their head (ptilinum) is used to force their way through soil. Once emerged, the ptilinum is retracted into the head but retains fluorescent dye particles. These particles fluoresce under an epifluorescent microscope, and fluorescence is used to separate and count sterile and wild flies.

In the 1989 response, sterile Qfly were released from the ground in bins at 1,500 release points. In the 2020 response, flies were chilled to reduce activity prior to release. Approximately 19.7 million 5-day-old Qfly were released over 2,053 hectares from 12 November 2020 to 27 January 2021. Flies were released from the ground using a release machine attached to a specially marked vehicle, which travelled slowly (20 km/hour), following a predetermined route in the QA. Releases took place once per week, at the rate of 3,000 Qfly per hectare. The first test releases occurred on 2 October 2020. The number of sterile Qfly trapped per week varied from 2 in November 2020 to 2,983 in January 2021 (Figure 16.3). The last sterile Qfly was trapped on 20 April 2021, 84 days after the last release of SIT Qfly.

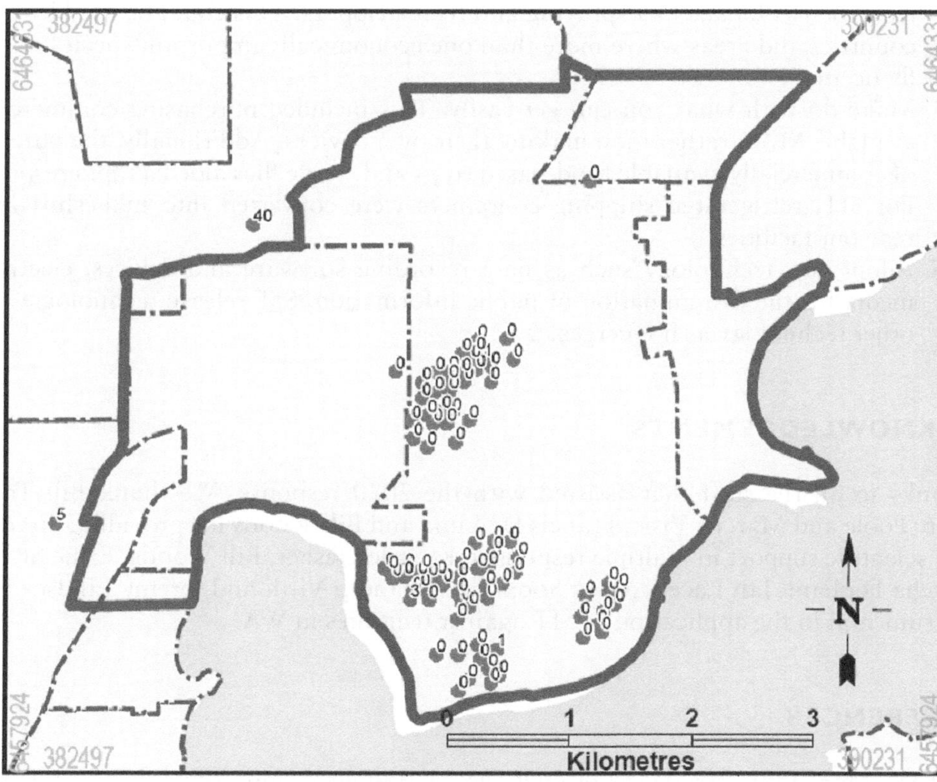

Figure 16.3 Map of the Quarantine Area Notice (thick black line) in the 2020 response. Black dots indicate the location of Lynfield traps baited with Capilure®. Trap catches of Medfly are used to evaluate the effectiveness of eradication techniques such as foliage baiting and fruit stripping.

16.13 REINSTATEMENT OF AREA FREEDOM

National guidelines require that reinstatement of pest-free area status occurs one generation and 28 days, or 12 weeks, after the last wild fly is captured in the traps, or the last larva is found, whichever is the longer period. Generation times are calculated using an accumulated day-degree model. At the time of reinstatement, surveillance levels return to normal. Area freedom was reinstated for the 2020 response in May 2021.

16.14 LEARNING OUTCOMES

During the decades of responses, there were several primary lessons learnt.

1. A permanent surveillance array is key to early detection and hence lower costs of eradication.
2. Most Qfly outbreaks in Western Australia can be eradicated using foliage baiting and fruit stripping if the population is detected early enough.
3. Medfly can be used as an indicator species to evaluate the efficacy of fruit fly controls such as foliage bait spraying and fruit stripping. This could be used by other countries and areas where more than one economically important species of fruit fly occurs.
4. Make do with what you can get easily. This included purchasing commercially available MAT, rather than making their own devices. Additionally, the purchase of commercially available food-based traps and sterile flies aided a rapid response. For SIT, refrigerated shipping containers were converted into makeshift adult rear-out facilities.
5. Adopt new technology such as data recording software and devices, electronic media for the dissemination of public information, SIT release technologies and other technology as it emerges.

ACKNOWLEDGEMENTS

Thanks to all the staff that assisted with the 2020 response. We thank Bill Trend, Marc Poole and Marcus Visic, Francis De Lima and Bill Woods for providing direction and scientific support in multiple responses. Kingsley Fisher, Bill Woods, Ernie Steiner, Roselia Fogliani, Ian Lacey, Alven Soopaya, Amandip Virdi and Jeremy Lindsey were instrumental in the application of SIT against fruit flies in WA.

REFERENCES

Bateman, M. A. 1991. *The Impact of Fruit Flies on Australian Horticulture (No. 3)*. Canberra: Horticultural Policy Council.

Blacket, M. J., A. Agarwal, L. J. Zheng, et al. 2020. A LAMP assay for the detection of *Bactrocera tryoni* Queensland fruit fly (Diptera: Tephritidae). *Scientific Reports* 10:9554.

Bradshaw, J. M., and J. J. Scott. 1992. Field operations. In *Queensland Fruit Fly Eradication Campaign*, ed. A. N. Sproul, S. Broughton, and N. Monzu, 29–48. Department of Agriculture Western Australia.

Cook, D. C. 2000. *Benefit Cost Analysis. Queensland Fruit Fly, March 2020 Incident*. Department of Primary Industries and Regional Development.

Cowley, J. M., F. D. Page, P. R. Nimmo, et al. 1990. Comparison of the effectiveness of two traps for *Bactrocera tryoni* (Froggatt) (Diptera: Tephritidae) and implications for quarantine surveillance systems. *Australian Journal of Entomology* 29:171–176.

Dadour, I. R., D. K. Yeates, and A. C. Postle. 1992. Two rapid diagnostic techniques for distinguishing Mediterranean fruit fly from *Bactrocera tryoni* (Froggatt) (Diptera: Tephritidae). *Journal of Economic Entomology* 85(1):208–211.

Dominiak, B. C., C. Bain, N. Sharma, et al. 2023. Evaluating trap and lure combinations using Biotraps and Lynfield traps for the surveillance of Queensland fruit fly *Bactrocera tryoni* (Froggatt) (Diptera: Tephritidae) and other tephritids in southern New South Wales, Australia. *International Journal of Tropical Insect Science* 43: 2083–2093.

Dominiak, B. C., D. Daniels, and R. Mapson. 2011. Review of the outbreak threshold of Queensland fruit fly (*Bactrocera tryoni* Froggatt). *Plant Protection Quarterly* 26:141–147.

Dominiak, B. C., and B. G. Fanson. 2020. Current quarantine and suspension distances are excessive for incipient populations of Queensland fruit fly (*Bactrocera tryoni* (Froggatt)) (Diptera: Tephritidae) in southern New South Wales, Australia. *Crop Protection* 138:105341.

Dominiak, B. C., and R. Mapson. 2017. Revised distribution of *Bactrocera tryoni* in Eastern Australia and effect on possible incursions of Mediterranean fruit fly: Development of Australia's Eastern Trading Block. *Journal of Economic Entomology* 110:2459–2465.

Dominiak, B. C., A. E. Westcott, and I. M. Barchia. 2003. Release of sterile Queensland fruit fly, *Bactrocera tryoni* (Froggatt) (Diptera: Tephritidae), at Sydney, Australia. *Australian Journal of Experimental Agriculture* 43:519–528.

Fanson, B. G., S. Sundaralingam, L. Jiang, et al. 2014. A review of 16 years of quality control parameters at a mass-rearing facility producing Queensland fruit fly, *Bactrocera tryoni*. *Entomologia Experimentalis et Applicata* 151:152–159.

FAO (Food and Agriculture Organization of the United Nations). 2015. *Establishment of Pest Free Areas for Fruit Flies (Tephritidae). International Standard for Phytosanitary Measures No.* 26. Rome: FAO on behalf of the Secretariat of the International Plant Protection Convention.

FAO (Food and Agriculture Organization of the United Nations). 2017. *Requirements for the Establishment of Pest Free Areas. International Standard for Phytosanitary Measures No.* 4. Rome: FAO on behalf of the Secretariat of the International Plant Protection Convention.

Fisher, K. 1996. Queensland fruit fly (*Bactrocera tryoni*): Eradication from Western Australia. In *Fruit Fly Pests. A World Assessment of Their Biology and Management*, ed. B. A. McPheron, and G. J. Steck, 535–541. Delray Beach, FL: St Lucie Press.

Fletcher, B. S. 1987. The biology of Dacine fruit flies. *Annual Review of Entomology* 32:155–144.

Guillén-Aguilar, J. C., L. López-Muñoz, E. Pérez-Espinoza, et al. 2016. *Manual to Differentiate Wild Mediterranean Fruit Flies* Ceratitis capitata *(Wied.) from non-irradiated (Fertile) and Irradiated (Sterile) VIENNA Temperature Sensitive Lethal Strain Flies*, ed. J. Reyes-Flores. Vienna, Austria: FAO/IAEA Programme of Nuclear Techniques in Food and Agriculture.

Horticulture Innovation. 2024. Australian Horticulture Statistics Handbook 2021–22 to 2023–24 (MT21006). https://www.horticulture.com.au/growers/help-your-business-grow/research-reports-publications-fact-sheets-and-more/australian-horticulture-statistics-handbook/ (accessed September 3, 2024).

Meats, A., and J. E. Edgerton. 2008. Short- and long-range dispersal of the Queensland fruit fly, *Bactrocera tryoni* and its relevance to invasive potential, sterile insect technique and surveillance trapping. *Australian Journal of Experimental Agriculture* 18:1237–1245.

Sproul, A. N., S. Broughton, and N. Monzu. 1992. *Queensland Fruit Fly Eradication Campaign. Internal Report*. Perth, Australia: Western Australia Department of Agriculture.

Vargas, R. I., J. D. Stark, B. Mackey, et al. 2005. Weathering trials of Amulet cue-lure and Amulet methyl eugenol "Attract-and-Kill" stations with male melon flies and Oriental fruit flies (Diptera: Tephritidae) in Hawaii. *Journal of Economic Entomology* 95:1551–1559.

Weldon, W. C. 2005. *Dispersal and mating behaviour of Queensland fruit fly, Bactrocera tryoni (Froggatt) (Diptera: Tephritidae): Implications for population establishment and control.* PhD Thesis, University of Sydney, Australia.

Weldon, C., and A. Meats. 2010. Dispersal of mass-reared sterile, laboratory-domesticated, and wild male Queensland fruit flies. *Journal Applied Entomology* 134:16–25.

Yeates, D. K., P. B. Yeoh, and S. N. Rechich. 1992. Trap monitoring for Qfly. In *Queensland Fruit Fly Eradication Campaign*, ed. A. N. Sproul, S. Broughton, and N. Monzu, 97–139. Department of Agriculture Western Australia.

Chapter 17

Area-wide fruit fly control with environmentally friendly techniques in Mauritius

Nausheen Azhaar Patel, Preeaduth Sookar, and Aaishah Ollee

17.1 INTRODUCTION

Fruit flies (Diptera: Tephritidae) cause large losses to fruits throughout the world and are recognized today as major pests of the horticulture industry. A growing international trade has further increased their significance (Drew and Allwood 1997; Blaser et al. 2018; Papadopoulos et al. 2024). In Mauritius, fruit fly problems date back to the beginning of the nineteenth century (Orian and Moutia 1960). Damage is caused when the female flies insert their ovipositor through the skin of the fruit and deposit their eggs just beneath. The larval stage of the fruit fly is the most devastating as they destroy a large proportion of the flesh of fruits and vegetables, rendering them unfit for human consumption. Infested fruits drop to the ground and decay. Fruit flies show rapid reproductive potential with short and overlapping generations and sudden outbreaks (Bateman 1972; Dukas 2020). The high fruit fly populations and the abundance of both cultivated and wild fruits throughout the year combine to maintain the status of the fruit flies as the major pests of cultivated fruits.

The Entomology Division has reviewed its adult fruit fly surveillance system in 2015, and traps were relocated and placed on a geographic information system grid, within a 2 km radius. To date, there are 137 points, each having a series of three independent dry traps with the following lures: methyl eugenol, cuelure and trimedlure. On 68 points, there are wet traps baited with either modified waste brewery yeast (WBY) or torula yeast (Figure 17.1). Larval surveillance consists of the collection of fruits and vegetables from different localities and incubation in a laboratory to assess host plants and fruit fly infestation levels.

Recent surveillance reports have identified two major flies of economic importance in Mauritius: the oriental fruit fly *Bactrocera dorsalis* (Hendel) and the melon fly *Zeugodacus cucurbitae* (Coquillett) (Sookar et al. 2021).

Bactrocera dorsalis was first intercepted in 1996 and a second time in 2013. The pest was successfully eradicated on both occasions (De Meyer et al. 2014; Sookar et al. 2021). The third incursion of *B. dorsalis* was in 2015 followed by its rapid establishment as a result of the presence of its favored host, i.e., mango (Mwatawala et al. 2006) at that time. *Zeugodacus cucurbitae* is present in Mauritius since 1940 and affects mainly the local cucurbits, namely snake gourd (*Trichosantes cucumerina*), sponge gourd (*Luffa acutangulata*), cucumber (*Cucumis sativus*), pumpkin (*Cucurbita*

DOI: 10.1201/9781003644330-22

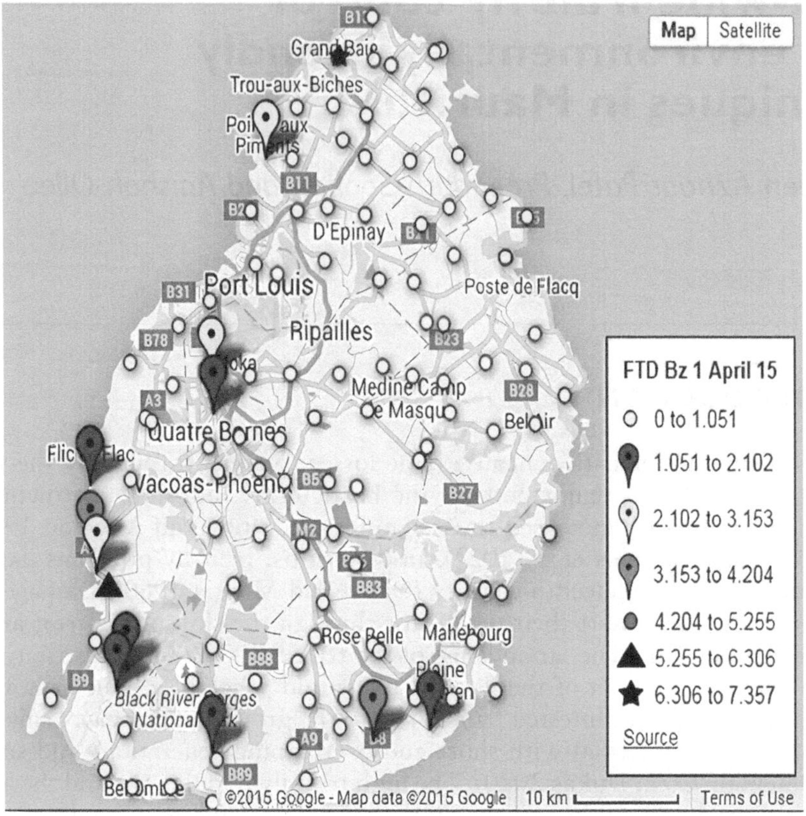

Figure 17.1 Map of Mauritius illustrating adult surveillance through trappings since 2015. Light gray is a treated region and dark gray is an untreated region.

maxima), bitter gourd (*Momordica charantia*), zucchini/squash (*Cucurbita pepo*) and melon (*Cucumis melo*) (Sookar et al. 2006).

Strategies for the fruit fly management include physical, cultural, behavioral, genetic, chemical and biological control (Kapatos 1989). However, no single technique can be used but a combined strategy is recommended, using the basics of the Integrated Management Approach (IPM) (Mau et al. 2007). Hence, the objective of the study was to demonstrate the efficacy of the integrated use of sterile insect technique (SIT) for the management of fruit flies in farmers' fields.

17.2 MATERIALS AND METHODS

17.2.1 Site location

Two sites were selected, one cucurbit plantation of 3 ha in the Northern Part of the Island and a mango orchard of approximately 2 ha in the North-West region, as treated regions, while regions outside the treated zones were considered control zones (untreated regions).

17.2.2 Bait application technique (BAT)

Protein bait was produced at the Entomology Division using a pre-established protocol by the modification of WBY (Patel et al. 2019), which is a by-product of the beer industry. A concentration of 10.0% v/v of the modified WBY, and a toxicant, Malathion 57% Emulsifiable Concentrate (EC) was used (<8 g/ha), at the rate of less than 2 L mixture per hectare (Sookar et al. 2006). The mixture was applied at the rate of 40 mL as a coarse spot spray on the undersides of foliage in the fruit orchard, and 10–20 mL/spot were sprayed on trap crops (maize) in the cucurbit plantation. The spraying activity was repeated every 7–10 days and carried out before 9.00 a.m., when the flies are more actively looking for food (Patel et al. 2019).

17.2.3 Monitoring (adult and larval monitoring)

Pheromone-baited traps (dry traps) were placed in the cucurbit field and in the orchard to monitor the population of the male flies. EDMA (Entomology Division of the Ministry of Agriculture) traps were used, which are simple yellow plastic containers of capacity 1 L, locally purchased, modified and adapted for use as low-cost traps. Entry holes were perforated on the sides of the containers using a drill machine, and a cotton wick was fixed in the inner bottom side of each container. Then, the latter were lured with cuelure or methyl eugenol, respectively, mixed with a toxicant (Lambda cyalothrin 10 CS) in a ratio of 1:1. Finally, traps were fixed at a height of <1 m from the ground level, with the lid facing downwards.

The traps were serviced on a fortnightly basis, i.e., the dead flies were collected from the lid, and the cotton wicks were relured accordingly using 0.1 mL of lure/malathion 57 EC. Wet traps were prepared using 10.0% v/v WBY and placed in Tephri traps at the rate of 300 mL/trap, to monitor the female population. Borax was added at 2% w/v in all wet traps to prevent the disintegration of the fruit flies. Servicing of the wet traps was done weekly. All flies were collected in properly labeled collection tubes and brought to the laboratory for identification. Data was analyzed and represented as flies per trap per day (FTD).

Regular collection of fruit and vegetables was done in treated and nontreated regions to assess the percentage infestation level (larval monitoring) throughout the trial. Samples (10–25 units) of infested zucchini fruits 3–5 cm long were collected on a weekly basis, while random fallen fruits (10–25 units) were collected in the fruit orchard. The collected fruits were counted, weighed and placed in a tray holding sand as a pupating medium. The trays were covered with a thin breathable cotton cloth to prevent contamination with *Drosophila*, and fruits were incubated for two weeks under controlled conditions (25°C, 75% RH). Larvae were allowed to pop in the sand. Resulting pupae were sieved and placed in Petri dishes for adult emergence. Adult flies were counted and identified.

17.2.4 Male annihilation technique (MAT)

The MAT involves the use of a high density of trapping stations consisting of a male lure combined with an insecticide. The MAT blocks, made from a 12 mm plywood sheet, were cut in sizes of 50 mm×50 mm. Then, the blocks were soaked for 8–10 days in a mixture of lures (cuelure targeting the melon fly and methyl eugenol targeting the oriental fruit fly) and malathion ULV, in a ratio of 4:1. The MAT blocks were placed at the rate of 10 blocks per hectare at flowering in the orchard and in the cucurbit field, respectively.

17.2.5 Sanitation

Sanitation involves removing fallen fruit, rotting vegetation, and other organic matter that serve as breeding sites for fruit flies. The planters and fruit growers were provided with an augmentorium, each for placement in the treated areas with the objective of trapping the fruit fly pests and also allowing beneficial insects, notably the parasitoids, to escape through the wire mesh on top of the cage. All infested fruits and vegetables were disposed of in the special disposal cage.

17.2.6 Sterile insect technique

Zeugodacus cucurbitae and *B. dorsalis* were reared under laboratory conditions in the mass-rearing facility of the Entomology Division. Pupae, collected from the F10 generations of *B. dorsalis* and F16 generations of *Z. cucurbitae*, were irradiated at 70 Gys and 80 Gys, respectively, using a cesium irradiator two days before emergence, after observing hypoxia. Irradiated flies (100 mL) were placed in paper bags with an agar block (60 g) and adult food (20 g) consisting of sugar and enzymatic yeast (3:1).

The irradiated flies were released eight days after emergence at the rate of 1,000 male flies/ha in both cucurbit fields and orchards. Releases were carried out on a weekly basis when the wild population had dropped below 0.5 flies/trap/day (FTD). However, all MAT blocks were removed prior to releases, and no bait spray was done during the release period to ensure that no sterile flies were caught and killed.

17.2.7 Data analysis

Assessment of the fruit fly population in treated and untreated regions was done by converting trap catches into FTD using the formula below. FTD is obtained by dividing the total number of captured fruit flies by the product obtained from multiplying the total number of inspected traps by the average number of days the traps were exposed. The formula is as follows:

$$\text{FTD} = F/(T \times D)$$

where F is the total number of fruit flies, T is the number of inspected traps, and D is the average number of days the traps were exposed in the field between two inspections.

Infestation by *B. dorsalis* and *Z. cucurbitae* followed the methodology of Cowley et al. (1992), and it was calculated as the ratio of the number of adults per kilogram of fruit collected (infestation index).

17.3 RESULTS

Figure 17.2 shows that the number of melon flies caught in traps in the treated regions was significantly different from the number caught in the traps in the untreated regions. Peak fly trap catches in the nontreated region varied from 15.0 to 20.0 flies/trap/day,

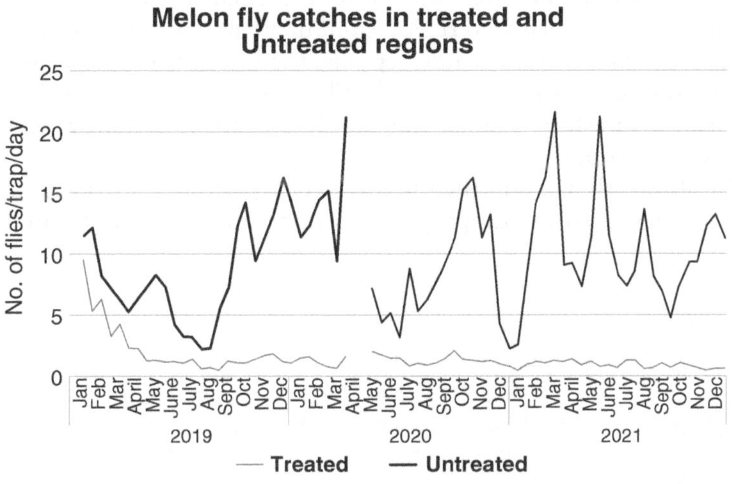

Figure 17.2 Population of melon fly *Zeugodacus cucurbitae* (Coquillett) recorded from monitoring of cuelure-baited traps for the period January 2019 to December 2021.

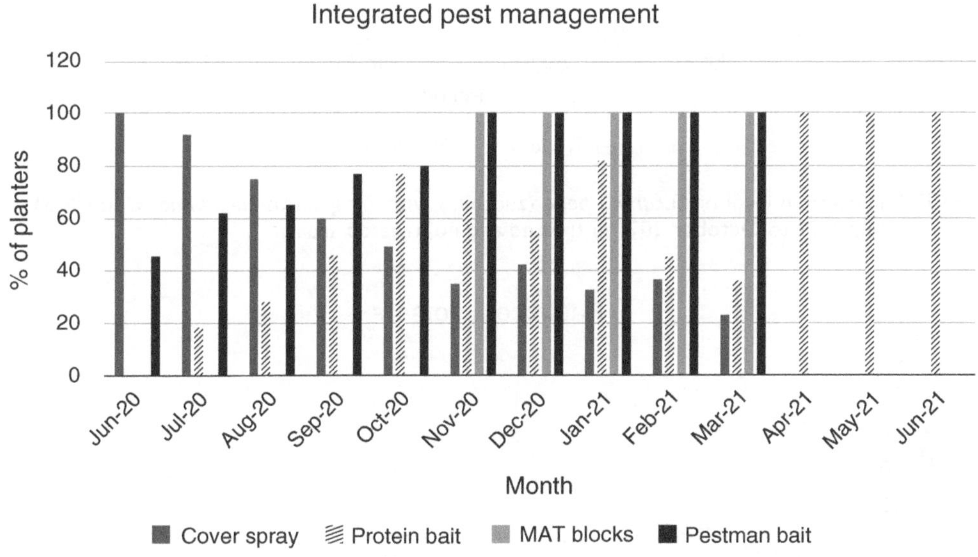

Figure 17.3 Survey of IPM strategies used by selected planters from June 2020 to June 2021.

mostly observed during the summer months. In the treated regions, the flies/trap/day were >5.0 during the first few months, after which a decline in the fruit fly population was seen, assuming an FTD of <2.0 thereon.

Figure 17.3 shows that at the start of the project the planters were using mostly cover sprays. However, during the study period, there was a shift from the use of cover sprays to the use of protein baits only (WBY and a commercial protein bait called 'Pestman') at the end of the project. From November 2020 to March 2021, cuelure MAT blocks were placed in the field on two occasions to knock down the population of male melon fruit flies.

Figure 17.4 shows that the infestation level recorded in *Cucurbita pepo* fruits collected from treated and nontreated regions and incubated in the laboratory was significantly different ($p<0.05$). In untreated regions, the infestation index ranged from 17.3 flies/kg fruit in July 2021 to up to 53.2 flies/kg fruit in October 2021. Conversely, in treated regions, the infestation index was 3.0 flies/kg fruit, and dropped to 0.3 flies/kg fruit for the same period.

Figure 17.5 shows that a significant difference ($p<0.05$) in the population of *B. dorsalis* was observed from trappings done in treated and nontreated regions. An

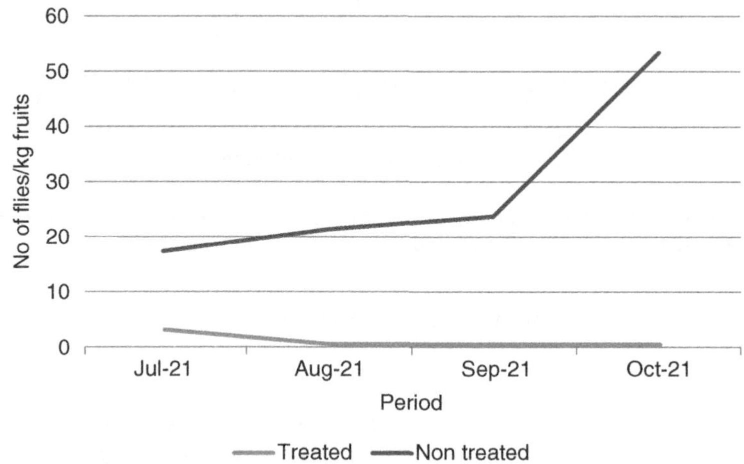

Figure 17.4 Infestation level of *Cucurbita pepo* (zucchini) with *Zeugodacus cucurbitae* (Coquillett) from July 2021 to October 2021 in treated and nontreated areas.

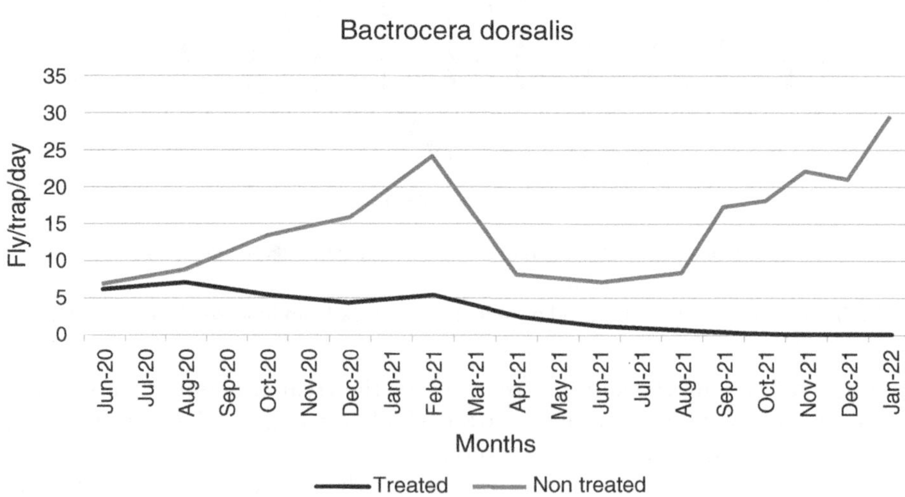

Figure 17.5 Population of Oriental fruit fly *Bactrocera dorsalis* (Hendel) recorded from methyl eugenol-baited traps in treated and untreated regions observed from June 2020 to January 2022.

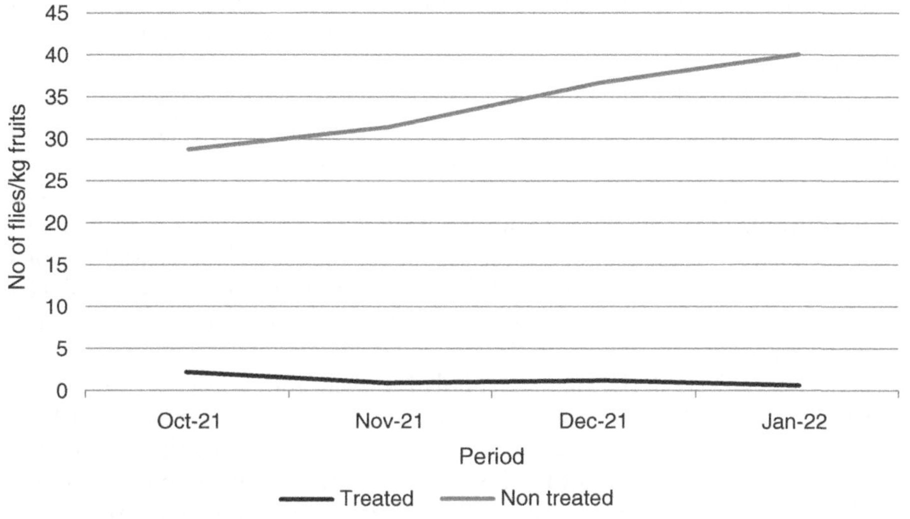

Figure 17.6 Infestation level of Mango *Mangifera indica* with *Bactrocera dorsalis* (Hendel) from October 2021 to January 2022 in treated and nontreated areas.

increase in fly/trap/day was recorded from June 2020, with 6.9 FTD in winter, to January 2022, assuming peaks of 24.1 and 29.3 FTD in the summer months. In treated regions, the FTD was 6.3 at the start of the project and declined to 0.1 flies/trap/day by January 2022.

Figure 17.6 shows a significant difference ($p<0.05$) in the infestation level of mangoes collected from treated and nontreated regions. The infestation index ranged from 28.7 flies/kg fruit in October 2021 to 40.2 flies/kg fruit in January 2022 in untreated regions. Conversely, in the treated regions, the infestation index ranged from 2.3 to 0.5 flies/kg fruit for the same period.

17.4 DISCUSSION AND CONCLUSION

Fruit fly populations are normally higher during the summer season due to the favorable environmental conditions that facilitate the proliferation of the pests. High oviposition rates and high egg-to-adult development success have been reported when temperature and humidity are favorable (Winkler et al. 2020). This explains the peaks observed from November to April (summer months) in the population of both fruit fly species (*B. dorsalis* and *Z. cucurbitae*).

The fruit fly control methods, i.e., area-wide application of protein bait sprays, mass trapping of males, sanitation, and release of sterile flies, reduced the trap catches for both *B. dorsalis* and *Z. cucurbitae* in the treated regions as compared to the untreated ones.

The application of the BAT as spot sprays under the leaves of plants on a weekly basis helped continuously kill both the emerging male and female flies. Females need the ammonia from the protein for sexual development and for ovipositing and are thus more attracted (Thomas et al. 2008). This method of bait spraying is relatively safe for nontarget insects and is less costly as less insecticide is used compared to cover spray and thus more environment-friendly, since it produces very little spray drift and very little chemical residue.

The application of the MAT by placing pheromone blocks reduced the male population. MAT disrupts the fruit fly life cycle by reducing the probability of fruit flies being fertilized because the male fly population declines. Due to the difference in the behaviors of the two fruit fly species, MAT placement targeting *Z. cucurbitae* is done on borders. In the orchard context and targeting *B. dorsalis*, MAT is placed within the orchard on the host trees. From previous unpublished research (Preeaduth Sookar, personal communication), it was observed that the effectiveness of the MAT blocks lasted approximately 2.5 months. Thus, the pheromone blocks were replaced accordingly on the respective trial sites.

Sanitation helped break the vicious fruit fly life cycle and reduce the population of the pests. In cultivated fields where sanitation measures are implemented, fruit fly populations decreased significantly (Verghese et al. 2004). The use of the augmentorium in this study was to allow escape of the beneficial parasitoids through the mesh on top of the cage, for instance, *Fopius arisanus*, an egg parasitoid that parasitizes *B. dorsalis* (Vargas et al. 2007).

The ultimate strategy was the release of the sterile *Z. cucurbitae* in the zucchini field and sterile *B. dorsalis* in the mango orchard. However, the MAT blocks initially placed had to be removed and bait application stopped during the release period to ensure that no released flies were lured and killed. When the released sterile males mate with wild female counterparts, there is no production of fertile eggs. Thus, SIT acts as a birth control, resulting in no offspring. The success of the control program requires provision of adequate overflooding ratios of sterile to wild males (Pereira et al. 2013). That is why BAT and MAT are prerequisites to SIT to reduce the initial wild population.

The results of the combined strategies were apparent from the lower infestation index observed when fruit and vegetables collected from the treated areas were incubated. A very high infestation index was observed from collections in the nontreated areas corresponding to the high fruit fly population level. These findings support the assumption that concurrent use of the applied techniques has a desirable complementary effect, and therefore this strategy could be used advantageously in efforts to suppress, contain or eradicate populations of fruit fly pests. The specificity of each technique and the complementary effect have a significant role in efficiently managing the fruit fly population. By integrating these strategies, pest management becomes more comprehensive, addressing multiple aspects of fruit fly behavior and lifecycle, resulting in reduced pest populations.

Timely implementation of each strategy (BAT, MAT and SIT) is essential. When BAT is applied at the right time, i.e., during peak fruit fly activity, the attractants efficiently reduce the adult fly population. The probability of the flies ingesting the protein bait while most active is somehow enhanced, ensuring that the insects ingest the lethal product, reducing their lifespan and limiting their ability to reproduce. The effectiveness of the bait decreases if it is applied outside of the active flight or feeding period of the flies. Mating disruption by placement of MAT blocks is another crucial component in this integrated strategy. The blocks release semiochemicals (sex pheromone) that cause confusion around male fruit flies and reduce their ability to find mates. The wild males are more prone to be attracted to the MAT blocks and get killed upon ingestion of the toxicant in the killer blocks. Thus, the strategic positioning of these blocks within orchards or around field borders reduces the mating success with wild female fruit flies. Regular sanitation aided in the elimination of potential fly larvae and pupae, thereby reducing the number of new adults emerging. When combined with BAT and MAT blocks, sanitation further weakened the fruit fly population by decreasing the wild fly population and reducing the available breeding sites, enhancing the overall effectiveness of the control strategy (Nagel and Peveling 2021; Zida et al. 2023).

Finally, SIT, which involves releasing sterile male flies into the environment, complements these methods by reducing the number of fertile females. When combined with the timely application of BAT, MAT blocks, and sanitation, SIT offers a powerful approach to fruit fly control.

No Peach fruit flies, *Bactrocera zonata* (Saunders), were recovered through the lab incubations of infested mangoes or from the traps throughout the study period, while very few (negligible) *Ceratitis quilici* (De Meyer, Mwatawala & Virgilio sp.) were observed. This demonstrates the likely competitiveness of *B. dorsalis* (Ekesi et al. 2006; Vargas et al. 2007). From these observations, it is deduced that the Oriental fruit fly has completely displaced *B. zonata* in Mauritius. Similar results have been reported for *Ceratitis quilici* and *Ceratitis capitata* (Wiedemann) in several fruit species. This more competitive fruit fly pest with a reportedly higher reproductive potential resulted in elevated infestation indices, which was consistent with Sookar et al. (2021). *Zeugodacus cucurbitae* was the only species recovered from the incubation of infested zucchini indicating the negligible population of *Dacus ciliatus* (Loew), which has been previously reported in cucurbits in Mauritius (Sookar et al. 2006).

This research has confirmed the devastating potential of the two fruit fly species (*Z. cucurbitae* and *B. dorsalis*) in Mauritius. These fruit flies are rightly categorized as economically important pests causing major harm to the horticulture industry, if not properly managed with appropriate strategic timely controls (APHIS/USDA 2016).

The way forward is to (i) replicate this best practice in other promising localities; (ii) increase the production of WBY (protein bait) for broader use in other areas; (iii) increase the production of cuelure and methyl eugenol MAT blocks; (iv) increase the breeding of *B. dorsalis* and *Z. cucurbitae* in the mass-rearing facility for release; (v) rear and release *Fopius arisanus* and reintroduce the larval parasitoid *Psyttalia fletcheri* to combat *Z. cucurbitae*, and finally (vi) release sterile flies pre-exposed to semiochemicals.

ACKNOWLEDGMENTS

This study is part of a Coordinated Research Project (CRP) with the International Atomic Energy Agency (IAEA). The authors are grateful to the organization and extend a special thanks to the Head of the Insect Pest Control Laboratory, Rui Cardoso Pereira, for his trust in our capabilities and constant support and assistance.

REFERENCES

APHIS/USDA. 2016. Oriental fruit fly, *Bactrocera dorsalis* host list. https://www.aphis.usda.gov/planthealth/plantpestinfo/fruitflies/downloads/host-list/off-hostlists.pdf (accessed December 24, 2023).

Bateman, M. A. 1972. The ecology of fruit flies. *Annual Review of Entomology* 17:493–518.

Blaser, S., C. Heusser, H. Diem, et al. 2018. Dispersal of harmful fruit fly pests by international trade and a loop-mediated isothermal amplification assay to prevent their introduction. *Geospatial Health* 13 (2). https://doi.org/10.4081/gh.2018.726

Cowley, J. M., R. T. Baker, and D. S. Harte. 1992. Definition and determination of host status for multivoltine fruit fly (Diptera: Tephritidae) species. *Journal of Economic Entomology* 85:312–317.

De Meyer, M., S. Mohamed, and I. M. White. 2014. Invasive fruit fly pests in Africa: A diagnostic tool and information reference for the four Asian species of fruit fly (Diptera: Tephritidae) that have become accidentally established as pests in Africa, including the Indian Ocean. https://www.africamuseum.be/fruitfly/AfroAsia.htm (accessed December 20, 2023).

Drew, D., and A. Allwood. 1997. Fruit fly control: Arrest of a new pest. *Partners* 10:25–33.

Dukas, R. 2020. Natural history of social and sexual behavior in fruit flies. *Science Reports* 10:21932.

Ekesi, S., P. W. Nderitu, and I. Rwomushana. 2006. Field infestation, life history and demographic parameters of the fruit fly Bactrocera invadens (Diptera: Tephritidae) in Africa. *Bulletin of Entomological Research* 96:379–386.

Kapatos, E. T. 1989. Integrated pest management systems of *Dacus oleae*. In *Fruit Flies, Their Biology, Natural Enemies and Control, World crop pests, Vol.* 3B, ed. W. Helle, A. S. Robinson, and G. Hooper, 391–398. Amsterdam, the Netherlands: Elsevier.

Mau, R. F. L., E. B. Jang, and R. I. Vargas. 2007. The Hawaii area-wide fruit fly pest management programme: influence of partnerships and a good education programme. In *Area-Wide Control of Insect Pests*, ed. M. J. B. Vreysen, A. S. Robinson, and J. Hendrichs, 671–683. Dordrecht, the Netherlands: Springer.

Mwatawala, M. W., M. De Meyer, R. H. Makundi, et al. 2006. Seasonality and host utilization of the invasive fruit fly *Bactrocera invadens* (Dipt., Tephritidae) in Central Tanzania. *Journal of Applied Entomology* 130:530–537.

Nagel, P., and R. Peveling. 2021. Environment and the sterile insect technique. In *Sterile Insect Technique. Principles and Practice in Area-Wide Integrated Pest Management*, ed. V. A. Dyck, J. Hendrichs, and A. S. Robinson, 753–780. Boca Raton, FL: CRC Press.

Orian, A. J. E., and L. A. Moutia. 1960. Fruit flies (Trypetidae) of economic importance in Mauritius. *Revue Agricole et Sucriere de l'Ile Maurice* 39 (3):142–150.

Papadopoulos, N. T., M. De Meyer, J. S. Terblanche, et al. 2024. Fruit flies: Challenges and opportunities to stem the tide of global invasions. *Annual Review of Entomology* 69:355–373.

Patel, N. A., S. Facknath, and P. Sookar. 2019. Assessment of modified waste brewery yeast as an attractant for fruit flies of economic importance in Mauritius. In *Area-Wide Management of Fruit Flies Pests*, ed. D. Perez-Staples, F. Diaz-Fleisher, P. Montaya, and M. T. Vera, 142–156. London: Taylor and Francis.

Pereira, R., B. Yuval, P. Liedo, et al. 2013. Improving sterile male performance in support of programmes integrating the sterile insect technique against fruit flies. *Journal of Applied Entomology* 137(S1):178–190.

Sookar, P., N. Patel, and P. Ramkalawon. 2021. *Bactrocera dorsalis*, an invasive fruit fly species in Mauritius. *Fruits* 76(6):269–275.

Sookar, P., S. Permalloo, B. Gungah, et al. 2006. An area wide control of fruit flies in Mauritius. In *Proceedings of the 7th International Symposium on Fruit Flies of Economic Importance*, 10–15 September 2006, Salvador, Brazil, 261–269.

Thomas, D. B., N. D. Epsky, P. Kendra, et al. 2008. Ammonia formulations and capture of *Anastrepha* fruit flies (Diptera: Tephritidae). *Journal of Entomological Science* 43(1):76–85.

Vargas, R. I., L. Leblanc, R. Putoa, et al. 2007. Impact of introduction of *Bactrocera dorsalis* (Diptera: Tephritidae) and classical biological control releases of *Fopius arisanus* (Hymenoptera: Braconidae) on economically important fruit flies in French Polynesia. *Journal of Economic Entomology* 100:670–679.

Verghese, A., P. L. Tandon, and J. M. Stonehouse. 2004. Economic evaluation of the integrated management of the oriental fruit fly *Bactrocera dorsalis* (Diptera: Tephritidae) in mango in India. *Crop Protection* 23(1):61–63.

Winkler, A., J. Jung, B. Kleinhenz, et al. 2020. A review on temperature and humidity effects on *Drosophila suzukii* population dynamics. *Agricultural and Forest Entomology* 22(3):179–192.

Zida, I., K. Nebie, A. Sawadogo, et al. 2023. Effectiveness of four integrated pest management approaches in the control of fruit flies (Diptera: Tephritidae) in mango agro-ecosystems in the South-Sudanian zone of Burkina Faso. *Advances in Entomology* 11(3):124–142.

Chapter 18

A tiered approach to the implementation of integrated fruit fly management programmes in the Americas

Emilio Arévalo Peñaranda and Walther Enkerlin

18.1 INTRODUCTION

According to statistics on global fruit and vegetable production, during 2022, a total of 933 million tons of fruits and 42 million tons of fresh vegetables were produced worldwide. This figure meant an increase of more than ten million tons compared to the previous year (Orús 2024). During the same year, U.S. fruit exports reached a commercial value of approximately US$14.7 billion, making it the world's leading exporter of this agricultural product (Rodríguez 2023). In the case of Latin America, the region is considered the largest exporter of tropical fruits in the world according to the 2019 FAO statistics (FAO/IAEA 2019), which represents around 25% of total global fruit and vegetable exports (Mello Garcia 2024a). In 2021, in Brazil alone, the volume of fruit production exceeded 39 million metric tons, making it the country with the highest production of such crops in Latin America. During the same year, the second place was occupied by Mexico with approximately 23.6 million metric tons produced (Statista 2024). As can be seen, fruit and vegetable production represents an important item in agricultural production throughout the American continent.

18.2 IMPORTANCE OF FRUIT FLIES (DIPTERA: TEPHRITIDAE) IN THE AMERICAS

In relation to fruit and vegetable production and the levels of productivity achieved in different countries, one of the main problems affecting it is related to attacks by fruit flies (Diptera: Tephritidae). Their attack causes direct damage due to the impact that the larvae produce during their feeding process and the entry of microorganisms that deteriorate the quality of the fruit, reducing income and increasing production costs. Moreover, there are adverse impacts due to the quarantine restrictions for international trade of these products, in addition to the environmental impact generated by conventional control measures used in most countries.

In the Americas, the presence of 977 species of tephritids is recorded, of which 73% occur in the Neotropical region (Hernández-Ortiz et al. 2020). Of all these species, only 21 species (belonging to the genera *Anastrepha* Schiner, *Bactrocera* Macquart, *Ceratitis* MacLeay, and *Rhagoletis* Loew) have economic importance (Mello Garcia 2024a). Most of the economically important fruit fly species in the Americas belong to the genus *Anastrepha* (11 species): *Anastrepha curvicauda* (Gerstaecker),

DOI: 10.1201/9781003644330-23

A. fraterculus (Wied.), *A. grandis* (Macquart), *A. ludens* (Loew), *A. obliqua* (Macquart), *A. serpentina* (Wied.), *A. sororcula* Zucchi, *A. striata* Schiner, *A. suspensa* (Loew), *A. pseudoparallela* Loew, and *A. zenildae* Zucchi (Mello Garcia 2024a). The second genus with the highest number of pest species is *Rhagoletis* with seven referenced species: *Rhagoletis cerasi* Loew, *R. cingulata* (Loew), *R. completa* (Cresson), *R. fausta* (Osten-Sacken), *R. indifferens* Curran, *R. mendax* Curran, and *R. pomonella* Walsh (Mello Garcia 2024a). The third group in order of importance is the genus *Bactrocera*, native to Asia and Europe, which has species with a high capacity for invasion. This genus is represented by three species in the Americas: *Bactrocera carambolae* Drew and Hancock, *B. dorsalis* (Hendel) (only present in Hawaii), and *B. oleae* (Rossi) (Mello Garcia 2024a). In the case of the genus *Ceratitis*, Mediterranean fruit fly (Medfly) *C. capitata* (Wied.), a single species native to tropical areas of Africa is recorded, which has been established in the Americas since the early 1900s (Mello Garcia 2024a). These are species that in some cases have a generalized distribution in the different countries of the Americas; in other cases, they have a distribution restricted to certain areas.

There is another important number of species of fruit flies (Diptera: Tephritidae), not present in the American continent, which represent a high risk for fruit growing in the countries, such as *B. scutellata, B. tau, B. latifons, B. frauenfeldi, B. cucurbitae, B. zonata,* and *B. albistrigata* (APHIS 2024). Some of them were detected in sporadic incursions that were controlled by the phytosanitary authorities of the affected countries.

18.3 ESTABLISHMENT OF NATIONAL FRUIT FLY PROGRAMMES IN THE AMERICAS

The countries of the Americas have been making efforts for the integrated management of these species of fruit flies. Many countries have implemented national programmes with different levels of action, according to their economic and technical resources, and with the importance of fruit growing for the countries and the economic impact that these pests may cause to their productive areas.

Countries such as Argentina, Brazil, Chile, Guatemala, Mexico, Peru, and the USA have national programmes that include prevention activities for non-native fruit flies of quarantine importance through the operation of epidemiological surveillance networks. These countries have biofactories and sterilization plants for the suppression, eradication, and prevention of fruit flies with the use of the sterile insect technique (SIT), mainly focused on the control of the Medfly. In some cases, these programmes have allowed countries to declare themselves as free of fruit flies and, in other countries, programmes have allowed the establishment of free areas or of low pest prevalence areas of these pests. Also, SIT has been used against the Mexican fruit fly (*A. ludens*) and the West Indies fruit fly (*A. obliqua*) to eradicate populations and establish areas of low fruit fly prevalence and free areas. Recently, pilot projects have been carried out to demonstrate its technical and economic feasibility with the South American fruit fly (*A. fraterculus*) in some of its morphotypes (Brazil, Argentina, and Peru) (Aluja et al. 2024).

In other cases, countries have managed to establish their National Fruit Fly Programmes, with progress in the implementation of surveillance activities against non-native flies of quarantine importance and the establishment of areas of low pest prevalence (ALPP) with a systems approach that does not include the SIT. Additionally, countries developed the application of some post-harvest phytosanitary treatments or the maintenance of small areas which are naturally free without the use of SIT (Reyes et al. 2007).

Despite these advances, it is important to mention the situation in which some countries do not have fruit fly programmes. The primary reason is that fruit flies are not considered a phytosanitary problem of high national priority or because their economy does not depend on agriculture but on tourism. This is the case for some Antillean and Caribbean countries (van Sauers-Muller and Goldsmith 2025).

In an analysis of the conditions of National Fruit Fly Programmes (personal communication with national programme coordinators and some plant health directors), the countries that have made the greatest progress are those in which fruit fly programmes are considered to be of public interest. These programmes are supported by a legal basis or phytosanitary regulation, and therefore they have their own budget and an operational structure.

18.3.1 North America

For North American countries, the Canadian Inspection Agency (CFIA) is part of the plant protection mandate and the Plant Protection Agency and has adopted regulatory measures under the Plant Protection Act (PPA) to prevent the introduction and spread of fruit flies into Canada (Department of Justice Canada, PPA). They have the Blueberry Protection Programme to prevent the spread of the blueberry maggot, *Rhagoletis mendax*, within Canada; the British Columbia Apple Maggot (*R. pomonella*) Programme, and the Programme to prevent the introduction and spread within Canada of the European Cherry Fly (*R. cerasi*) (Julie Laplante, personal communication). In the USA, general regulatory authority is provided by the PPA of June 20, 2000, as amended, Section 412(a), 7U.S.C. 7712(a). This act authorizes the U.S. Secretary of Agriculture to prohibit or restrict the movement in interstate commerce of any plant, plant part, or article if the secretary determines that the prohibition or restriction is necessary to prevent the spread of a plant pest within the USA. The U.S. Fruit Fly Regulation is detailed in the Code of Federal Regulations (CFR) 7 301.32 (CFR 2025).

Mexico has a Federal Law on Plant Health published in 1994 through which plant health is regulated and promoted and generally its measures are observed throughout the territory. Also, Mexico has regulations (Norma Oficial Mexicana), which have a budget to attend all activities related to fruit flies. This budget allows Mexico to maintain Medfly, as well as Mexican fruit fly and West Indies fruit fly, production and sterilization facilities and to successfully keep the country free of Medfly and to advance in the National Campaign against endemic fruit fly native species (Gutiérrez-Ruelas et al. 2013). Medfly was eradicated from Mexico in 1982. This goal was achieved through the Moscamed Programme, which was established as a tripartite international commission by the governments of Mexico, Guatemala, and the USA (Enkerlin et al. 2015). Since 1982, the programme has maintained a biological barrier between the borders of Guatemala and Mexico, preventing the introduction and northward spread of the pest.

18.3.2 Central America

Also, the Moscamed Programme has protected the region of Peten in the north of Guatemala from the introduction and spread of the pest and has kept Belice as a Medfly-free country (Enkerlin et al. 2017). The programme was fundamental to creating capacity for fruit fly surveillance and control in the Central American countries. This has been done with the assistance of the International Atomic Energy Agency (IAEA) technical cooperation projects (Aluja et al. 2024). Through one of these regional projects, at least one fruit fly-free area and low-prevalence areas were established and are being maintained in each of the Central American countries with the production of fruits and vegetables including tomatoes, bell peppers, and papayas that are being exported to the USA (Reyes et al. 2007).

18.3.3 South America

In the case of South America, Argentina is another country that has made significant progress in fruit fly research and development and in the establishment and maintenance of Medfly-free areas. This country has Law 27233 (SENASA 2015) that declares animal and plant health to be of national interest, which was regulated through Decree 776/2019 (SENASA 2019). This regulation strengthens the legal status of the National Service of Agrifood Health and Quality (SENASA), formalizes its competencies, and establishes the appropriate mechanisms for its financing, allowing it to deepen the levels of efficiency of public management to face the challenges. This law allows SENASA to work with better rules for the management of health issues (Garavelli et al. 2025).

Chile is the only country in the Americas which is free of fruit flies. Medfly was eradicated in 1995 using SIT. This has been possible because of the long-standing commitment from the National Agriculture and Livestock Service (SAG) and strong alliance with the grower and exporter associations supporting the national fruit fly programme. This has allowed Chile to diversify its export markets and the horticultural industry to become a key contributor to the country's gross domestic product (Yevenes Flores et al. 2025).

In 2016, Ecuador launched a National Fruit Fly Programme under the framework of AGROCALIDAD (Agencia de Regulación y Control Fito y Zoosanitario). Under this programme and with the support of IAEA's technical cooperation projects, areas where nontraditional fruits are cultivated in the highlands have been established as fruit fly low-prevalence areas and as fruit fly-free places of production. This has allowed exports of fruits including dragon fruit, golden berries, and tree tomato to China, Peru, and the USA, with a value of over US$ 80 million per year (IAEA 2024).

Peru has a Supreme Decree No. 009-2000-AG, which approved the regulations for the control, suppression, and eradication of fruit flies. The regulations are of general observance in Peruvian territory and their provisions are of public order and social interest (Carlos Rivera, personal communication and Liz Karen Villanueva, personal communication). The same legislative support occurs in other countries that have well-structured fruit fly programmes, with which they keep their countries free or have areas free of this pest within their territories. The activities developed in the programmes of some countries are based on the experiences of other countries where greater development has been achieved in the control of these pests. Also, these countries are very much aligned with the approved standards in the International Plant Protection Convention (IPPC) in relation to the problem of fruit flies. However, in all programmes, regardless of their level, the quality of the actions of the programmes varies, which exposes countries to a greater risk of introduction and establishment of non-native fruit flies of quarantine importance.

18.4 PLANT PROTECTION REGIONAL ORGANIZATIONS

A revision of the situation of fruit flies at the regional level of the countries that make up the Regional Plant Protection Organizations (RPPOs) with respect to regulatory issues showed the following situation:

18.4.1 NAPPO

In the case of the North American Plant Protection Organization (NAPPO), all member countries have established National Fruit Fly Programmes with legal and financial support and have fruit fly-free and low-prevalence areas.

18.4.2 OIRSA

In the case of the International Regional Organization for Agricultural Health (OIRSA), all member countries have National Fruit Fly Programmes. However, not all countries have a regulation at the country level that allows them to have the sustainability of their programme, independent of the governments in power and the directors of the organizations in charge of the programme. In some of the member countries, areas of low prevalence and freedom of some species of fruit flies have been established and maintained. OIRSA provides a forum, for technical work teams on phytosanitary issues to help with economic resources to address outbreaks of non-native fruit flies, through phytosanitary emergency mechanisms, which allows them to carry out prevention or eradication actions (Carlos Urias, personal communication).

18.4.3 CAN

In relation to the member countries of the Andean Community (CAN), not all countries have legislation, and therefore the sustainability of their fruit fly programmes depends on the policies of the governing administrations. Only Peru, through SENASA, has a national programme for the suppression and eradication of fruit flies. Ecuador has a national programme through AGROCALIDAD for the establishment of fruit fly-free sites of production in the highlands. Ecuador does not have a regulation related to fruit flies, even though it has established technical working groups trying to articulate actions for the surveillance and management of fruit flies as quarantine pests. In the other countries, there are some pilot suppression projects for the establishment of low-prevalence and free areas.

Despite CAN being the only RPPO in the Americas that issues decisions and resolutions that are binding and superior in hierarchy to the national standards of its member countries (McAllister Tafur, personal communication; Claudia Cabezas, personal communication; Plata López and Yepes Ceballos 2009), it does not have a regulation related to fruit flies, even though it has established technical working groups trying to articulate actions for the surveillance and management of fruit flies as quarantine pests. Yet, it has many examples of the integration that can be achieved in the actions carried out against an invasive pest by the member countries, which can be seen in the resolutions issued by this RPPO. For example, Resolution No. 2234 adopted the "Phytosanitary surveillance actions for *Fusarium oxysporum* f.s. *cubense* tropical race 4 (syn *Fusaarium odoratissimum*)" (Secretaria General CAN 2021) and Resolution 1850 adopted the "Andean Plan for Prevention and Contingency for Huanglongbing Citrus Disease" (Secretaria General CAN 2016). Something similar would be interesting to have in the region in the case of surveillance and control of outbreaks of non-native flies of quarantine importance.

18.4.4 COSAVE

At the level of COSAVE (Technical Committee on Plant Health of the Southern Cone), some countries have legislation at the country level, through which they declare fruit flies as pests of public interest. This legislation allowed those countries to install biofactories for the mass rearing of insects of interest and irradiation plants for the production of sterile insects and the establishment of suppression and eradication programmes for some species of fruit flies of public interest (Ricardo Rodríguez personal communication).

18.4.5 CAHFSA and other Caribbean PPOS

In the case of the recently formed CAHFSA (Caribbean Agricultural Health and Food Safety Agency) (CAHFSA 2024), not all of its countries have fruit fly programmes at the national level, nor do they have legislation at the country level declaring fruit flies to be of public interest. Additionally, they do not have a recurring budget for the implementation of their activities. This may be due to the source of income of their economies, in which tourism is prioritized and not agriculture (Alies van Sauers-Muller, personal communication). On the other hand, not all CARICOM (Caribbean Community) countries are part of the CAHFSA, because not all of them are independent, but their territories are part of other countries, which belong to other Regional PPOs, a situation that makes it difficult to integrate their actions.

Notwithstanding the above, in the Caribbean, there is an organization that allows them to coordinate their actions in the field of plant health and in turn serves as a technical committee of the CAHFSA. This is the Caribbean Plant Health Director Forum, in which aspects related to the phytosanitary risks common to them are discussed (Alies van Sauers-Muller, personal communication). Some countries have surveillance programmes for non-native flies of quarantine importance for the region and work on exclusion programmes, financially supported by the United States Department of Agriculture (USDA) of Trinidad and Tobago. Unfortunately, in the area of phytosanitary legislation, the situation is quite precarious in the Caribbean, where there are also problems with the availability of technicians for surveillance work. Only a few Caribbean islands have implemented phytosanitary legislation regarding this problem, such as Antigua, Barbados, and Jamaica. Also, Grenada had implemented its legislation earlier. However, the sustainability of the programmes in the countries where they have been established is uncertain, because donated resources are becoming less frequent (Nelson Laville, personal communication). CAHFSA (Committee of Plant Health Directors) has formed technical working groups, including the fruit fly group (Diptera: Tephritidae). The main objective is to protect the Caribbean from the threats and/or effects of the introduction, economic damage, and spread of fruit flies through research, advice, and training on this topic (CAHFSA 2024).

It is necessary to highlight here the importance of legislation at the country level and not only at the level of institutions, to achieve the sustainability of the programmes over time and meet the objective of keeping our territories free of non-native fruit flies of quarantine importance. Many examples from different countries can be found in Garcia Mello (2024b).

18.5 REGIONAL AND INTERNATIONAL SUPPORT FOR FRUIT FLY PROGRAMMES IN THE AMERICAS

Fruit flies are a group of insect pests for which the most advances have been recorded at the level of research in different areas of knowledge such as biology, ecology, taxonomy, genetics, physiology, morphology, and surveillance and control, with significant advances made in the management of these pests. The development and use of various phytosanitary measures and risk mitigation schemes include the systems approach and the establishment of free areas using integrated pest management with SIT. Regarding SIT, some species use genetic sexing strains for the release of sterile males only, which has increased the efficiency of the technology and made it more sustainable.

18.5.1 TWWH

The transfer of all this knowledge and technology development showed the need to bring together researchers and technical personnel in charge of programmes against fruit flies. It was in the 1980s, when the Tephritid Workers of the Western Hemisphere (TWWH) was established to achieve this goal, particularly to address issues associated with the programmes against the Medfly in Guatemala, Mexico, and the USA. The objective was to analyse and discuss research needs and priorities to more effectively address the problems of these programmes. Subsequently, a similar group was formed to address the research needs of *Anastrepha* fruit fly programmes in the Americas. In the 1990s, after the agreement to unite these two groups, the TWWH was created. In the last three decades, the number of programmes implemented against fruit flies in the Western Hemisphere has increased significantly.

18.5.2 IAEA

The IAEA (through the Joint FAO/IAEA Centre for Nuclear Techniques in Food and Agriculture) is one of the international organizations that has been a pillar for research and development on fruit flies and has supported the establishment of national and regional fruit fly programmes in the Americas. The FAO/IAEA Centre promotes and supports (among FAO and IAEA Member States) the safe and correct use of nuclear and related technologies in food and agriculture to contribute to global food security and sustainable agricultural development in the world. This is achieved through adaptable research and development activities in its own laboratories in Seibersdorf. Additionally, the centre supports and coordinates more than 25 research project networks worldwide involving some 400 research institutions and experimental stations annually. Further, the centre supports capacity development and technology transfer to more than 200 national and regional technical cooperation projects; and technical and policy advice to phytosanitary policymakers (IAEA 2025). Nuclear technologies promoted by the Joint FAO/IAEA Centre, through the Insect Pest Control Section, include SIT. This technique is integrated with other sustainable and environmentally friendly surveillance and control methods to control the main insect pests of importance to agriculture, livestock, and humans. SIT focuses on the IPM over wide areas, consisting of the application of SIT where feasible, to promote food security and international food trade (IAEA 2025).

Additionally, the IAEA has supported the development of International Standards for Phytosanitary Measures (ISPMs), which are discussed and approved through the IPPC and become international standards that are adopted by the Commission on Phytosanitary Measures. These standards are used by the National Plant Protection Organizations (NPPOs) of the member countries to harmonize and adopt these standards for the benefit of international trade in plant products (IPPC 2025). In the specific case of fruit flies, there are several regulations related to this group of pests, such as:

- ISPM No. 26 "Establishment of pest-free areas for fruit flies (Tephritidae)".
- ISPM No. 28 "Phytosanitary treatments for regulated pests", which includes 29 of the 46 annexes, phytosanitary treatments for different species of fruit flies of the Tephritidae family and in different types of fruits.
- ISPM No. 35 "Systems approach to the risk management of fruit fly pests (Tephritidae)". It combines monitoring and control practices to reduce pest risk that may or may not include postharvest treatment.
- ISPM No. 37 "Determination of the Status of a Fruit as a Host for Fruit Flies (Tephritidae)".

The lack of national programmes, with their own operational structure and budget, is one of the main challenges faced by the IAEA's national and regional technical cooperation projects. This deficiency limits the implementation in member countries because these countries lack their own operational structure and budget to adequately deliver their programmes. Additionally, the lack of phytosanitary legislation (that supports the implementation of programmes and emergency actions in the event of incursions of the pest, noncompliance with commitments by NPPOs) limits the implementation of work plans. Furthermore, programmes are also affected by the wrong choice of national counterparts responsible for project implementation or frequent changes of National Counterparts for political reasons.

18.5.3 IICA

Another international organization that has been supporting the area of fruit flies is the Inter-American Institute for Cooperation on Agriculture (IICA). This organization has contributed with its member countries to organize their phytosanitary schemes and to strengthen fruit fly programmes. This support included its participation in different activities and support for the Inter-American Coordination Group on Plant Health (GICSV), through its technical secretariat, for the development of the different operational plans prepared by representatives of the RPPOs (Aluja et al. 2024).

18.5.4 Other organizations

Similarly, other development and financial organizations that have contributed to the establishment and strengthening of National Fruit Fly Programmes include the Standards and Trade Development Facility, the IDB, and the World Bank (Aluja et al. 2024). In the case of RPPOs, there is support from the GICSV and its Fruit Fly Technical Group (GTMF) created to support the different actions around this pest.

The GICSV is a collaborative arrangement of RPPOs from the countries of the Western Hemisphere (America and the Caribbean). GICSV was created to promote joint actions aimed at preventing the introduction and spread of pests in plants, plant products, and other regulated items, as well as encouraging the adoption of appropriate measures for their control. This group has a long history, but it was not until 1998 that its constitution was formalized through a memorandum of understanding signed by the representatives of the RPPOs of the Americas (GICSV 2025).

Despite all this support and all this progress, the lack of national programmes with legal and budgetary stability in several countries of the Americas, the scarce availability of emergency protocols in some countries and the absence of quality certification schemes for each of the basic activities of the programmes, put at risk progress in the establishment and maintenance of areas free or of low prevalence of some species of fruit flies of quarantine importance. Unfortunately, these limitations open the doors to the possible incursion and establishment of non-native fruit flies of quarantine importance for the continent.

18.6 TIERED APPROACH

18.6.1 Why

Below are some reflections taken from Chapter 16 of the book *Management of Fruit Flies in the Americas* (Mello Garcia 2024b), of which one of the authors of this publication is also one of its authors, entitled "The management of fruit flies in Colombia: a long road to travel" (Canal and Arevalo 2024). Also, these reflections are based on conversations with some directors of plant or fruit fly health programmes in several countries of the American continent. As has been observed, there has been international support with the participation of different organizations, which has been increasing and becoming more constant. It is time for countries to strengthen or establish their national fruit fly programmes, to make a transboundary front to more effectively control endemic fruit fly pests and prevent the introduction and establishment of non-native fruit flies of quarantine importance for the American Continent and for the territories.

To this end, it is essential that fruit fly programmes are on the institutional agendas of each country and have a regulation at the country level that allows them to have a budget for the development of their activities. Several countries in the Americas have regulations at the level of laws or decrees that establish priority phytosanitary programmes against pests that impact regional economies and have, as a priority, the establishment of pest eradication programmes for the development of free areas, declaring these programmes of public interest. Therefore, it is an unavoidable priority to establish the legal framework that gives sustainability to the programme over time, independent of the governments and managers in power.

What is important is the commitment at the country level to establish the programme in a tiered manner with quality parameters, advancing according to the objectives, the availability of financial resources and according to the impact that this pest may cause to agriculture or the economic and environmental sector of each country. The tiered approach consists of using a different intervention strategy in each country that ranges from programmes with the most basic structure to well-structured programmes, according to the particular situation in which countries find themselves in the management of fruit flies, starting with the fundamental pillars that give sustainability to the programmes for the control of endemic species and the prevention of non-native flies of quarantine importance.

18.6.2 First pillar

The most basic condition in the tiered approach can be found in those countries that do not have a fruit fly programme and for which it is considered essential to establish a National Fruit Fly Programme with three strategic pillars. The first pillar is to prevent the introduction of non-native fruit fly species at high-risk points of entry (ports, international airports, and border sites) through the enforcement of quarantine regulations and a surveillance programme for the timely detection of these species that are a threat to the continent or to the territory of each country. A surveillance programme should follow international standards, so that it can be integrated into a network that is established in the Western Hemisphere for the surveillance of quarantine fruit flies.

18.6.3 Second pillar

Similarly, the second pillar is the development of capacities for the control of endemic species. These strategies include other basic control methods such as a basic IPM where phytosanitary measures are used, population monitoring using traps and sampling of fruit hosts, and population control using insecticide baits and the collection and destruction of fruits. Harmonized guidelines for fruit fly trapping and fruit sampling and phytosanitary procedures for their control from the IAEA and FAO are key to achieving this goal, of course adapted to the conditions of each territory (FAO/IAEA 2018, 2019, 2023). In this regard, the IAEA and FAO have already made important efforts through national and regional cooperation projects where capacities have been developed through training of staff, expert visits and the provision of trapping, diagnostic and control materials, as well as the development of a regional database.

18.6.4 Third pillar

A third pillar is to be able to have contingency plans for the timely response to possible outbreaks that may occur, with harmonized strategies for their eradication. The purpose is to prevent the establishment of non-native fruit flies of quarantine importance in the free areas of the Western Hemisphere. Countries should strengthen this activity with harmonized and standardized procedures. In this regard, international organizations such as the IAEA and FAO and regional organizations, such as OIRSA and COSAVE, have provided intensive training to their member countries that include field drills through technical cooperation projects.

18.6.5 Tiered approach implementation

The development and/or strengthening of these three pillars is the basis for sustainable fruit fly programmes, which contribute positively to local production and strengthen protection against non-native fruit flies of quarantine importance. The purpose is "To prevent the introduction of more species of fruit flies to a territory and to achieve more effective control of established species". In a tiered approach, the three pillars are established in the first step of the **Basic Level** of the fruit fly programmes.

The second step of the three-tiered approach, or **Intermediate Level**, of the programme is for countries that have already established a National Fruit Fly Programme. The Intermediate Level is where, through their activities, they have achieved admissibility in the international market of their fruit and vegetable production, through the implementation of mitigation measures with phytosanitary treatments or the establishment of small pest-free areas (PFAs), ALPP, or pest-free sites of production within a systems approach. This activity requires that fruit fly monitoring and control procedures be used in the defined production areas and for prioritized fruit products. The aim is that countries that are already in this condition strengthen their procedures and activities, both for monitoring and for control in the production sites where it is important that the phytosanitary procedures are in line with harmonized procedure manuals of the IAEA and FAO and with fruit fly ISPMs. The subscription of partnerships with beneficiaries and other stakeholders is expected. Apart from the international quarantine regulations at points of entry, national quarantine regulations need to be in place to protect the PFA. An active communication plan is necessarily focused on beneficiaries and the public at large, in support of the National Fruit Fly Programme.

At the third step, or **Advanced Level**, we could include countries that have been using an IPM approach on an area-wide basis, that is, countries that have mass rearing and sterilization facilities for the application of SIT for eradication and the establishment of PFA and ALPP in large-scale interventions as well as sterile fly release programmes as a preventive method to the introduction and establishment of non-native flies of quarantine importance (Enkerlin 2021). Generally, countries in this condition apply surveillance and IPM measures in accordance with harmonized procedures manuals of IAEA and FAO and with fruit fly ISPMs. The advanced level requires the strengthening and expansion of international and national quarantine networks. Also, it requires expanding the communication and outreach plan to have the support of stakeholders and the public at large, in particular, for the smooth implementation of area-wide SIT actions (Figure 18.1).

Regardless of the level at which a country stands in the tiered approach, programme activities and plant resources should always be protected from the introduction of other species of quarantine importance, ensuring effective surveillance and in all cases having contingency plans following the international standards available in ISPMs. An essential element is to establish extension activities aimed at producers in the integrated management of endemic fruit flies, which significantly affect fruit production in each country.

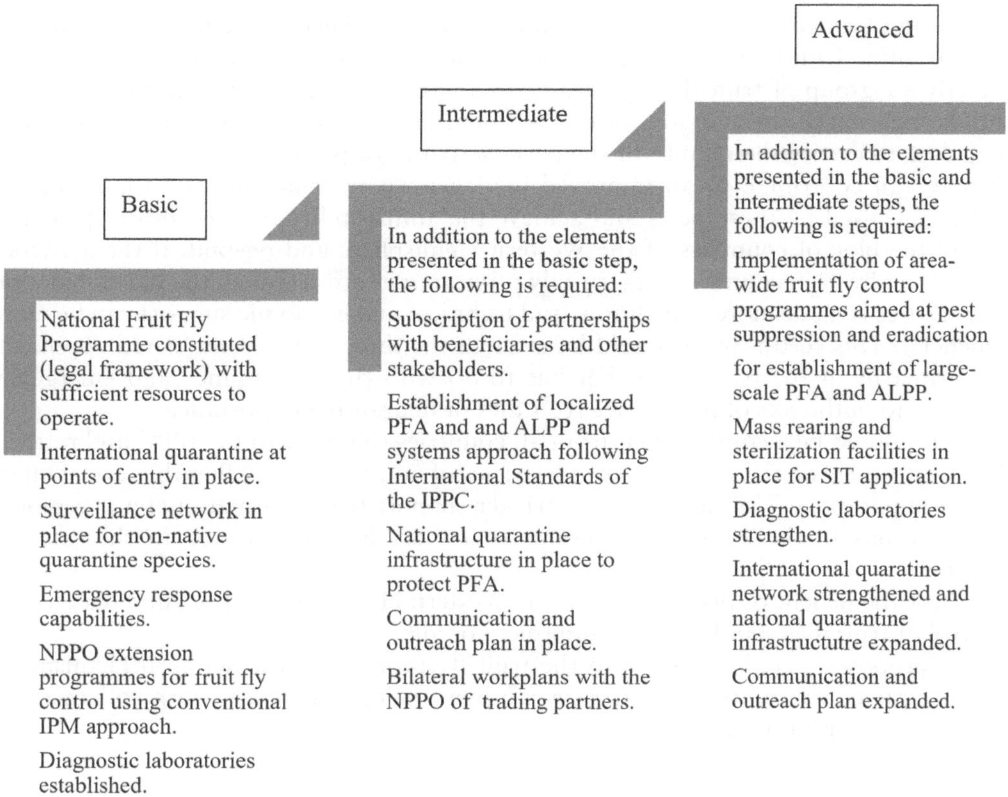

Figure 18.1 Tiered approach levels for the implementation of fruit fly control programmes in Latin America and the Caribbean.

This activity would improve the quality of local production and strengthen the articulation of farmers with the actions of the National Fruit Fly Programme. Additionally, this action promotes public-private partnerships that give sustainability to the programmes. Several countries have implemented important programmes in this regard.

18.7 FINAL RECOMMENDATIONS

At the Western Hemisphere level, the formation of a group of experts (called the Technical Advisory Panel – TAP) should be formalized to support national or regional programmes. TAPs should be called to action in the event of outbreaks of flies of quarantine importance that require their eradication, with the support of the IAEA, FAO, OIRSA, IICA, the RPPOs of the Western Hemisphere and the GICSV. An example of a TAP-supporting eradication actions has been seen in recent years with the outbreaks of *C. capitata* that have occurred in the Dominican Republic and have been successfully eradicated (Zavala et al. 2021).

Additionally, TAP would be called into action in the case of ongoing programmes to review and provide advice on specific activities of the programmes. So far, the TAPs have been implemented at the request of countries and under the leadership of the FAO/IAEA Joint Center of Nuclear Techniques in Food and Agriculture mostly in response to quarantine fruit fly outbreaks. A more permanent TAP could be established under the GICSV group of fruit fly experts, as an advisory group for the improvement and standardization of the activities of the National Fruit Fly Programme of each country including to certify the quality of the basic activities of programmes. In this way, this activity can contribute to an improved management of fruit flies and the reduction of losses caused by these pests and achieve the admissibility of agricultural products within the bloc of countries of the Western Hemisphere and beyond. If the activities carried out by each country in its fruit fly programmes are certified, the harmonization of measures is made more expeditious and technical and economic support can be more beneficial. This certification should be given at the three levels of the tiered approach, but primarily at the basic level that has to do with the surveillance and emergency response to outbreaks of non-native fruit flies of quarantine importance.

There should be a commitment from all countries to report to the IPPC and to their trading partners on the detection of any species of non-native fruit flies that are of quarantine importance. This action would trigger support from international and regional organizations to the eradication actions that must be implemented in the affected country.

All for the benefit of plant health in the Western Hemisphere "The surveillance and control of fruit flies should be a commitment of all".

The proposed tiered approach to the fruit fly problem in the Western Hemisphere can be applied to any other region or continent facing challenges with endemic and non-native quarantine species.

ACKNOWLEDGEMENTS

The information obtained by personal communication with several persons is gratefully acknowledged: Ricardo Rodríguez (Former Director of the Chile Fruit Flies Program – retired); Alies Muller (Fruit Fly Programme – Suriname); Wilda Ramírez (Former Director of Plant Health – SENASA Argentina – retired); Carlos Rivera (Fruit Fly Programme – Peru); Francisco Ramirez y Ramírez (General Director of Plant Health – Senasica – Mexico); Gerardo Ortiz (Fruit Fly Consultant – Mexico); Pedro Morales (INIA Researcher – Venezuela); Carlos Urias (Regional Director of Plant Health OIRSA- El Salvador); Nancy Villegas (OIRSA Risk Analysis – El Salvador); Stephanie Bloem (former Director NAPPO-USA); Richard Johnson (APHIS-USA National Policy Manager); Julie Laplante (Horticulture Section – CFIA – Canada), Nelson Laville (President CPHD – Dominica), Claudia Cabezas (former CAN official), McAllister Tafur (former CAN official), and Katty Rojas (CAN official).

REFERENCES

Aluja, M., S. M. Ovruski, F. R. Mello Garcia, et al. 2024. Fruit fly (Tephritidae) management in the neotropical region: History, state of the art, and perspectives. In *Management of Fruit Flies in the Americas*, ed. F. R. Mello Garcia, 11–66. Cham, Switzerland: Springer.

APHIS. 2024. Exotic Fruit Flies. https://www.aphis.usda.gov/plant-pests-diseases/fruit-flies (accessed May 20, 2024).

CAHFSA (Caribbean Agricultural Health and Food Safety Agency). 2024. https://www.cphdforum.org/ (accessed September 12, 2024).

Canal, N.A., and E. Arévalo-Peñaranda. 2024. The management of fruit flies in Colombia: A long road to travel. In *Management of Fruit Flies in the Americas*, ed. F. R. Mello Garcia, 289–316. Cham, Switzerland: Springer. https://doi.org/10.1007/978-3-031-48608-1_1

CFR (Code of Federal Regulations). 2025. eCFR :: 7 CFR Part 301 Subpart C -- Fruit Flies (accessed March 30, 2025).

Enkerlin, W. R. 2021. Impact of fruit fly control programmes using the sterile insect technique. In *Sterile Insect Technique: Principles and Practice in Area-Wide Integrated Pest Management*, ed. V. A. Dyck, J. Hendrichs, and A. S. Robinson, 979–1006. Boca Raton, FL: CRC Press.

Enkerlin, W. R., J. M. Gutiérrez Ruelas, C. Cáceres Barrios, et al. 2017. The Moscamed Regional Programme: A success story of area-wide sterile insect technique application. *Entomologia Experimentalis et Applicata* 164(3):1–16.

Enkerlin W., J. M. Gutiérrez-Ruelas, A. Villaseñor Cortes, et al. 2015. Area Freedom in Mexico from Mediterranean fruit fly (Diptera: Tephritidae): A review of over 30 years of a successful containment using an integrated area-wide SIT approach. *Florida Entomologist* 98(2):665–681.

FAO/IAEA. 2018. *Trapping Guidelines for Area-Wide Fruit Fly Programmes*, ed. W. R. Enkerlin, and J. Reyes-Flores. Rome, Italy: Food and Agriculture Organization of the United Nations.

FAO/IAEA. 2019. *Fruit Sampling Guidelines for Area-Wide Fruit Fly Programmes*, ed. W. R. Enkerlin, J. Reyes-Flores, and G. Ortíz. Rome, Italy: Food and Agriculture Organization of the United Nations.

FAO/IAEA. 2023. *Guideline on Phytosanitary Procedures for Area-Wide Management of Fruit Fly Pests*, ed. J. Reyes-Flores, W. R. Enkerlin, J. Hendrichs, and R. Pereira. Vienna, Austria: Food and Agriculture Organization of the United Nations/International Atomic Energy Agency.

Garavelli, E., M. F. Vazquez, and P. Fedyszak. 2025. National fruit fly control and eradication program (procem) in Argentina. In *Fruit Flies of Economic Importance XI*, ed. Dominiak, B., D. Perez-Staples, M. T. Vera, and A. Bakri. Netherlands: CRC Press.

Grupo Interamericano de Coordinación en Sanidad Vegetal (GICSV). 2025. https://apps.iica.int/GICSV/programas/SanidadVegetal/pages/informacionGeneral.aspx.

Gutiérrez-Ruelas, J. M., G. S. Martínez, A. Villaseñor Cortes, et al. 2013. *Los Programas de Moscas de la Fruta en México. Su Historia Reciente*. México, D. F., México: Talleres de S y G Editores.

Hernández-Ortiz, V., M. Hernández-López, and G. J. Steck, G. J. 2020. Morfología y taxonomía de Tephritidae: Especies de importancia económica y cuarentenaria en América. In *Moscas de la fruta: fundamentos y procedimientos para su manejo*, ed. P. Montoya, J. Toledo, and E. Hernández. pp. 71–116, 71–116. Mexico: SyG Editores.

IAEA (International Atomic Energy Agency). 2024. Insect Pest Control Section (IPCS). Newsletter 102. https://www.iaea.org/publications/15624/insect-pest-control-newsletter-no-102-january-2024 (accessed March 30, 2025).

IAEA (International Atomic Energy Agency). 2025. Insect Pest Control Section (IPCS). https://www.iaea.org/about/insect-pest-control-section (accessed March 30, 2025).

IPPC (International Plant Protection Convention). 2025. International Standards for Phytosanitary Measures. https://www.ippc.int/es/core-activities/standards-setting/ispms/ (accessed March 30, 2025).

Mello Garcia, F. R. 2024a. Fruit flies: A threat to fruit production in the Americas. In *Management of Fruit Flies in the Americas*, ed. F. R. Mello Garcia, 3–9. Cham, Switzerland: Springer. https://doi.org/10.1007/978-3-031-48608-1_1.

Mello Garcia, F. R. (ed.). 2024b. *Management of Fruit Flies in the Americas*. Cham, Switzerland: Springer.

Orús, A. 2024. Volumen de producción global de fruta 2000–2022. https://es.statista.com/estadisticas/635091/produccion-de-fruta-fresca-a-nivel-mundial-de-1990-a/.

Plata López, L. C., and D. Yepes Ceballos. 2009. Naturaleza jurídica de las normas comunitarias andinas. Revista de Derecho 31:196–223.

Reyes, J., X. Carro, J. Hernandez, et al. 2007. A multi-institutional approach to create fruit fly low prevalence and fly-free areas in Central America. In *Area-Wide Control of Insect Pests*, ed. M. J. B. Vreysen, A. S. Robinson, and J. Hendrichs, 627–640. Vienna, Austria: IAEA.

Rodríguez, D. P. 2023. EE.UU. es el principal exportador de fruta en todo el mundo. https://www.larepublica.co/economia/estados-unidos-es-el-principal-exportador-de-fruta-en-todo-el-mundo-3609856 (accessed May 20, 2024).

van Sauers-Muller, A. and J. Goldsmith. 2025. Fruit flies and their monitoring in the Caribbean Region. In *Fruit Flies of Economic Importance XI*, ed. Dominiak, B., D. Perez-Staples, M. T. Vera, and A. Bakri. Netherlands: CRC Press.

SENASA (Servicio Nacional de Sanidad y Calidad Agroalimentaria). 2015. Ley 27233. https://www.boletinoficial.gob.ar/detalleAviso/primera/11499884/20160724 (accessed March 30, 2025).

SENASA (Servicio Nacional de Sanidad y Calidad Agroalimentaria). 2019. Decreto 776/2019. https://www.boletinoficial.gob.ar/detalleAviso/primera/221429/20191120?busqueda=2 (accessed March 30, 2025).

SGCAN (Secretaría General CAN). 2016. Resolución No. 1850. Adopción del Plan Andino de prevención y contingencia para la enfermedad de los cítricos Huanglongbing. https://www.comunidadandina.org/DocOficialesFiles/resoluciones/RESO1850.pdf (accessed March 30, 2025).

SGCAN (Secretaría General CAN). 2021. Resolución No. 2234. Acciones de Vigilancia Fitosanitaria para *Fusarium oxysporum* f. sp. *cubense* raza 4 Tropical Foc R4T (syn. *Fusarium odoratissimum*). https://www.comunidadandina.org/DocOficialesFiles/Gacetas/Gaceta%204367.pdf (accessed March 30, 2025).

Statista. 2024. América Latina: principales países productores de fruta 2021. https://es.statista.com/estadisticas/580269/volumen-de-produccion-frutas-america-latina-por-pais/ (accessed March 30, 2025).

Yevenes Flores, J., J. Nitor Carrillo, N. Álvarez Ugalde, et al. 2025. Current status of the national fruit fly detection programme in Chile. In *Fruit Flies of Economic Importance XI*, ed. Dominiak, B., D. Perez-Staples, M. T. Vera, and A. Bakri. Netherlands: CRC Press.

Zavala-Lopez, J., G. Marte-Diaz, and F. Martinez-Pujols. 2021. Successful area-wide eradication of the invading Mediterranean fruit fly in the Dominican Republic. In *Area-Wide Integrated Pest Management: Development and Field Application*, ed. J. Hendrichs, R. Pereira, and M. J. B. Vreysen, 519–537. Boca Raton, FL: CRC Press.

Chapter 19

Fruit flies and their monitoring in the Caribbean region

Alies van Sauers-Muller and Juliet Goldsmith

19.1 INTRODUCTION

The Caribbean region is a vibrant and diverse area comprising numerous islands and several mainland countries in Central and South America. This region is characterized by a rich tapestry of ecosystems, agricultural practices, and cultural heritage. The Caribbean Community (CARICOM) serves as the region's primary economic and political union, consisting of 15 Member States (Antigua and Barbuda, The Bahamas, Barbados, Belize, Dominica, Grenada, Guyana, Haiti, Jamaica, Montserrat, Saint Kitts and Navis, Saint Lucia, Saint Vincent and the Grenadines, Suriname, and Trinidad and Tobago), and 5 Associate Members (Anguilla, Bermuda, The British Virgin Islands, Curacao, and Turks and Caicos Islands). All CARICOM members, except for Belize, Guyana, and Suriname, are island states. Additionally, CARICOM includes observer countries, which are neighboring nations outside the Caribbean collaborating on various initiatives. Additional information on CARICOM composition can be found in CARICOM web site.

The primary objective of CARICOM is to promote economic integration and cooperation among its members, ensuring that the benefits of integration are equitably shared. Agriculture is central to this mission, as it is critical for food provision, employment, and economic development. However, the region faces significant challenges in maintaining agricultural productivity and competitiveness. One of CARICOM's key initiatives is the "25 by 25" goal, which aims to increase regional food production by 25% by 2025. This ambitious target is essential for reducing dependency on food imports, enhancing food security, and fostering sustainable development.

A major obstacle to achieving this goal is the threat posed by agricultural pests and diseases, particularly fruit flies. These pests not only reduce crop yields but also hinder international trade by triggering quarantine restrictions. Addressing these challenges requires a coordinated regional approach, leveraging the expertise and resources of CARICOM's specialized agencies, such as the Caribbean Agricultural Health and Food Safety Agency (CAHFSA), and collaborative platforms like the Caribbean Plant Health Directors (CPHD) Forum.

DOI: 10.1201/9781003644330-24

19.2 THE IMPORTANCE OF AGRICULTURE IN THE CARIBBEAN

Agriculture is a cornerstone of Caribbean economies, providing livelihoods for millions of people and contributing significantly to food security and rural development. The sector accounts for approximately 16% of overall employment in the region, with higher percentages in some countries, such as Guyana (30%) and Dominica (25%) (Goldsmith, 2024). In addition to its economic importance, agriculture plays a vital role in preserving the region's cultural heritage and biodiversity.

Despite its significance, Caribbean agriculture faces numerous challenges threatening its sustainability and competitiveness. These challenges include the following:

- **Climate change**: The Caribbean is one of the regions which is most vulnerable to climate change. Rising temperatures, changing rainfall patterns, and the increasing frequency and intensity of hurricanes and tropical storms disrupt agricultural production and damage infrastructure. For example, Hurricane Maria in 2017 caused widespread devastation to agricultural systems in Dominica and Puerto Rico, resulting in billions of dollars in losses. The impact of climate change is expected to intensify, leading to further disruptions in food production and increased vulnerability for farmers.
- **Natural disasters**: In addition to hurricanes, the region is prone to other natural disasters, such as floods, droughts, and volcanic eruptions. These events can destroy crops, erode soils, and displace farming communities. Often, the economic costs of these disasters are compounded by the loss of agricultural knowledge and skills, as communities struggle to rebuild.
- **Pests and diseases**: The introduction and spread of invasive pests and diseases pose a significant threat to Caribbean agriculture. Fruit flies, in particular, are a major concern due to their ability to infest a wide range of fruit crops, reducing yields and triggering trade restrictions. The economic impact of fruit fly infestations can be devastating, leading to losses in both local markets and international trade.
- **Tourism-related risks**: Tourism, a major economic driver in many Caribbean islands, increases the risk of pest and disease introduction through the movement of people, goods, and vehicles. For example, the Giant African Snail (*Lissachatina fulica*), an invasive species, was introduced to the region through human activity and has since become a significant agricultural pest. The movement of tourists can inadvertently facilitate the spread of pests, making it essential to implement stringent biosecurity measures.
- **Limited resources**: Many Caribbean countries face resource constraints, including limited access to modern agricultural technologies, insufficient funding for research and development, and inadequate infrastructure for pest surveillance and management. These limitations hinder the ability of farmers to adopt best practices and respond effectively to pest threats.

These challenges underscore the need for regional collaboration and coordinated action to protect Caribbean agriculture and ensure its long-term sustainability.

19.3 CARICOM'S ROLE IN AGRICULTURAL HEALTH

To address the challenges facing Caribbean agriculture, CARICOM has established several specialized agencies and collaborative platforms. One of the most important of these is the CAHFSA, which serves as the Regional Plant Protection Organization for the Caribbean. CAHFSA works closely with the CPHD forum, a platform that brings together plant health directors from CARICOM Member States, Associate Members, and non-CARICOM territories, such as the Dutch, French, and English-speaking islands.

The CPHD was established in 2007 and plays a critical role in safeguarding the region's agriculture and biodiversity. Its inclusive structure is essential, as pests and diseases do not respect national borders. The CPHD meets annually to discuss pest and disease management, share information, and coordinate regional efforts. Also, it forms specialized working groups to address specific challenges, such as the Musa group (focused on banana and plantain pests) and the Mollusc group (established in response to the introduction of the giant African snail).

Key activities of the CPHD include the following:

- **Publishing information**: The CPHD publishes information on pests and diseases on its website (https://www.cphdforum.org/), providing valuable resources for farmers, researchers, and policymakers.
- **Conducting simulation exercises**: These exercises prepare stakeholders for potential pest outbreaks, enhancing the region's readiness to respond effectively.
- **Organizing training programs**: The CPHD organizes training programs and workshops to build capacity in pest surveillance and management and ensure that national plant protection organizations are equipped with the necessary skills and knowledge.
- **Developing technical materials**: The CPHD develops technical materials and fact sheets to support national plant protection organizations in managing pests and diseases.

19.4 FRUIT FLIES: A MAJOR THREAT TO CARIBBEAN AGRICULTURE

Fruit flies, particularly those in the family Tephritidae, are among the most significant agricultural pests in the Caribbean. These insects infest a wide range of fruit crops, causing extensive damage and rendering the produce unmarketable. The economic impact of fruit fly infestations is twofold:

- **Local impact**: Infested fruits are unsuitable for human consumption or sale, reducing farmers' incomes and increasing food insecurity. The loss of income can have cascading effects on rural communities, affecting access to education, healthcare, and other essential services.
- **Trade restrictions**: The presence of fruit flies can trigger quarantine restrictions or outright bans on the export of affected crops, limiting access to international markets or increasing the costs due to the need to implement postharvest measures.

This affects farmers and has broader implications for national economies that rely on agricultural exports.

19.4.1 Fruit fly species in the Caribbean

The majority of fruit fly species in the Caribbean belong to the genus *Anastrepha*. While the number of economically significant species is relatively low, their impact is substantial. Key pest species include the following:

- **West Indian fruit fly (*Anastrepha obliqua*)**: Present in all Caribbean countries except Antigua and Barbuda and the Turks and Caicos Islands.
- **Papaya fruit fly (*A. curvicauda*)**: Found in Belize and Trinidad and Tobago.
- **Star apple fruit fly (*A. ocresia*)**: Found in Jamaica and occasionally in the Cayman Islands.
- **Sapodilla fruit fly (*A. serpentina*)**: Found in Belize, Guyana, Suriname, and Trinidad and Tobago.
- **Caribbean fruit fly (*A. suspensa*)**: Found in the Bahamas, Cayman Islands (eradicated), Jamaica, and Montserrat.
- **Guava fruit fly (*A. striata*)**: Found in Belize, Guyana, Suriname, and Trinidad and Tobago.
- **Carambola fruit fly (*Bactrocera carambolae*)**: Found in Guyana and Suriname.
- **Mediterranean fruit fly (*Ceratitis capitata*)**: Found in Aruba and the Dominican Republic (eradicated twice in the latter).

Additionally, recent research has revealed the presence of many nonpest species of *Anastrepha* in the region, particularly in Trinidad and Suriname. For example, studies conducted in Suriname between 2016 and 2023 identified at least 70 *Anastrepha* species, most of which were found in the forested interior (Muller et al. 2020a, 2020b). These findings highlight the need for continued research to better understand the diversity and distribution of fruit fly species in the Caribbean.

19.4.2 Regional efforts to combat fruit flies

The CPHD has been instrumental in coordinating regional efforts to combat fruit fly threats. A dedicated working group was established to address the issue, and significant progress has been made since 2013. Key achievements include the following:

- **Training and capacity building**: Sixteen countries received extensive training in fruit fly surveillance and management techniques, including identification, trapping, fruit collection, data management, and control methods. These training programs have strengthened the capacity of national plant protection organizations to monitor and manage fruit fly populations effectively. While fruit collection is regularly conducted in several countries, a general reporting system for sharing information has not been established, aside from published papers. Most national plant protection organizations prioritize monitoring over publishing scientific papers.

- **Development of the Fruit Fly Monitoring System (FFMS)**: In 2014, the FFMS was introduced to centralize trapping data and improve data management. This secure online database allows countries to enter and manage their data, ensuring consistency and accuracy. The FFMS includes features such as:
  - Dropdown lists of host tree species, including local and scientific names. Local names of host trees differ in the various countries in the region, and combining them with the scientific names solved the problem of not being able to find "their" host tree.
  - Geopositioning coordinates for trap locations.
  - Visualization of trapping data on Google Maps, enabling stakeholders to assess trap distribution and identify areas of concern.
  - Production of graphs with F/T/D per fruit fly for certain time periods.
  - Reports on trapping intervals.

  Countries primarily use Jackson and McPhail traps, with lures such as methyl-eugenol, TriMed lure, Cue-lure, and Torula for McPhail traps. Trap and lure placement is determined by each country based on the specific fruit fly species and threats they face. For example, countries with direct flights to Venezuela or Aruba must closely monitor for Medfly. Also, it is beneficial to target specific areas, such as regions with certain fruit crops or districts with populations from neighboring countries where particular fruit flies pose a threat.
- **Open sharing platform**: Since 2020, a push for greater transparency has led to the development of the FFMS Open Sharing Platform. Six countries (Cayman Islands, Guyana, Jamaica, Nevis, St. Lucia, and Suriname) currently share their data, with plans to include Flies per Trap per Day (FTD) data in the future. This platform facilitates regional collaboration and allows countries to take preemptive measures to safeguard their agriculture. The FFMS and Open Sharing Platform are centrally maintained online. However, sustaining these systems presents a significant challenge due to the involvement of numerous small countries with limited resources. Examples of the information that can be retrieved from the FFMS Open Sharing Platform are presented in Figures 19.1 and 19.2.
- **Provision of resources**: The CPHD has provided trapping materials and supplies to support national surveillance programs. These resources have been critical for maintaining effective monitoring systems and ensuring that farmers have the tools they need to manage fruit fly populations.
- **Research and innovation**: The CPHD has supported research on fruit fly biology, ecology, and management, contributing to the development of new control methods and technologies. For example, research on using pheromone traps and baiting techniques has shown promise in reducing fruit fly populations in targeted areas.
- **Public awareness campaigns**: To enhance community engagement and awareness, the CPHD has initiated public awareness campaigns to educate farmers and the general public about the importance of pest management and biosecurity measures. These campaigns emphasize the role of local communities in preventing the introduction and spread of invasive species.

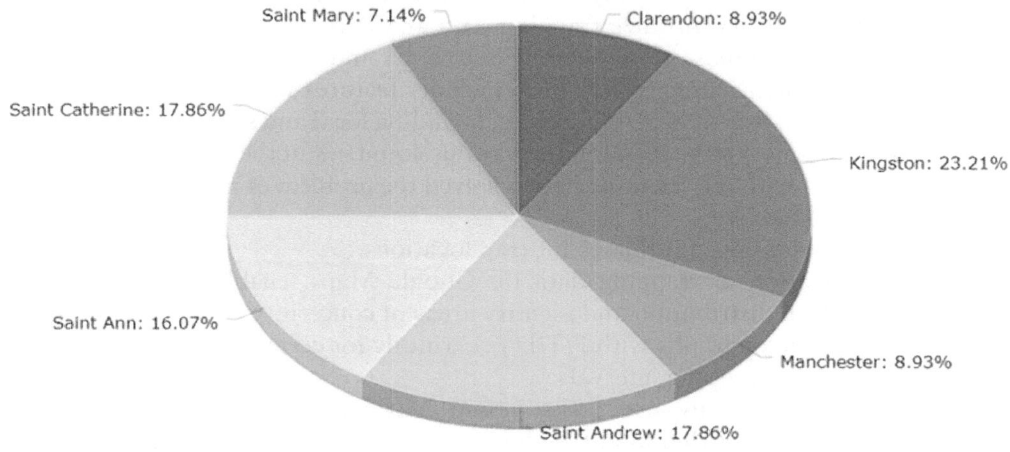

Figure 19.1 Percentage of traps installed in each of the six participating countries in the Open Sharing Platform. Caution with "reading" this graph: Guyana is slightly bigger than Suriname, but has a higher amount of traps per square km and a bigger area in which monitoring takes place. Cayman Islands are three small islands, which is almost totally accessible, while Suriname and Guyana have vast areas with forest in which no traps are located.

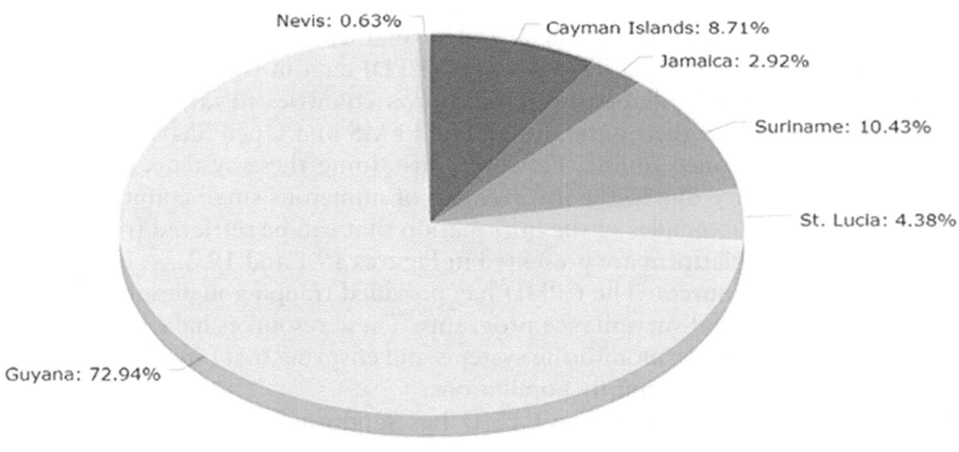

Figure 19.2 Example on how the FFMS Open Sharing Platform shows the distribution of traps within a given country: in this case, Jamaica.

19.4.3 Achievements and future directions

The CPHD Fruit Fly Working Group has made significant contributions to regional pest management, including:

- National training programs and short-term attachments for in-depth training.
- Development of fact sheets and technical materials to support pest management efforts.
- Maintenance of a comprehensive list of Tephritid species and host plants.
- Establishment and maintenance of the FFMS database.

Looking ahead, several priorities have been identified to strengthen regional pest management efforts:

- **Expansion of the FFMS Open Sharing Platform**: Encouraging more countries to participate in the platform and share their data will enhance regional preparedness and facilitate early detection of invasive species. Also, increased data sharing will allow for better analysis of pest trends and the effectiveness of management strategies.
- **Development of biological control methods**: Research on natural enemies, such as parasitoids and predators, could provide sustainable alternatives to chemical control methods. For instance, introducing parasitoids that specifically target fruit fly larvae could help reduce populations without harming beneficial insects.
- **Climate change adaptation**: Developing strategies to address the impacts of climate change on fruit fly populations and their host plants will be critical for long-term pest management. This may include identifying climate-resilient crop varieties and implementing adaptive management practices considering changing environmental conditions.
- **Policy and advocacy**: Strengthening regional policies and advocating for increased investment in agricultural health will ensure the sustainability of pest management programs. Policymakers should prioritize research, training, and infrastructure development funding to support effective pest management.
- **Strengthening biosecurity measures**: Enhancing biosecurity measures at ports of entry and within agricultural systems is essential for preventing the introduction of new pests. This includes implementing stricter inspection protocols, promoting the use of certified pest-free planting materials, and educating stakeholders about the importance of biosecurity.
- **Collaboration with international organizations**: Engaging with international organizations, such as the Food and Agriculture Organization and the International Plant Protection Convention, can provide valuable resources and expertise for managing fruit fly threats. Collaborative research initiatives and knowledge-sharing platforms can enhance regional capacity to address pest challenges.

19.5 CONCLUSION

The challenges posed by fruit flies and other agricultural pests underscore the importance of regional collaboration in the Caribbean. Through initiatives such as the CPHD and the FFMS, CARICOM has made significant strides in protecting agriculture and ensuring food security. However, sustained efforts and continued investment will be necessary to address the evolving threats to the region's agriculture and biodiversity. By building on its achievements and embracing innovation, the Caribbean can safeguard its agricultural heritage and secure a sustainable future for its people.

REFERENCES

Goldsmith, J. 2024. Fruit flies in the Caribbean. *Presented at the 11th Tephritid Workers of the Western Hemisphere (TWWH) Meeting*, June 3–7, Jamaica. Montego Bay. https://nucleus.iaea.org/sites/naipc/twd/Lists/Symposia/Attachments/37/11th%20TWWH%20Book%20of%20Abstracts.pdf (accessed March 30, 2025).

Muller, A., A. Gangadin, B. Sutton, et al. 2020a. New species and records of *Anastrepha* from Suriname and the Guianan region. *Poster at 10th Meeting of Tephritid Workers of the Western Hemisphere*, November 2–6. Colombia.

Muller, A., A. Gangadin, B. Sutton, et al. 2020b. Host plants of *Anastrepha* species in Suriname. *Poster at 10th Meeting of Tephritid Workers of the Western Hemisphere*, November 2–6, Colombia.

Index

Note: **Bold** page numbers refer to tables and *italic* page numbers refer to figures.

A

M

N

Q

R

S

T

U

V

W

X

Z

For Product Safety Concerns and Information please contact our
EU representative GPSR@taylorandfrancis.com Taylor & Francis
Verlag GmbH, Kaufingerstraße 24, 80331 München, Germany